Nanotubes and Nanowires

RSC Nanoscience & Nanotechnology

Series Editors

Professor Paul O'Brien, *University of Manchester, UK*
Professor Sir Harry Kroto, FRS, *University of Sussex, UK*
Professor Harold Craighead, *Cornell University, USA*

This series will cover the wide ranging areas of Nanoscience and Nanotechnology. In particular, the series will provide a comprehensive source of information on research associated with nanostructured materials and miniaturised lab on a chip technologies.

Topics covered will include the characterisation, performance and properties of materials and technologies associated with miniaturised lab on a chip systems. The books will also focus on potential applications and future developments of the materials and devices discussed.

Ideal as an accessible reference and guide to investigations at the interface of chemistry with subjects such as materials science, engineering, biology, physics and electronics for professionals and researchers in academia and industry.

Titles in the series:

Nanotubes and Nanowires
C.N.R. Rao, FRS and A. Govindaraj

Visit our website on www.rsc.org/nanoscience

For further information please contact:

Sales and Customer Services
Royal Society of Chemistry
Thomas Graham House
Science Park, Milton Road
Cambridge CB4 0WF
Telephone +44 (0)1223 432360, Fax +44 (0)1223 426017, Email sales@rsc.org

Nanotubes and Nanowires

C.N.R. Rao, FRS

A. Govindaraj

Jawaharlal Nehru Centre for Advanced Scientific Research, Bangalore, India

RSCPublishing

ISBN 0-85404-832-4

A catalogue record for this book is available from the British Library

Published by The Royal Society of Chemistry,
Thomas Graham House, Science Park, Milton Road,
Cambridge CB4 0WF, UK

Registered Charity Number 207890

For further information see our web site at www.rsc.org

Typeset by Alden Bookset, Northampton, UK
Printed by Biddles Ltd, King's Lynn, Norfolk, UK

Preface

The science of nanomaterials has become the flavour of the day, with research being driven both by academic curiosity and the promise of useful applications. Amongst the nanomaterials, nanocrystals, nanowires and nanotubes constitute three major categories, the last two being one-dimensional. Since the discovery of the carbon nanotubes in 1991, interest in one-dimensional nanomaterials has grown remarkably, and a phenomenal number of research articles has been published on nanotubes as well as nanowires. The nanotubes are not only those of carbon but also of inorganic materials. Several strategies have been developed for the synthesis of these materials and a range of interesting properties reported. Thus, the electronic and mechanical properties of carbon nanotubes have been studied extensively, and several of them directly relate to potential applications. Typical of the important properties of carbon nanotubes are high mechanical strength, good electrical and thermal conductivity and excellent electron emission characteristics. The electronic and Raman spectra of carbon nanotubes have helped immensely in characterization as well as in understanding some of the intrinsic structural characteristics.

While nanotubes of several inorganic materials, many of which possess layered structures, have been synthesized and characterized, the literature on inorganic nanowires is much more extensive. Every conceivable inorganic material seems to have been prepared in nanowire form. Properties and possible applications of these inorganic one-dimensional materials have been investigated to some extent, but there seems to be ample scope for study.

This monograph provides an up-to-date survey of various aspects of carbon nanotubes, inorganic nanotubes and nanowires. Nanotubes of lipids, peptides, polymers and DNA are known, but they have not been discussed in this monograph due to its limited scope. We have found it difficult to cover the entire gamut of properties and applications of the nanotubes and nanowires in detail in view of the immense literature that has accumulated in the last three to four years. We have been selective, emphasizing more the chemical aspects of nanotubes and nanowires, especially those related to synthesis and characterization to a greater extent. We have provided an extensive list of references to enable those who would like more complete information on the properties and other aspects of these

v

materials. It is possible that we have failed to cite some important references by oversight or error in judgement, and we would like to be excused for such omissions. We have done our best to make the monograph contemporary and we hope that students, teachers and practitioners of nanoscience will find it useful.

C.N.R. Rao
A. Govindaraj

About the Authors

C.N.R. Rao obtained his PhD degree from Purdue University and DSc degree from the University of Mysore. He is the Linus Pauling Research Professor and Honorary President of the Jawaharlal Nehru Centre for Advanced Scientific Research and Honorary Professor at the Indian Institute of Science (both at Bangalore). His research interests are mainly in the chemistry of materials (e.g., transition-metal oxides, open-framework structures, and nanomaterials). He has authored nearly 1000 research papers and edited or written 30 books in materials chemistry. A member of several academies including the Royal Society, US National Academy of Sciences, Japan Academy, French Academy of Sciences, and the Pontifical Academy of Sciences, he is also Distinguished Visiting Professor at the University of California, Santa Barbara. He was awarded the Einstein Gold Medal by UNESCO, the Hughes Medal by the Royal Society, and the Somiya Award of the International Union of Materials Research Societies (IUMRS). In 2005, he received the Dan David Prize for materials research from Israel and is the first recipient of the India Science Prize.

A. Govindaraj obtained his M.Sc and Ph.D degrees from the University of Mysore. His main research interests are in fullerenes, nanowires and nanotubes in which areas he has published extensively. He is a Senior Scientific Officer at the Indian Institute of Science and Honorary Faculty Fellow at the Jawaharlal Nehru Centre for Advanced Scientific Research.

Contents

Chapter 2   Inorganic Nanotubes

Chapter 3   Inorganic Nanowires

Abbreviations

AAM	Anodic Alumina Membrane
AFM	Atomic Force Microscope (Microscopy)
BNNTs	Boron Nitride Nanotubes
c.m.c.	Critical Micelle Concentration
μCP	Microcontact Printing
CNTs (or CN)	Carbon Nanotubes
CSR	Carrot-shaped Rods
CTAB	Cetyltrimethylammonium Bromide
CVD	Chemical Vapour Deposition
DOS	Density of States
DWNTs	Double-walled Carbon Nanotubes
EDX	Energy Dispersive X-ray Spectroscopy
FE	Field Emission
FEED	Field-emitted Electron Energy Distribution
FET	Field-effect Transistor
F-N	Fowler–Nordheim (plot)
F-SWNTs	Fluorinated Single-wall Nanotubes
HFCVD	Hot-filament Chemical Vapour Deposition
HiPco	High-pressure CO Disproportionation Process
HOPG	Highly Oriented Pyrolytic Graphite
HREM	High-resolution Electron Microscope (Microscopy)
IF	Inorganic Fullerenes
IMJs	Intramolecular Junctions
ITO	Indium Tin Oxide
MFP	Mean-free Path
MR	Magnetoresistance
MWNTs	Multi-wall Nanotubes
NR	Nanoribbons
NSP	Nebulized Spray Pyrolysis
NT-FETs	Nanotube Field-effect Transistors
NW	Nanowires (often prefixed by an elemental symbol, *e.g.* BiNW for Bismuth Nanowires)

PANI	Polyaniline
PDMS	Polydimethylsiloxane
PEG	Poly(ethylene glycol)
PEI	Polyethyleneimine
PFO	Poly(9,9-di-*n*-octylfluorenyl-2,7-diyl)
PL	Photoluminescence
PMMA	Poly(methyl methacrylate)
PPV	Poly(*p*-phenylene vinylene)
PVD	Physical Vapour Deposition
PVP	Poly(vinylpyrrolidone)
SAED	Selected-area Electron Diffraction
SDS	Sodium Dodecyl Sulphate
SEM	Scanning Electron Microscope (Microscopy)
SET	Single Electron Tunnelling
SFLS	Supercritical fluid–Liquid–Solid
SLS	Solution–Liquid–Solid
SHG	Second-harmonic Generation
sscm	Standard Cubic Centimetres per Minute
STM	Scanning Tunnelling Microscopy
STS	Scanning Tunnelling Spectroscopy
SWNTs	Single-wall Nanotubes
s-SWNTs	Shortened Single-wall Nanotubes
TB-DFT	Tight-binding Density Functional Theory
TEM	Transmission Electron Microscope (Microscopy)
TEP	Thermoelectric Power
THG	Third-harmonic Generation
VLS	Vapour-Liquid–Solid
VS	Vapour–Solid
XRD	X-ray Diffraction

CHAPTER 1

Carbon Nanotubes

1 Introduction

Diamond and graphite are the two well-known forms of crystalline carbon. Diamond has four-coordinate sp^3 carbon atoms that form an extended three-dimensional network, whose motif is the chair conformation of cyclohexane. Graphite has three-coordinate sp^2 carbons that form planar sheets, whose motif is the flat six-membered benzene ring. The new carbon allotropes, the fullerenes, are closed-cage carbon molecules with three-coordinate carbon atoms tiling the spherical or nearly-spherical surfaces, the best known example being C_{60}, with a truncated icosahedral structure formed by twelve pentagonal rings and twenty hexagonal rings (Figure 1.1a). Fullerenes were discovered by Kroto et al.[1] in 1985 while investigating the nature of carbon present in interstellar space. The coordination at every carbon atom in fullerenes is not planar, but slightly pyramidalized, with some sp^3 character present in the essentially sp^2 carbons. The key feature is the presence of five-membered rings, which provide the curvature necessary for forming a closed-cage structure. In 1990, Krätschmer et al.[2] found that the soot produced by arcing graphite electrodes contained C_{60} and other fullerenes. The ability to generate fullerenes in gram quantities in the laboratory, using a relatively simple apparatus, gave rise to intense research activity on these molecules and caused a renaissance in the study of carbon. Iijima[3] observed, in 1991, that nanotubules of graphite were deposited on the negative electrode during the direct current arcing of graphite for the preparation of fullerenes. These nanotubes are concentric graphitic cylinders closed at either end due to the presence of five-membered rings. Nanotubes can be multi-walled with a central tubule of nanometric diameter surrounded by graphitic layers separated by ~3.4 Å. Unlike the multi-walled nanotubes (MWNTs), in single-walled nanotubes (SWNTs), there is only the tubule and no graphitic layers. A transmission electron microscope (TEM) image of a MWNT is shown in Figure 1.1(b). In this nanotube, graphite layers surround the central tubule. Figure 1.1(c) shows the structure of a nanotube formed by two concentric graphitic cylinders, obtained by force-field calculations. A single-walled nanotube can be visualized by cutting C_{60} along the centre and spacing apart the hemispherical corannulene end-caps by a cylinder of graphite of the same diameter. Carbon nanotubes are the only form of carbon with extended

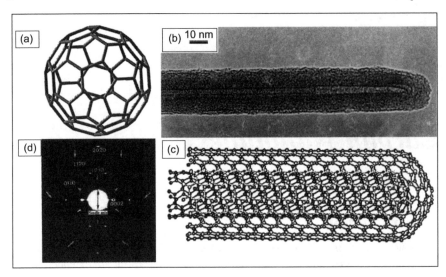

Figure 1.1   (a) *Schematic diagram of a C₆₀ molecule; (b) A TEM image of a multi-walled carbon nanotube; (c) Minimum energy structure of a double-walled carbon nanotube; (d) Electron diffraction pattern of a multi-walled carbon nanotube* (Parts (a–c) reproduced from ref. 20a; part (d) reproduced from ref. 20c)

bonding and yet with no dangling bonds. Since carbon nanotubes are derived from fullerenes, they are referred to as tubular fullerenes or bucky tubes.

Ever since the discovery of the carbon nanotubes,[4,5] several ways of preparing them have been explored. Besides MWNTs, SWNTs have been prepared by various methods, including electrochemical synthesis[6] and pyrolysis of precursor organic molecules.[7] The structure of carbon nanotubes has been extensively investigated by high-resolution electron microscopy.[8–10] The nanotubes, prepared by arc vaporization of graphite, are closed at both ends, but can be opened by various oxidants.[11,12] There has been considerable success in filling nanotubes with various materials.[13] Apart from opening and filling, carbon nanotubes have been doped with boron and nitrogen, giving rise to p-type and n-type materials, respectively. By employing carbon nanotubes as removable templates, oxidic, carbidic and other nanostructures have been prepared. One of the developments is the synthesis of aligned nanotube bundles for specific applications. Various properties and phenomena as well as several possible and likely applications of carbon nanotubes have been reported. Unsurprisingly, therefore, these nanomaterials have elicited great interest. Several review articles, special issues of journals and conference proceedings[14–20] have dealt with carbon nanotubes. Some of the reviews present possible technological applications, with focus on the electronic properties,[19,20] the recent book of Reich *et al.*[21] being devoted to a detailed presentation of the basic physics of carbon nanotubes. There are several other reviews and books as well, some of which are cited as references.[22–27]

Since the discovery of the carbon nanotubes, there has been considerable work on inorganic layered materials such as MoS₂, WS₂ and BN to explore the formation of nanotubes of these materials. Indeed several have been synthesized and characterized.[28–31] Inorganic nanotubes are discussed at length in Chapter 2. Here, we shall present several aspects of carbon nanotubes, such as their preparation, structure, mechanism of formation, chemical substitution, properties and applications. We briefly examine the three fundamental aspects of CNTs, namely, their electronic structure and related properties, their vibrational and thermal characteristics and their mechanical properties. These aspects are interrelated, since both thermal and mechanical properties reflect the chemical bonding in the carbon network, which controls their electronic structure as well.

2 Synthesis

Multi-walled Nanotubes

Carbon nanotubes are readily prepared by striking an arc between graphite electrodes in ~0.7 atm (~500 Torr) of helium, which is considerably larger than the helium pressure used to produce fullerene soot. The schematic diagram of the apparatus is shown in Figure 1.2. A current of 60–100 A across a potential drop of about 25 V gives high yields of carbon nanotubes. The arcing process can be optimized such that the major portion of the carbon anode gets deposited on the cathode as carbon nanotubes and graphitic nanoparticles.[32a] Arc evaporation of graphite has been carried out in various kinds of ambient gases (He, Ar, and CH$_4$).[32b] Hydrogen appears to be effective in producing MWNTs of high crystallinity. Arc-produced MWNTs in hydrogen also contain very few carbon

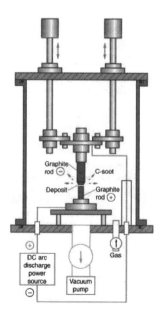

Figure 1.2 *Schematic diagram of an arc discharge apparatus*
(Reproduced from ref. 32b)

nanoparticles. Carbon nanotubes have been produced in large quantities by using plasma arc-jets by optimizing the quenching process in an arc between a graphite anode and a cooled copper electrode.[33,34] If both the electrodes are of graphite, MWNTs are the main products, along with side products such as fullerenes, amorphous carbon, and graphite sheets.

A route to highly crystalline MWNTs is the arc-discharge method in liquid nitrogen.[35] In this method, vacuum is replaced with liquid nitrogen in the arc discharge chamber. Typically, direct current was supplied to the apparatus using a power supply. The anode is a pure carbon rod (8 mm diameter) and the cathode is a pure carbon rod (10 mm diameter). The Dewar flask is filled with liquid nitrogen and the electrode assembly immersed in nitrogen. Arc discharge occurs as the distance between the electrodes became less than 1 mm, and a current of ~80 A flows between them. When the arc discharge is over, carbon deposits near the cathode are recovered for analysis. Liquid nitrogen prevents the electrodes from contamination with unwanted gases and also lowers the temperature of the electrodes. Furthermore, CNTs do not stick to the chamber wall. The content of the MWNTs can be as high as 70% of the reaction product. Analysis with Auger-spectroscopy revealed that no nitrogen was incorporated in the MWNTs. Synthesis in a magnetic field gives defect-free and high purity (>95%) MWNTs, which can be used as nanosized electric wires for device fabrication.[36] Here, the arc discharge is controlled by a magnetic field around the arc plasma, created by using extremely pure graphite (purity >99.999%) electrodes (Figure 1.3a and 1.3b). MWNTs can be mass produced economically by

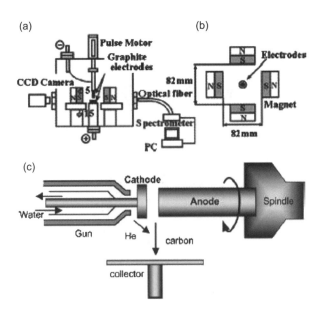

Figure 1.3   (a) *and* (b) *Schematic diagrams of the system for the synthesis of MWNTs in a magnetic field* (c) *Schematic diagram of a plasma rotating electrode system* (Parts (a) and (b) reproduced from ref. 36; part (c) Reproduced from ref. 37)

the plasma rotating arc discharge technique.[37] The centrifugal force caused by the rotation generates turbulence and accelerates the carbon vapour perpendicular to the anode (Figure 1.3c). Rotation also distributes the micro discharges uniformly and generates a stable plasma. Consequently, it increases the plasma volume and raises the plasma temperature. At a rotations of 5000 rpm, a yield of 60% is obtained at 1025 °C (without the use of a catalyst). The yield increases up to 90% if the rotation speed is increased at 1150 °C. The MWNTs obtained generally have an inner diameter of 1–3 nm and an outer diameter of ~10 nm.

Deposition of carbon vapour on cooled substrates of highly oriented pyrolytic graphite affords tube-like structures.[38] Carbon nanotubes are also produced by electrolysis in molten halide salts with carbon electrodes under argon.[39a] MWNTs with well-ordered graphitic structures have also been obtained under hydrothermal conditions around 800 °C, under 60–100 MPa pressure, using a polyethylene–water mixture in the presence of a nickel catalyst.[39b] Besides the conventional arc-evaporation technique, carbon nanotubes are produced by chemical vapour deposition (CVD), by the decomposition of hydrocarbons such as C_2H_2 under inert conditions around 700 °C over Fe/graphite,[40] Co/graphite[41] or Fe/silica[42] catalysts. Transition metal particles are essential for the formation of nanotubes by the CVD or pyrolysis process, and the diameter of the nanotube is generally determined by the size of the metal particles.[43]

Chemical Vapour Deposition (CVD)

The chemical vapour deposition (CVD) method uses a carbon source in the gas phase and a plasma or a resistively heated coil, to transfer the energy to the gaseous carbon molecule. Commonly used carbon sources are methane, carbon monoxide and acetylene. The energy source cracks the molecule into atomic carbon. The carbon then diffuses towards the substrate, which is heated and coated with a catalyst (usually a first row transition metal such as Ni, Fe or Co) and binds to it. Carbon nanotubes are formed in this procedure if the proper parameters are maintained. Good alignment[44] as well as positional control on a nanometric scale[45] are achieved by using CVD. Control over the diameter, as well as the growth rate of the nanotubes is also achieved. Use of an appropriate metal catalyst permits preferential growth of single-walled rather than multi-walled nanotubes.[46]

CVD synthesis of nanotubes is essentially a two-step process, consisting of a catalyst preparation step followed by synthesis of the nanotube. The catalyst is generally prepared by sputtering a transition metal onto a substrate, followed by etching by chemicals such as ammonia, or thermal annealing, to induce the nucleation of catalyst particles. Thermal annealing results in metal cluster formation on the substrate, from which the nanotubes grow. The temperature for the synthesis of nanotubes by CVD is generally in the 650–900 °C range.[44–47] Typical nanotube yields from CVD are around 30%. Various CVD processes have been used for carbon nanotubes synthesis, including plasma-enhanced CVD, thermal chemical CVD, alcohol catalytic CVD, aerogel-supported CVD and laser-assisted CVD.

Plasma-enhanced Chemical Vapour Deposition

The plasma-enhanced CVD method involves a glow discharge in a chamber or a reaction furnace through a high-frequency voltage applied to both the electrodes. Figure 1.4 shows a schematic diagram of a typical plasma CVD apparatus with a parallel plate electrode structure. A substrate is placed on the grounded electrode. To form a uniform film, the reaction gas is supplied from the opposite plate. Catalytic metals such as Fe, Ni and Co are deposited on a Si, SiO_2, or glass substrate using thermal CVD or sputtering. After the nanoscopic fine metal particles are formed, the carbon nanotubes grow on the metal particles on the substrate by the glow discharge generated from a high frequency power source. A carbon-containing gas, such as C_2H_2, CH_4, C_2H_4, C_2H_6 or CO is supplied to the chamber during discharge.[48] The catalyst has a strong effect on the nanotube diameter, growth rate, wall thickness, morphology and microstructure. Nickel seems to be the most suitable catalyst for the growth of aligned MWNTs by this technique.[49] The diameter of the MWNTs is around 15 nm. The highest yield of carbon nanotubes achieved by Chen *et al.* was about 50%, at a relatively low temperature (<330 °C).[48]

Thermal Chemical Vapour Deposition

In this method, Fe, Ni, Co or an alloy of these metals is initially deposited on a substrate. After the substrate is etched by a dilute HF solution, it is placed in a quartz boat, positioned in a CVD reaction furnace. Nanometre-sized catalytic metal

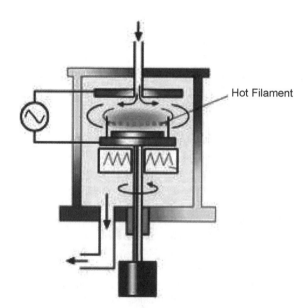

Figure 1.4 *Schematic diagram of a plasma CVD apparatus*
(Reproduced from ref. 48)

particles are formed after an additional etching of the catalytic metal film using NH_3 gas at 750–1050 °C. Nanotubes grow on the fine catalytic metal particles by the CVD process. When growing MWNTs on a Fe catalytic film by thermal CVD, the diameter range of the carbon nanotubes depends on the thickness of the catalytic film. When the film thickness was 13 nm, the diameter was between 30 and 40 nm. When a thickness of 27 nm was used, the diameter range was between 100–200 nm.[50]

Vapour Phase Growth

In the vapour phase growth, pyrolysis or the floating catalyst method the carbon vapour and the catalytic metal particles are both deposited in the reaction chamber without a substrate. The diameter of the carbon nanotubes by vapour phase growth is in the range 2–4 nm for SWNTs and 70 and 100 nm for MWNTs. Using this technique, Sen *et al.*[51,52] prepared MWNTs as well as metal-filled onion-like and nanotube structures by the pyrolysis of metallocenes such as ferrocene, cobaltocene, and nickelocene under reducing conditions, wherein the precursor acts as the source of the metal as well as carbon. The pyrolysis set-up consists of stainless steel gas flow lines and a two-stage furnace system fitted with a quartz tube (Figure 1.5), the flow rate of the gases being controlled by the use of mass flow controllers. In a typical preparation, a known quantity (100 mg) of the metallocene (presublimed 99.99% purity) is taken in a quartz boat and placed at the centre of the first furnace, and a mixture of Ar and H_2 of the desired composition is passed through the quartz tube. The metallocene is sublimed by heating the first furnace to 200 °C at a controlled heating rate (20 °C/min^{-1}). The metallocene vapour generated is carried by the Ar–H_2 gas stream into the second furnace, maintained at 900 °C, where pyrolysis occurs. The main variables in the experiment are the heating rate of ferrocene, the flow rate of Ar, and the pyrolysis temperature. Figure 1.6(a) shows a TEM image of MWNTs obtained by the pyrolysis of mixture of C_2H_2 and ferrocene. Ferrocene vapour carried by a 75% Ar + 25% H_2 mixture at 900 sccm (sccm = standard cubic centimetre per minute) into the furnace yields large quantities of carbon deposits, mainly containing carbon nanotubes. Under similar conditions, nickelocene with benzene gave MWNTs (Figure 1.6b). This procedure has been employed for large-scale production of carbon nanotubes.[53]

Nebulized spray pyrolysis has been successfully employed for the synthesis of MWNTs.[54] Figure 1.7 shows the schematic diagram of the experimental set-up. Silicon substrates were placed in the regions I–IV of the reactor to collect the product. In a typical synthesis, 2 g of metallocene was dissolved in 100 mL of a hydrocarbon and the solution nebulized using a 1.54 MHz ultrasonic beam carried into a 25 mm quartz tube, in a SiC furnace maintained at the required temperature (800–1000 °C). Pure argon was used as the carrier gas and the gas flow rate was controlled using UNIT mass flow controllers. In a typical procedure, the carrier gas flow rate was kept at 1000 sccm and pyrolysis was carried for 30 min. After the reaction, the furnace was allowed to cool. Products were collected after the tube cooled to room temperature. The average droplet size for the various solvents for 1.54 MHz frequency is ~2.2 μm. Ferrocene, cobaltocene, nickelocene and iron

(a)

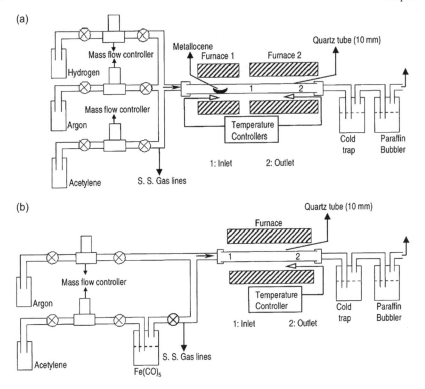

(b)

Figure 1.5 *Apparatus employed for the synthesis of carbon nanotubes by the pyrolysis of mixtures of (a) metallocene + C_2H_2 and (b) $Fe(CO)_5 + C_2H_2$. Numbers 1 and 2 in the figure represent the inlet and outlet, respectively*
(Reproduced from refs. 62 and 63)

pentacarbonyl were used as both catalysts and carbon sources. Acetylene, benzene, toluene, xylene, mesitylene and *n*-hexane, used as solvents for the catalysts, act as additional carbon sources. The TEM images in Figure 1.8 show the MWNTs obtained by this method. Random or aligned bundles of MWNTs with fairly uniform diameters are obtained, depending on the flow rate and hydrocarbon solvent used for dissolving the metallocene. Well-graphitized MWNTs were obtained with a solution of ferrocene in xylene (inset in Figure 1.8d). The product quality depends on the pyrolysis temperature, carrier gas flow rate, additional carbon sources used and the catalyst precursor concentration. This procedure can be scaled up for continuous production of MWNTs.

Aligned Nanotube Bundles and Micropatterning

Some applications of carbon nanotubes, *e.g.* as electron emitters,[55] require that they are aligned or micropatterned. Aligned nanotube bundles have been obtained by CVD over transition metal catalysts embedded in the pores of mesoporous silica or the channels of alumina membranes.[56,57] Terrones *et al.*[58,59] prepared aligned

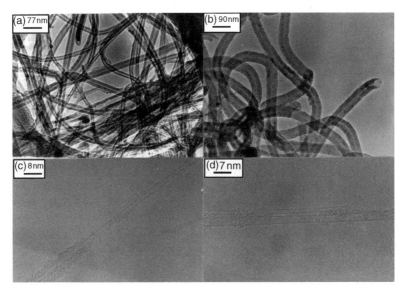

Figure 1.6 (a) *TEM image of MWNTs obtained by the pyrolysis of a mixture of C_2H_2 (25 sccm) and ferrocene at 1100 °C at 1000 sccm Ar flow.* (b) *TEM image of MWNTs obtained by the pyrolysis of a mixture of nickelocene and benzene at 900 °C in 85% Ar and 15% H_2 mixture at a flow rate of 1000 sccm.* (c) *HREM image of a SWNT obtained by the pyrolysis of nickelocene and C_2H_2 (50 sccm) at 1100 °C in a flow of Ar (1000 sccm).* (d) *HREM image of SWNTs obtained by the pyrolysis of ferrocene and CH_4 at 1100 °C in a flow of Ar (990 sccm) and CH_4 (10 sccm)*
(Reproduced from refs. 51 and 63)

nanotubes over silica substrates, laser-patterned with cobalt. Ren *et al.*[44] employed plasma-enhanced CVD on nickel-coated glass, using acetylene and ammonia mixtures, for this purpose. The mechanism of growth of nanotubes by this method and the exact role of the metal particles are not clear, although a nucleation process involving the metal particles is considered important. Fan *et al.*[60] obtained aligned nanotubes by employing CVD on porous silicon and plain silicon substrates patterned with Fe films. The role of the transition metal particles assumes significance since aligned nanotubes are obtained by the pyrolysis of acetylene over iron/silica catalyst surfaces.[61] Knowing that carbon nanotubes can be prepared by the pyrolysis of mixtures of organometallic precursors and hydrocarbons,[51,52] one would expect that transition metal nanoparticles produced *in situ* in the pyrolysis may not only nucleate the formation of carbon nanotubes but also align them. This has been examined by pyrolysing metallocenes along with additional hydrocarbon sources, in a suitably designed apparatus (Figure 1.5).[62,63] Figure 1.9 (a–c) shows scanning electron microscope (SEM) images of aligned nanotubes obtained by the pyrolysis of ferrocene. The image in Figure 1.9(a) shows large bundles of aligned nanotubes. Figure 1.9(b) shows the side-view whereas Figure 1.9(c) shows the top-view of the aligned nanotubes, wherein the nanotube tips are seen. A TEM image of a part of an aligned nanotube bundle obtained from the pyrolysis of an

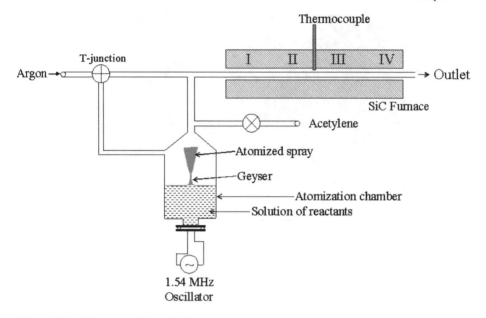

Figure 1.7 *Schematic of the nebulised spray pyrolysis set-up*
(Reproduced from ref. 54)

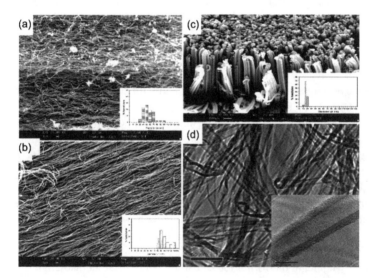

Figure 1.8 (a) *and* (b) *SEM images of aligned MWNT bundles obtained by the nebulized*
spray pyrolysis of ferrocene along with benzene and mesitylene, respectively, at
900 °C at an argon flow rate of 1000 sccm. (c) *and* (d) *SEM and TEM images,*
respectively of aligned MWNTs obtained by the pyrolysis of a toluene solution
of ferrocene (20 g l^{-1}) for 5 min *and acetylene flow of 100* sccm *for 30 min.*
Insets in (a)–(c) *show diameter histograms. Inset in* (d) *TEM of well-*
graphitized MWNTs obtained with a solution of ferrocene in xylene
(Reproduced from ref. 54)

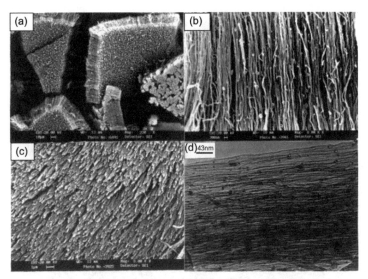

Figure 1.9 *SEM image* (a) *showing the bundles of aligned nanotubes obtained by pyrolysing ferrocene along with butane (50 sccm) at 1100 °C in a Ar flow of 950 sccm.* (b) *and* (c) *Show aligned nanotubes perpendicular and along to the axis of the nanotubes respectively.* (d) *TEM image of a part of an aligned nanotube bundle obtained from the pyrolysis of a mixture of acetylene (85 sccm) and ferrocene at 1100 °C in a Ar flow of 1000 sccm* (Reproduced from ref. 62a, 63)

acetylene–ferrocene mixture is shown in Figure 1.9(d). The average length of the nanotubes is generally around 60 μm with methane and acetylene. Andrews *et al.*[64a] have carried out the pyrolysis of ferrocene–xylene mixtures to obtain aligned carbon nanotubes. Pyrolysis of Fe(II)phthalocyanine also yields aligned nanotubes.[64b] Hexagonally ordered arrays of nanotubes are produced by using alumina templates with ordered pores.[65a] Quasi-aligned carbon nanotubes are obtained by using metal impregnated zeolite templates.[65b] This method offers a control of the diameter of nanotubes through the use of zeolite templates of known diameter. The advantage of the precursor method is that aligned nanotube bundles are produced in one step, at a relatively low cost, without prior preparation of substrates. Rao and Govindaraj have reviewed the precursor route to carbon nanotubes.[66a]

TEM observations of aligned nanotubes produced by ferrocene + hydrocarbon pyrolysis show iron nanorods encapsulated inside the carbon nanotubes, the proportion of nanorods depending on the proportion of ferrocene. Typical TEM images of such nanorods or nanowires are shown in Figure 1.10. The inset in Figure 1.10b shows the selected-area electron diffraction (SAED) pattern of the nanorods, with spots due to (010) and (011) planes of α-Fe. The high-resolution electron microscope (HREM) image of an iron nanorod shows well-resolved (011) planes of α-Fe in single-crystalline form. X-ray diffraction studies also show the presence of α-Fe, with a small portion of Fe_3C as the minor phase. In addition to the nanorods, iron nanoparticles (20–40 nm diameter) are encapsulated by graphite

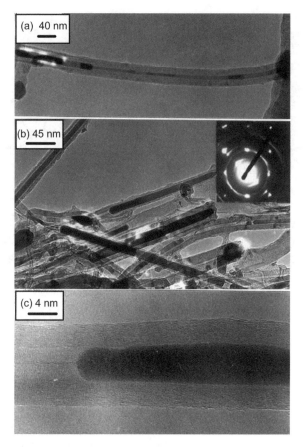

Figure 1.10   (a), (b) *TEM images of the iron nanorods encapsulated inside the carbon nanotubes from aligned nanotube bundles.* (c) *HREM image of a single-crystal iron nanorod encapsulated inside a carbon nanotube. Inset in* (b) *represents the selected area electron diffraction (SAED) pattern of an iron nanorod*
(Reproduced from ref. 63, 66(b))

layers. Both the iron nanorods and nanoparticles are well protected against oxidation by the graphitic layers. The iron nanorods exhibit complex magnetic behaviour.[66b] Such iron-filled carbon nanotubes could be useful as probes in magnetic force microscopy. By using the technique of nebulized spray pyrolysis mentioned earlier, aligned MWNT bundles are obtained. Figures 1.8(a) and 1.8(b) show SEM images of MWNTs prepared under these different conditions. The inclusion of acetylene to an atomized spray of ferrocene in toluene yield aligned MWNT bundles with a narrow diameter distribution. Nebulized spray pyrolysis of iron pentacarbonyl in the presence of acetylene also yields aligned MWNT bundles.

Dai *et al.*[67] have reviewed the formation of aligned and micropatterned carbon nanotubes. Various synthetic and microfabrication techniques have been developed for preparing aligned and micropatterned nanotubes with desirable structural/property characteristics and the details of some are given below. Horizontally aligned carbon nanotubes have been prepared either by slicing a nanotube-dispersed polymer composite or by rubbing a nanotube-deposited plastic surface with a thin sheet of Teflon or aluminium foil emitters as the cathode.[55] Ajayan[68] mixed nanotubes with an epoxy-based resin, hardened the mixture and cut it into slices ranging in thickness from 50 to 1000 nm with lateral dimensions of a few millimetres. TEM images of the slices showed that the tubes were preferentially oriented parallel during cutting. These results demonstrate the nature of rheology, on a nanometer scale, in composite media and flow-induced anisotropy.

Burghard *et al.*[69] carried out region-specific deposition of carbon nanotubes grafted with surfactants containing negatively charged groups (*e.g.* sodium dodecyl-sulphate) onto substrate surfaces prepatterned with positively charged functionalities (*e.g.* ammonium groups on silanized silica). Subsequently, Liu *et al.*[70] demonstrated that individual single-wall carbon nanotubes purified by refluxing in HNO_3 (2.6 M) could be region-specifically deposited onto surfaces prepatterned with a self-assembled monolayer (SAM) of NH_2 functionalities. In a separate study, Chen and Dai[71] have developed a versatile method for making micropatterns of either aligned or nonaligned carbon nanotubes. They first generated plasma-induced (either by nondepositing plasma treatment or by plasma polymerization) surface patterns of NH_2 groups onto a substrate (*e.g.* a quartz glass plate, mica sheet, or polymer film), and then performed region-specific adsorption of COOH-containing carbon nanotubes from an aqueous medium, through the polar–polar interaction between the COOH groups and the plasma-patterned NH_2 groups. The COOH-containing carbon nanotubes were prepared by acid treatment (HNO_3) of nanotubes generated by the pyrolysis of iron(II) phthalocyanine (FePc).[72] To construct a nanotube field emitter, de Heer *et al.*[55] made an ethanol dispersion of arc-produced carbon nanotubes, which was then passed through an aluminium oxide micropore filter to align the nanotubes perpendicularly onto the filter surface.

In a photolithographic approach, Yang *et al.*[73] patterned a positive photoresist film of diazonaphthoquinone (DNQ)-modified cresol novolak (Figure 1.11, Scheme 1a) onto a pristine quartz substrate. Upon UV irradiation through a photomask, the DNQ-novolak photoresist film in the exposed regions was rendered soluble in an aqueous solution of sodium hydroxide due to the photogeneration of the hydrophilic indenecarboxylic acid groups from the hydrophobic DNQ *via* a photochemical Wolff rearrangement (Scheme 1b in Figure 1.11). The photolithographically patterned photoresist film, after an appropriate carbonization process, acts as a shadow mask for the patterned growth of the aligned nanotubes. Figure 1.11(a) shows the steps of the photolithographic process. Pyrolysis of FePc onto the photoresist-prepatterned quartz plate after carbonization leads to region-specific growth of the aligned carbon nanotubes in the UV-exposed regions (Figure 1.11b). This method is compatible with existing photolithographic processes.

By means of a modified photolithographic method for the patterned pyrolysis of FePc, three-dimensional (3D) micropatterns of aligned carbon nanotubes normal to the substrate surface have been obtained, with region-specific tubular lengths and packing densities.[74] The photoresist system consisted of novolak/hexa-methoxymethylmelamine (HMMM) as the film former, phenothiazine as the

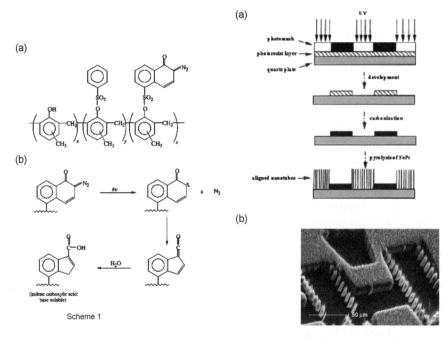

Figure 1.11 *Scheme 1 describes the photolithographic method: photolithographically patterning of positive photoresist film of diazonaphthoquinone (DNQ)-modified cresol novolak on a pristine quartz substrate. (a) Schematic representation of the micropattern formation of aligned carbon nanotubes by the photolithographic process. (b) SEM micrograph of patterned films of aligned nanotubes prepared by the pyrolysis of FePc onto a photolithographically prepatterned quartz substrate*
(Reproduced from ref. 73)

photosensitizer, and diphenyliodoniumhexafluorophosphate $(Ph_2I^+X^-)$ as a photoacid generator that can photochemically generate the acid, through a photomask, required for the region-specific crosslinking of the photoresist film. UV irradiation through a photomask creates a latent acid pattern, formed by the photolithographic generation of the acid from $Ph_2I^+X^-$. Post-exposure baking at 110 °C for 10 min caused an acid-induced crosslinking of the novolac resin and HMMM, rendering the photoresist film in the UV-exposed regions insoluble in an aqueous solution of sodium hydroxide (3 wt.%) and ethanol (10 wt.%). The photoresist film in the regions not exposed to the UV light was removed by immersing in developer solution for 10–20 s, affording a negative polymer pattern on the substrate. The crosslinking reactions between HMMM and the novolac resin were completed by immersing the photoresist-patterned substrate in an aqueous solution of *p*-toluenesulfonic acid (10 wt.%) for 30 min and baking at 150 °C for 30 min. The photoresist-prepatterned silica wafer or quartz plate was directly used as the substrate for the region-specific growth of aligned nanotubes. SEM images of the 3D aligned carbon nanotube micropatterns showed a region-specific packing density as well as tubular length. In contrast to the DNQ-novolak photoresist film, the HMMM-crosslinked novolac photoresist film supported the nanotube growth, probably due to surface characteristics, which allow the Fe nanoparticles to deposit onto this particular photoresist layer at the initial stage of FePc pyrolysis.[74] The 3D micropatterns of aligned carbon nanotubes so prepared offer the possibility of constructing advanced microdevices with multidimensional features.

DNQ-novolak photoresist patterns were carbonized into carbon black for the region-specific growth of aligned nanotubes in the polymer-free regions, by the pyrolysis of FePc in an Ar/H$_2$ atmosphere at 800–1100 °C, as with photolithographic patterning. The spatial resolution was limited by the resolution of the mask used. The ease with which micro-/nanopatterns of organic materials can be made, even on curved surfaces by soft lithographic techniques,[75] may provide additional benefits, especially in the construction of flexible devices. Both the photolithographic and soft-lithographic patterning methods involve a tedious carbonization process prior to aligned nanotube growth. To eliminate the carbonization process, high-resolution carbon nanotube arrays aligned in a direction normal to the substrate surface have been prepared by radio frequency glow-discharge plasma polymerization of a thin polymer pattern onto a quartz substrate, followed by region-specific growth of aligned nanotubes in the plasma-polymer-free regions by pyrolysis of FePc.[72]

Huang *et al.*[64b] have used microcontact printing (μCP) and micromolding techniques to prepare micropatterns of carbon nanotubes aligned in a direction normal to the substrate surface. While the μCP process involves the region-specific transfer of self-assembling monolayers (SAMs) of alkylsiloxane onto a quartz substrate and the subsequent adsorption of polymer chains in the SAM-free regions, the micromolding method permits the formation of polymer patterns through solvent evaporation from a precoated thin layer of polymer solution confined between a quartz plate and a polydimethylsiloxane (PDMS) elastomer mold.

There have been other efforts to micropattern carbon nanotubes. For example, Zheng *et al.*[76] used micromolding in capillaries combined with catalyst-containing 3D cubic mesoporous silica films to fabricate microscopically ordered, aligned carbon nanotube patterns. In this case, PDMS mold having a patterned relief structure on its surface was placed on the surface of a substrate to form a network of channels between them.

Plasma-enhanced chemical vapour deposition (PECVD) is a promising growth technique for the selective positioning and vertical alignment of CNTs.[77] By combining photolithography with PECVD, regular arrays of freestanding single carbon nanotubes (CNTs) were prepared on Ni dot arrays by dc plasma-enhanced chemical vapour deposition. The size of the Ni dot was reduced for single CNT growth by means of conventional photolithography and a lateral wet-etch process. By adjusting the wet etch time and the solution ratio, the lateral wet-etched region could be controlled so as to obtain the optimum size of Ni nanoparticles for single CNTs. Substrates consisted of p-type silicon wafers with resistivities of 3–6 Ω cm with a 30 nm Ni film deposited by a sputtering process. Photoresist patterns were obtained by a conventional photolithography process that included spin-coating, soft-baking, exposure, and development. Hard-baking at 105 °C enhanced the adhesion between the Ni film and the photoresist (PR) prior to the wet-etch process.

The wet-etch process for the thin Ni layer was carried out using a mixture of $[H_3PO_4]:[HNO_3]:[CH_3COOH]:[H_2O] = 2:2:2:4$. The lateral etch rate was about 10 nm min^{-1}. The etch rate could be controlled by adjusting the proportion of H_2O. By dissolving the photoresist in acetone, Ni nanopatterns were obtained. The planar Si substrate with Ni nanopatterns was placed on a cathode in dc PECVD at 550 °C with a base pressure below 10^{-6} Torr. Acetylene and ammonia, respectively, were used as the carbon source and etchant gases. The flow rates of NH_3 and C_2H_2 were 180 and 60 sccm during the growth of CNTs. A dc plasma was created by a negative bias of 450 V applied to the substrate. The vertical alignment of a CNT was directly dependent on the location of the catalyst metals. Using this method, well-separated, well-defined arrays of freestanding CNTs were fabricated. The process can be scaled up at a lower cost than electron beam lithography.

Dai[78] has reviewed carbon nanotubes prepared by CVD over patterned catalyst arrays. Controlled synthesis of nanotubes opens up exciting opportunities in nanoscience and nanotechnology, including electrical, mechanical, and electromechanical devices, chemical functionalization, surface chemistry and photochemistry, molecular sensors, and interfacing with soft biological systems.

Single-walled Nanotubes

Carbon nanotubes generally obtained by the arc method or hydrocarbon pyrolysis are multi-walled, having several graphitic sheets or layers (Figure 1.1). Depending on the exact technique, it is possible to selectively grow SWNTs or MWNTs. Two distinct methods of synthesis can be employed with the same arc discharge apparatus. If SWNTs are required, the anode has to be doped with a metal catalyst based on Fe, Co, Ni, or Mo. Several elements and mixtures of elements have been tested and the results vary considerably.[79]

Single-walled nanotubes were first prepared by metal-catalysed dc arcing of graphite rods in He.[4,5] The graphite anode was filled with metal powders (Fe, Co or Ni) and the cathode was of pure graphite. SWNTs generally occur in the web-like material deposited behind the cathode. Various metal catalysts have been used to make SWNTs by this route. Dai et al.[80] prepared SWNTs by the disproportionation of CO at 1200 °C over Mo particles of a few nanometre diameter dispersed in a fumed alumina matrix. Saito et al.[81] compared SWNTs produced by using different catalysts and found that a Co or a Fe/Ni bimetallic catalyst gives rise to tubes forming a highway-junction pattern. SWNTs have also been prepared by using various oxides, Y_2O_3, La_2O_3, CeO_2, as catalysts.[82] The arc discharge technique, though cheap and easy to implement, gives low yields of SWNTs. Journet et al.[83] obtained ~80% yield of SWNTs in the arc, by using a mixture of 1 at% Y and 4.2 at% Ni as catalyst. Arc evaporation of graphite rods filled with Ni and Y_2O_3 in He (660 Torr) gives rise to web-like deposits on the chamber walls near the cathode, consisting of SWNT bundles.[63] HREM images reveal bundles consisting of 10–50 SWNTs forming highway junctions (Figure 1.12a). The average diameter of the SWNTs is around 1.4 nm and the length extends up to 10 microns. The quantity and quality of the nanotubes depend on various parameters, such as

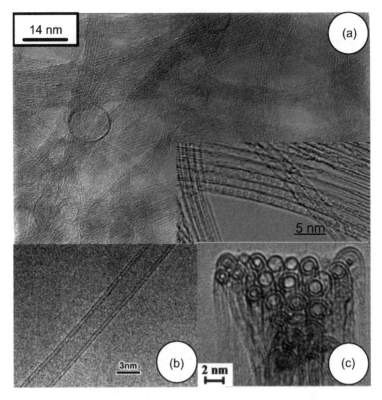

Figure 1.12 (a) *HREM image of SWNTs obtained by arcing graphite electrodes filled with Ni and Y_2O_3 under He (660 Torr). (b) and (c) HREM images of DWNTs. Inset in (a): HREM image of fullerenes encapsulated inside the SWNTs (peapods); scale bar is 5 nm*
(Part (a) reproduced from ref. 63 (inset reproduced from ref. 118b; parts (b) and (c) reproduced from ref. 123d)

the metal concentration, inert gas pressure, kind of gas, the current and system geometry. Some ways to improve the process of arc discharge are now discussed.

Inert Gas

The most common problem with SWNT synthesis is that the product contains metal catalyst particles and defects, rendering purification difficult. However, the diameter can be controlled by changing the thermal transfer and diffusion, and hence the condensation of atomic carbon. This has been demonstrated in an experiment where different mixtures of inert gases were used.[84] Argon, with a lower thermal conductivity and diffusion coefficient, gives SWNTs with a diameter of ~1.2 nm. A linear fit of the average nanotube diameter showed a 0.2 nm diameter decrease per 10% increase in argon:helium ratio, when nickel/yttrium was used (C:Ni:Y was 94.8:4.2:1) as catalyst.

Optical Plasma Control

Here, the anode to cathode distance is adjusted to obtain strong visible vortices around the cathode. This enhances anode vaporization and improves nanotube formation. Combined with suitable use of the argon–helium mixture, one can simultaneously control the macroscopic and microscopic parameters of the nanotubes formed.[85] With a nickel–yttrium catalyst (C:Ni:Y is 94.8:4.2:1) the optimum nanotube yield is found around 660 mbar for pure helium and 100 mbar for pure argon. The nanotube diameter ranges from 1.27 to 1.37 nm. Since CVD gives SWNTs with a diameter of 0.6–1.2 nm, different groups have used the same catalyst as in CVD in the arc discharge method. There seems to be a relation between the diameter of SWNTs synthesized by CVD and arc discharge. As a result, the diameter could be lowered to a range of 0.6–1.2 nm with arc-discharge, using a mixture of Co and Mo catalyst in high concentration. These diameters are considerably smaller than the 1.2–1.4 nm found normally.[79] Green and co-workers[86] have studied the correlation of the diameter of SWNTs, using the same catalyst by the CVD and the arc discharge methods.

Improvement of Oxidation Resistance

Some progress has been made in improving the oxidation resistance of the SWNTs. A strong oxidation resistance is needed if nanotubes are to be used for applications such as field emission displays. A modified arc-discharge method, using a bowl-like cathode, is found to decrease the defects and yield cleaner nanotubes, with improved oxidation resistance. The anode rod contained the Ni-and-Y catalyst.[87] Open-air synthesis of SWNTs has been carried out with a modified welding arc torch method.[88] A plate target made of graphite containing the Ni-Y catalyst (Ni:Y is 3.6:0.8 at%) was fixed at the sidewall of a water-cooled, steel-based electrode. The torch arc was aimed at the edge of the target and the soot was deposited on the substrate behind the target. The arc was operated at a dc of 100 A and shielding argon gas flowed through the torch, enhancing arc jet formation beyond the target. In the soot, carbon nanohorns and bundles of SWNTs with an average diameter of 1.32 nm were found. However, the yield was lower than in the conventional low-pressure arc discharge method. This is because the lighter soot escapes into the atmosphere in the open air synthesis. Secondly, the carbon vapour might be oxidized and emitted as carbon dioxide.

Laser Vapourization

Smalley's group reported the synthesis of carbon nanotubes by laser vapourization in 1995.[89] The laser vapourization apparatus used by the Smalley group is shown in Figure 1.13(a). A pulsed[90,91] or continuous[92] laser is used to vapourize a graphite target in an oven at 1200 °C. The oven is filled with helium or argon gas to keep the pressure at 500 Torr. A hot vapour plume forms, then expands and cools rapidly. As the vapourized species cool, small carbon molecules and atoms quickly condense to form larger clusters, possibly including fullerenes. The catalysts also

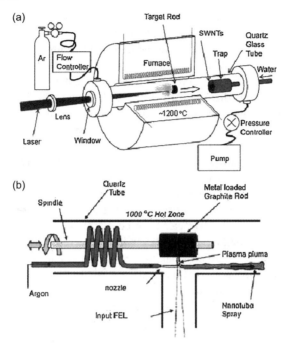

Figure 1.13 *Schematic drawings of* (a) *a Smalley-type laser ablation apparatus* (b) *Ultra-fast-pulsed laser ablation apparatus*
(Part (a) reproduced from refs. 32b and 89; Part (b) reproduced from ref. 91)

begin to condense, but more slowly at first, and attach to carbon clusters and prevent their closing into cage structures.[93] Catalysts may even open cage structures when they attach to them. From these initial clusters, tubular molecules grow into single-wall carbon nanotubes until the catalyst particles become too large, or until conditions have cooled sufficiently that carbon no longer can diffuse through or over the surface of the catalyst particles. It is also possible that the particles become so coated with a carbon layer that they cannot absorb more and the nanotube stops growing. The SWNTs formed are bundled due to van der Waals forces. SWNTs have been produced in more than 70% yield by the condensation of a laser-vaporized carbon–nickel–cobalt mixture at 1200 °C.[94]

There are some similarities in the spectral emission of the excited species in the laser ablation of a composite graphite target and in the laser-irradiated C_{60} vapour. This suggests that fullerenes may also be produced by laser ablation of catalyst-filled graphite, just as when no catalysts are present in the target. However, subsequent laser pulses excite the fullerenes to emit C_2 that adsorbs on the catalyst particles and feeds SWNT growth. There is not sufficient evidence to support this conclusion. Laser ablation is similar to arc discharge, since the optimum background gas and the catalyst mix are the same in the two processes.

Condensates obtained by laser ablation are contaminated with MWNTs and carbon nanoparticles. With pure graphite electrodes, MWNTs are formed, but uniform SWNTs are obtained if a mixture of graphite with Co, Ni, Fe or Y is used instead of pure graphite. SWNTs so-synthesized exist as ropes. The Ni-Y mixture catalyst (Ni:Y is 4.2:1) appears to give the best yield. With a Ni-Co catalyst and a pulsed laser at 1470 °C, SWNTs with a diameter of 1.3–1.4 nm are obtained. When a continuous laser is used at 1200 °C with the Ni:Y catalyst (Ni:Y is 2:0.5 at%), SWNTs with an average diameter of 1.4 nm are formed in 20–30% yield. Because of the good quality of nanotubes produced, efforts have been made to scale up laser ablation. However, the results are not yet as good as from the arc-discharge method. Laser vaporization results in a higher yield for SWNTs and the nanotubes have better properties and a narrower size distribution than the SWNTs produced by arc-discharge. Nanotubes produced by laser ablation have greater purity (up to *ca.* 90% purity) than those formed by the arc discharge process.

Ultrafast pulses from a free electron laser (FEL; pulse width is ~400 fs) with a pulse repetition rate of 75 MHz can be used effectively for SWNT synthesis.[91] Here, the intensity of the laser bundle behind the lens reaches ~5×10^{11} W:cm^{-2}, which is about 1000 times greater than in the Nd:YAG systems. A jet of preheated (1000 °C) argon through a nozzle tip is situated close to the rotating graphite target, which contains the catalyst. The argon deflects the ablation plume approximately 90° away from the incident FEL beam direction, clearing away the carbon vapour from the region in front of the target. The SWNT soot is collected in a cold finger. This process can be seen in Figure 1.13(b). The yield is 1.5 g h^{-1} at 20% of the maximum power. With this method, the maximum achievable yield with current lasers is 45 g h^{-1} (with a NiCo or NiY catalyst) in an argon atmosphere at 1000 °C at a wavelength of ~3000 nm. The SWNTs are in bundles. The number of tubes within a typical bundle varies from 8–200, with a length of 5–20 microns and a diameter of 1–1.4 nm.

A continuous-wave laser-powder method has also been employed for SWNT synthesis, based on the laser ablation of mixed graphite and metallic catalyst powders by a 2 kW continuous wave CO_2 laser in an argon or nitrogen stream.[92b] Because of the use of micron-size particles, thermal conductivity losses are significantly decreased compared with laser heating of the bulk solid targets in known laser techniques. As a result, the absorbed laser power is more effectively used for material evaporation. The yield in this technique is 5 g h^{-1} with a Ni/Co catalyst (Ni/Co is 1:1) at 1100 °C. The yield in SWNTs is 20–40%, with a mean diameter of 1.2–1.3 nm.

Arepalli[95] has reviewed progress in the laser ablation process for SWNT production. Different types of lasers are now routinely used to prepare single-walled carbon nanotubes. The original method developed by the Smalley group used a double-pulse laser oven process. Several researchers have used variations of the lasers to include one-laser pulse (green or infrared), different pulse widths (ns to ms as well as continuous wave), and different laser wavelengths (*e.g.*, CO_2, or free electron lasers in the near-to far-infrared). Some of these variations have been used with different combinations and concentrations of metal catalysts, buffer gases (*e.g.*, helium), oven temperatures, flow conditions, and different porosities of the graphite targets.

Pyrolysis or Vapour Phase Deposition

Under controlled conditions of pyrolysis, dilute hydrocarbon–organometallic mixtures yield SWNTs.[63,96a] Pyrolysis of a metallocene–acetylene mixtures at 1100 °C yields SWNTs,[96a] shown in the TEM image in Figure 1.6(c). The diameter of the SWNT in Figure 1.6(c) is 1.4 nm. Figure 1.6(d) shows the SWNTs obtained similarly by the pyrolysis of ferrocene–CH_4 at 1100 °C. It may be recalled that the pyrolysis of nickelocene in an admixture with benzene under similar conditions primarily yields MWNTs. The bottom portion of the SWNT in Figure 1.6(c) shows an amorphous carbon coating around the tube, common with such preparations. This can be avoided by reducing the proportion of the hydrocarbon C_2H_2 and mixing hydrogen in the Ar stream. Pyrolysis of binary mixtures of metallocenes and acetylene also gives SWNTs, due to the beneficial effect of such binary alloys.[96b] Pyrolysis of acetylene in mixture with $Fe(CO)_5$ at 1100 °C gives good yields of SWNTs. Pyrolysis of ferrocene–thiophene mixtures also affords SWNTs, but the yield appears to be somewhat low. Pyrolysis of benzene and thiophene along with ferrocene gives a high yield of SWNTs.[97]

Chemical Vapour Deposition (CVD)

Compared with arc and laser methods, the main advantage of CVD is that it is a straightforward way to scale up production to industrial levels. Colomer *et al.*[98] obtained SWNTs in high yield by the decomposition of methane over transition metal supported MgO substrates. Flahaut *et al.*[99] have synthesized SWNTs by passing a H_2–CH_4 mixture over transition metal containing oxide spinels, obtained by the combustion route. The quality of SWNTs has been characterized on the basis of adsorption measurements.

Zeolites containing 1D channels have been employed to synthesize monosized SWNTs.[100] Endo *et al.*[101] have described a possible route for the large-scale synthesis of SWNTs by a combination of a substrate (template) and floating catalyst. The template (nanosize zeolites) has exposed pores that act as anchoring sites for metal nanoparticles, prohibiting the metal particle aggregation. Such optimized seeding results in high purity SWNTs *via* suppression of impurity formation such as MWNTs and carbon nanoparticles. The three-dimensionally floated template in the reaction chamber makes it possible to produce nanotubes in the semicontinuous system, with a large variations in texture and a wide range of diameters (0.4–4 nm).

Hata *et al.*[102] have demonstrated an efficient CVD synthesis of SWNTs where the activity and lifetime of the catalysts are enhanced by water. Water-stimulated enhanced catalytic activity results in massive growth of superdense and vertically aligned nanotube forests, up to 2.5 mm high, that can be easily separated from the catalysts, providing nanotube material with carbon purity above 99.98%. Moreover, patterned, organized intrinsic nanotube structures are successfully fabricated. The water-assisted synthesis addresses some of the problems faced in carbon nanotube synthesis. Currently, complete control of selectively growing metallic or semiconducting SWNTs has not been achieved. More recently, preferential growth of semiconducting SWNTs (>85%) has been observed in a plasma-enhanced CVD (PECVD) process at 600 °C.[103] Preferential growth combined with postgrowth treatment and separation may possibly solve the problem of obtaining metallic *vs.* semiconducting nanotubes.

Alcohol Catalytic Chemical Vapour Deposition

Alcohol catalytic CVD has been developed for possible of large-scale production of high quality SWNTs at low cost. In this technique, evaporated alcohols (methanol and ethanol) are passed over iron and cobalt catalyst particles supported on zeolites at the relatively low temperature of 550 °C. It seems that hydroxyl radicals of the alcohol eliminate carbon atoms with dangling bonds, which are obstacles in forming high-purity SWNTs. The diameter of the SWNTs obtained by this method is ~1 nm.[104] The low reaction temperature and high-purity of the product suggest possibility of upscaling production. Furthermore, the low reaction temperature ensures that the technique is applicable for the direct growth of SWNTs on semiconductor devices patterned with aluminium.

Aerogel-supported Chemical Vapour Deposition

In this method, SWNTs are synthesized by the decomposition of carbon monoxide on an aerogel-supported Fe/Mo catalyst around 860 °C. Several factors affect the yield and quality of SWNTs, including the surface area of the supporting material, reaction temperature and feed gas. Because of the high surface area, porosity and ultralight density of the aero gels, the productivity of the catalyst is much higher than in other methods.[105] After a simple acid treatment and oxidation, high purity (>99%) SWNTs are obtained. When using CO as the feed gas, the yield of the nanotubes is low but the overall purity of the material is good. The diameter distribution of the nanotubes is between 1.0 and 1.5 nm.[106]

Laser-assisted Thermal Chemical Vapour Deposition

In the laser-assisted thermal CVD, a medium-power, continuous wave CO_2 laser, which is perpendicularly directed onto a substrate, pyrolyses sensitized mixtures of $Fe(CO)_5$ vapour and acetylene in a flow reactor. Carbon nanotubes are formed by the catalysing action of the small iron particles.[107] By using a reactant gas mixture of iron pentacarbonyl vapour, ethylene and acetylene, both single- and multi-walled carbon nanotubes are produced. Silica is used as the substrate. Diameters of the SWNTs range from 0.7 to 2.5 nm. The diameter range of the MWNTs is 30–80 nm.

CoMoCat Process

Here, SWNTs are grown by the disproportionation of CO at 700–950 °C.[108] This technique was developed at the University of Oklahoma and it is being commercialized by SouthWest Nanotechnologies (SweNT) Inc. The technique is based on a Co-Mo catalyst formulation that inhibits the sintering of Co particles, thereby inhibiting the formation of undesired forms of carbon (Figure 1.14(a)). During the reaction, cobalt is progressively reduced from the oxidic state to the metallic form. Simultaneously, molybdenum is converted into the carbidic form (Mo_2C). Cobalt acts as the active species in the activation of CO while the role of the Mo may be dual. It may stabilize Co as well-dispersed Co^{2+}, avoiding its reduction, and act as a carbon sink to moderate the growth of carbon, inhibiting the formation of undesirable forms of carbon. Critically, for an effective reactor operation, the space velocity has to be high enough to keep the CO conversion as low as possible. The advantage of fluidized bed reactors is that they permit continuous addition and removal of solid particles from the reactor without stopping the operation. The method can be scaled-up without loss in SWNT quality. By varying the operative conditions, SWNTs of different diameters can be produced. The CoMoCat catalyst has a high selectivity towards SWNTs, namely 80–90%.

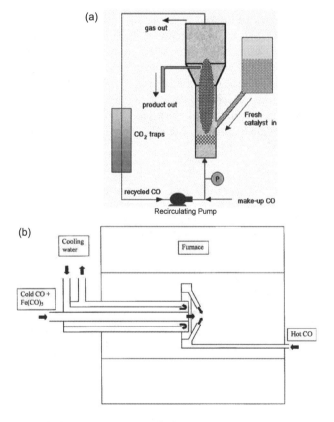

Figure 1.14 (a) *Schematic diagram of a CoMoCat apparatus* (b) *Layout of a CO flow-tube reactor, showing water-cooled injector and 'showerhead' mixer*
(Part (a) reproduced from ref. 108; part (b) reproduced from ref. 109)

High-pressure CO Disproportionation Process

Nikolaev *et al.*[109] obtained SWNTs using a gas-phase catalytic method involving the pyrolysis of $Fe(CO)_5$ and CO. Decomposition of CO on a silica-supported Co-Mo catalyst also yields SWNTs.[110] The improvised technique "High pressure CO" disproportionation process (HiPco) is a technique for catalytic production of SWNTs in a continuous-flow gas phase using CO as the carbon feedstock and $Fe(CO)_5$ as the iron-containing catalyst precursor. SWNTs are produced by flowing CO mixed with a small amount of $Fe(CO)_5$, through a heated reactor. Figure 1.14(b) shows the layout of CO flow-tube reactor. The size and diameter distribution of the nanotubes can be roughly selected by controlling the pressure of CO. This process is promising for bulk production of carbon nanotubes.[109] Nanotubes with a diameter as small as 0.7 nm, considered to be the smallest achievable chemically stable SWNTs, have been produced by this method.[111] The average diameter of HiPco SWNTs is ~1.1 nm. The yield that could be achieved is around 70%. SWNTs material with 97% purity can be produced at up to 450 mg h^{-1}.[112] Nikolaev[113] has recently conducted a parametric study on the gas-phase production of SWNTs by the HiPco process. This latest process for producing large quantities of single-walled carbon nanotubes is living up to its promise.[112] In the HiPco process, nanotubes grow in high-pressure, high-temperature flowing CO on catalytic clusters of iron. The catalyst is formed *in situ* by the thermal decomposition of $Fe(CO)_5$, which is delivered intact within a cold CO flow and then rapidly mixed with hot CO in the reaction zone. Upon heating, the $Fe(CO)_5$ decomposes into atoms, which condense into larger clusters. SWNTs nucleate and grow on these particles in the gas phase *via* CO disproportionation (Boudouard reaction) catalysed by the Fe surface. The rate of the Boudouard reaction scales as a square of CO pressure, hence the importance of high-pressure for efficient SWNT production.

$$2CO \xrightarrow{Fe} CO_2 + C \text{ (SWNT)}$$

The concentration of CO_2 produced in this reaction is equal to that of carbon and can, therefore, serve as a useful real-time feedback parameter. Present production rates approach 450 mg h^{-1} (or 10 g day^{-1}), and nanotubes typically have no more than 7 mol% of iron impurities. The second-generation HiPco apparatus can run continuously for 7–10 days. The HiPco process is an attractive alternative gas-phase process that does not use premade catalyst particles, like other CVD techniques. To improve the yield and to avoid the release of CO exhaust into the atmosphere, a closed-loop reactor has been developed. Here, CO is recirculated through the system using a compressor rather than being released into the atmosphere.

Flame Synthesis

This method is based on the use of a controlled flame environment, wherein carbon atoms are formed from inexpensive hydrocarbon fuels along with aerosols of metal catalysts.[114] The SWNTs grow on metal islands in the same manner as in laser ablation and arc discharge. A submonolayer film (<10 nm) of metal (cobalt) catalyst was applied to the stainless steel mesh (SS-304) by physical vapour deposition (PVD). In this manner, metal islands resembling droplets form upon the mesh support to serve as catalyst particles. These small islands become the aerosol when exposed to a flame. Another way is to create aerosol type metal particles by burning a filter paper that is rinsed with a solution of metal ions (*e.g.*, iron nitrate).[115] A thermal evaporation technique in which the metal powder (*e.g.*, Fe or Ni) is inserted in a trough and heated can also be used. A fuel gas is partially burnt under controlled conditions to attain a temperature of ~800 °C to form the carbon atoms for the production of SWNTs. The optimization parameters are the fuel gas composition, the catalyst, catalyst carrier surface and temperature. The technique developed by Laplaze *et al.*[116] uses concentrated solar energy to vaporize graphite to synthesize SWNTs.

Sonochemical Route

Jeong *et al.*[117] have described a sonochemical route for producing high-purity SWNTs in liquid solution under ambient conditions. This method for SWNT growth begins, typically, with the preparation of a mixture solution of ferrocene and *p*-xylene. Silica powder is then added to this solution. Ultrasonication is performed under ambient conditions to grow high purity SWNTs. Finally, the resulting SWNTs powder is collected on a filter membrane. Ferrocene acts as the precursor of the Fe catalyst for nanotube growth. The carbon source for SWNT growth is provided by *p*-xylene as well as by ferrocene, while silica powder provides the nucleation site for SWNT growth. The average diameter of the SWNTs produced was in the range of 1.3–1.8 nm. This method also presents a facile route to high-purity SWNTs without complex purification processes.

Peapods and Double-walled Nanotubes

Smith *et al.*[118a] discovered fullerenes encased in carbon single-walled nanotubes (SWNTs) in acid-purified nanotubes prepared by the laser oven method[94] (inset of Figure 1.12a). This material is called a *peapod*, because its structure resembles miniature peas in a pod. The fullerenes of the peapod are arranged one-dimensionally with a constant intermolecular distance, which is slightly smaller (by 3%) than the nearest molecular distance of solid C_{60}. C_{60} peapods can be synthesized with high yields by the vapour-phase reaction of C_{60} with an acid-purified SWNT bundle.[119a] The reaction is carried out at 400 °C in a vacuum-sealed glass ampoule, the reaction time being normally a day or more to fill the tube with C_{60} (reaching a yield of almost 100%). Prior to taking the samples in the glass ampoule, the acid-purified SWNTs have to be heated in dry air at 420–480 °C for 1 h to burn off contaminated carbon. This burning also guarantees open-ended tubes. The burning temperature depends on the cleanness of the tube surface. That is, if the rinsing step in the purification procedure for SWNTs is not sufficient to remove impurities, the burning temperature must be increased. Intact C_{60} molecules, thus arranged in the nanotube, start to merge above ~800 °C in vacuum, and they finally form an inner tube. According to a TEM study of the merging process of C_{60} in the intermediate 800–1000 °C, the internal tubes form as the result of a structural relaxation at the "bottleneck" area of the C_{60} dimers, trimers, tetramers, and so on. Once the fullerenes are in the nanotube space, they become stabilized, and it is rather difficult to remove them. Electronic structure calculations show that an energy of ~0.5 eV would be gained by peapod formation.[120] The encased C_{60} molecules are energetically stable and the de-doping of fullerenes is difficult and may not occur by heating. Metallofullerenes (endohedral fullerenes) have also been encapsulated in SWNTs and their structures investigated by TEM and other techniques.[119b]

Double-walled carbon nanotubes (DWNTs) are obtained by the arc discharge technique in an of Ar–H_2 mixture (1:1,v/v) at 350 Torr.[121] The catalyst is prepared from a mixture of Ni, Co, Fe and S powders heated in an inert gas atmosphere at 500 °C for 1 h. The predominantly obtained DWNTs bundles have an outer diameter in the range 1.9–5 nm and inner tube diameters in the range 1.1–4.2 nm. High-quality DWNTs are obtained by the high-temperature pulsed arc discharge method with Y/Ni alloy catalysts at 1250 °C.[122] Optimum conditions for DWNT synthesis are almost the same as those of carbon SWNTs, indicating that the growth processes of DWNTs and SWNTs are closely related. Carbon DWNTs have been prepared quantitatively from the chains of C_{60} molecules generated inside carbon SWNTs.[119a] Heating at ~1200 °C induces the coalescence between C_{60} and, eventually, the C_{60} molecules transform into a tubular structure. Bandow *et al.* [119a] have reviewed the formation pathway to DWNTs from C_{60} encased within single-walled carbon nanotubes (peapods). High-quality carbon DWNTs have been produced by catalytic CVD of carbon-containing molecules such as propanol, THF, hexane, benzene, alcohol around 900–1000 °C over Fe-Mo embedded MgO or Al_2O_3 support material.[123a–c] Figure 1.12b and 1.12c show TEM images of an isolated and of a bundle of DWNTs (77% yield), respectively, synthesized by the CVD of methane over a $Mg_{1-x}Co_xO$ solid solution containing Mo oxide.[123d] The diameter of DWNTs so prepared is dependent on the process conditions and the carbon source. The DWNTs have inner tube diameters in the range 0.6–3.4 nm and outer tube diameters in the 1.4–4.5 nm range. The interlayer spacing between graphene layers ranges from 0.35 to 0.38 nm.

Mechanism of Formation

Several growth models have been proposed for the carbon nanotubes prepared by the pyrolysis of hydrocarbons on metal surfaces. Baker and Harris[124] suggested a four-step mechanism. In the first step, the hydrocarbon decomposes on the metal surface to release hydrogen and carbon, which dissolves in the particle. The second step involves diffusion of the carbon through the metal particle and its precipitation on the rear face to form the body of the filament. The supply of carbon onto the front face is faster than the diffusion through the bulk, causing an accumulation of carbon on the front face, which must be removed to prevent the physical blocking of the active surface. This is achieved by surface diffusion and the carbon forms a skin around the main filament body, in step three. In the fourth step, overcoating and deactivation of the catalyst and termination of tube growth takes place. Oberlin *et al.*[125] proposed a mechanism where bulk diffusion is insignificant and carbon is transported around the particle by surface diffusion. Dai *et al.*[80] proposed a mechanism wherein carbon forms a hemispherical graphene cap (yarmulke) on the catalyst particle and the nanotubes grow from the cap. The diameter of the nanotube is controlled by the size of the catalytic particle, nanometre-size particles yielding SWNTs. A crucial feature of this model is that it avoids dangling bonds at all stages of growth. SWNTs produced by arc vaporization may also be formed by the yarmulke mechanism.

Endo and Kroto,[126] based on the observation of C_2 ejection from C_{60} in mass spectrometry, suggested that the formation of MWNTs involved the formation of fullerenes. Smalley,[127] however, pointed out that only the growth of outer layers of multi-walled tubes would be permitted by such a mechanism. Iijima *et al.*[9] presented electron microscopy evidence for the open-ended growth of carbon nanotubes and suggested that the termination of incomplete layers of carbon seen on the tube surface may arise because of the extension and thickening of the nanotubes by the growth of graphite islands on the surfaces of existing tubes. The nucleation of pentagons and heptagons on the open tube ends results in a change in the direction of the growing tube and some novel morphologies, including one where the tube turns around 180° during growth, have been observed. The growth is self-similar and fractal-like with the inner tubules telescoping out of the larger ones, with logarithmic scaling of the size.

Isotope scrambling experiments of Ebbesen *et al.*[128] showed that, under the conditions of fullerene formation, the plasma has vaporized carbon atoms. Based on tube morphologies, a mechanism similar to that of Saito *et al.*,[129] wherein the carbonaceous material reaching the cathode anneals into polyhedral particles, was suggested. Given the right conditions, the tip might open and continue to grow. Such a growth could occur inwards from outside. Ebbesen *et al.* suggested the possibility of tubes forming directly from the closing of a large graphene sheet. Such a suggestion gains credence from the simulations of Robertson *et al.*[130] who examined the curling and closure of small graphitic ribbons. These workers found that the formation of cylinders is favoured by both entropy and enthalpy. This could possibly, serve to nucleate the growth of multilayers by cladding as in the mechanism of Iijima *et al.*

Amelinckx *et al.*[131] employ the concept of a spatial velocity hodograph to describe the extrusion of a carbon tubule from a catalytic particle. The model is consistent with the observed tubule shapes and explains how spontaneous plastic deformation of a tubule can occur. These workers propose a model in which the graphene sheets can form both concentric cylinders and scroll-type structures. The nanotubes nucleate from a large fullerene type dome. For SWNTs produced by the arc method, two types of catalytic growth are observed: (a) one wherein many nanotubes grow radially from a single catalyst particle that is much larger than the nanotube diameter and (b) another where the metal particles are of the same size or even smaller than the tube diameter and a single tube grows from a metal particle. SWNT growth by mode (b) can be explained by the yarmulke mechanism[80] mentioned earlier. Based on molecular dynamics and total energy calculations, Maiti *et al.*[132] propose a model for mode (a) growth wherein nanometre-sized protrusions on the metal particle surface lead to the nucleation of SWNTs. The metal particle is capped by a graphene sheet and nanometre size protrusions on the metal surface leave holes in the graphene sheet. Carbon atoms are added to the graphene sheet in the form of handles between a pair of nearest neighbour carbon atoms. These handles migrate along the graphene sheet till they reach the edge of the initial hole, where they form seven-membered rings to give rise to a geometry curving upwards. The seven-membered rings get confined at this high curvature region due to energetics and the hexagons grow further to form the body of the tube.

TEM examination of the carbonaceous products obtained in the pyrolysis of hydrocarbons and organometallic precursors suggests that the size of the catalyst particle plays a role, with regard to the nature of the product formed. The role of the size of the catalyst particle on the nature of the carbon nanostructure formed is shown schematically in Figure 1.15. This should not be taken as a rule since nanotubes of diameters >50 nm do occur. If the catalytic metal nanoparticle is around 1 nm in diameter, SWNTs are the predominant product.[63,96] The report of Dai *et al.*[80] that SWNTs are formed on Mo nanoparticles in the size range of

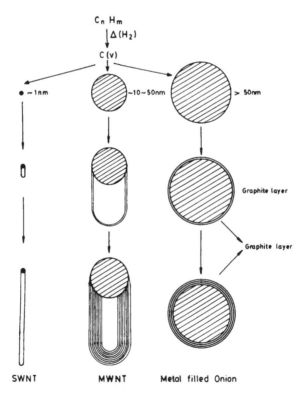

Figure 1.15 *Schematic showing the dependence of the carbon nanostructure obtained by hydrocarbon pyrolysis on the size of the metal nanoparticles*
(Reproduced from ref. 20a)

1–4 nm indicates that there can be a slight range in the size. With organometallic precursors, it seems that metal nanoclusters of ∼1 nm diameter are produced under controlled conditions. When the organometallic concentration is high, MWNTs are formed around the catalyst particles with sizes in the range of 5–20 nm. This is true of carbon nanotubes obtained by the metallocene route.[51,52] In the higher size range of ≥50 nm, mainly graphite-covered metal particles are formed.[63]

Moisala *et al.*[133] have reviewed the progress in CVD and aerosol synthesis of SWNTs with an emphasis on the role of catalytic metal nanoparticles. They discuss the evolution of the catalyst particle size distribution due to collision, sintering and evaporation of the metal during SWNT synthesis. The active catalyst has been demonstrated to be in a reduced metal form based on the experimental data and calculations on the equilibrium concentrations of carbon and oxygen in iron. The effect of the catalyst particle size on the melting temperature and on carbon solubility in the metal is discussed. Furthermore, the maximum length of SWNTs of 1 and 2.5 nm diameter as a function of carbon solubility has been estimated to determine whether carbon dissolution and precipitation are simultaneous or subsequent process steps. The form of the precipitated carbon is dictated by the catalyst particle size, carbon feed rate and reactor conditions. Also, the carbon precipitation rate affects the form of precipitated carbon in aerosol synthesis. In CVD, interactions between the catalyst particle and the support surface groups may be utilized in controlling the evolution of the particle size distribution. In aerosol synthesis, the velocity and temperature gradients in the flow reactor result in a distribution of particle sizes. The feed rate of carbon to the metal particles has to be controlled to selectively produce SWNTs and to avoid accumulation of amorphous carbon.

Purification and Separation

As-synthesized carbon nanotubes contain carbonaceous impurities, typically amorphous carbon and graphitic nanoparticles, which may enclose the transition-metal catalyst in the case of SWNTs. A simple and successful strategy for the purification of arc-grown MWNTs is high-temperature oxidation in air. Early attempts to apply such a method to the purification of SWNTs, however, failed. It is now clear that the problem lies with the catalyst metal particles present in the material. In the presence of oxygen and other oxidizing gases, the metal particles catalyse the low-temperature oxidation of carbons indiscriminately, destroying the SWNTs. The first-step before gas-phase oxidative purification is, therefore, the removal of the metal catalyst particles.

Most nanotube purification methods involve one or more of the following steps: gas- or vapour-phase oxidation, wet chemical oxidation/treatment, centrifugation, or filtration (including chromatographic methods). A combination of these steps is employed to achieve complete purification. Strong oxidation and acid refluxing have an effect on the structure of the tubes. Wet chemical purification methods (*e.g.* acid digestion of impurities) give rise reaction products on the surface of a nanotube that render it a natural surfactant. While the removal of the carbonaceous and catalyst impurities is necessary in purification, it is not sufficient. Amorphous carbon is readily burnt away by heating the nanotubes in air around 300 °C. Hydrothermal treatment can be used to remove the amorphous carbon from the nanotube surface.[134,135] Bandow *et al.*[136] have used microfiltration to rid SWNTs of the other contaminants. Martinez *et al.*[137] used a combination of air oxidation and microwave acid digestion to purify arc-discharge SWNTs. High-temperature annealing of the purified samples is carried out in some procedures to remove the surface functional groups.[138,139]

The need for the removal of metals was first pointed out by Chiang *et al.*[140,141] Their method begins with long (18 h), low-temperature (225 °C) oxidative cracking of the carbonaceous shells encapsulating the metal particles. For this purpose, wet oxygen (20% O_2 in argon passed through a water-filled bubbler) was passed over the nanotubes in the hot zone of a flow tube furnace. After this, the material was stirred in concentrated HCl to dissolve iron particles. After filtering the acid and drying, the oxidation and acid extraction cycles were repeated once more at 325 °C, followed by an oxidative bake at 425 °C. The 325 °C step may, however, be unnecessary.

Sen *et al.*[142] have examined the effect of oxidation conditions on the purity of the as-prepared SWNTs films prepared by the arc-discharge method. Purification was monitored by near-infrared spectroscopy. This technique also gives an estimate of the change in relative amounts of SWNTs and amorphous carbon impurities after each purification step, thereby providing control over the oxidation process. Oxidation was carried out by reaction in flowing oxygen at different temperatures and for different durations and the purity evaluated after each oxidation step. Optimization of the process allowed a three-fold improvement in purity with the quantitative retention of the SWNT portion of the sample. A purification efficiency is defined, which is expressed as the ratio of the fraction of SWNTs retained to the fraction of impurities retained in the sample after application of the purification process.

Vivekchand *et al.*[143] have developed a purification procedure involving high temperature hydrogen heat-treatment of SWNTs or by etching the appropriate acid treated carbon nanotubes with hydrogen at elevated temperatures [700–1000 °C]. The method has been effective in purifying both SWNTs and MWNTs. While acid washing dissolves the metal particles, the hydrogen treatment removes amorphous carbon as well as the carbon coating on the metal nanoparticles. Thus it is possible to eliminate amorphous carbon and other carbonaceous materials in the SWNTs by this technique without the use of microfiltration *etc.* This work needs to be compared with the other procedures reported in the literature. All the methods make use of acid washing to remove the metal particles present. In procedures involving air oxidation, SWNTs are subjected to heat treatment in the 350–500 °C range, depending on the method of synthesis. The method of Vivekchand *et al.*, however, employs hydrogen treatment around 1000 °C for most SWNTs and MWNTs, except for HiPco SWNTs which require a lower temperature. Whereas in air oxidation the amorphous carbon is converted into CO_2, it is converted into CH_4 on hydrogen treatment. Electron microscope images corresponding to the hydrogen purification method are shown in Figure 1.16. The TEM image in Figure 1.16(a) shows the as-synthesized arc SWNTs bundles (13 nm diametre) containing

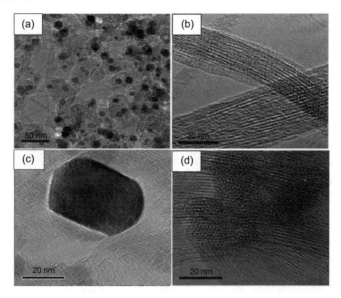

Figure 1.16 *TEM images of* (a) *as-synthesized SWNTs obtained by the arc-discharge method,* (b) *after first acid treatment,* (c) *after first hydrogen treatment and before second acid treatment and* (d) *after the second hydrogen treatment. After such purification only do we get proper electronic spectra of SWNTs* (Reproduced from ref. 143)

amorphous carbon and metal nanoparticles. Most of the metal nanoparticles are dissolved on acid washing, but the nanotubes are covered with amorphous carbon (TEM image in Figure 1.16b). The amorphous carbon is removed by high-temperature hydrogen treatment and the remaining small metal nanoparticles agglomerate into larger particles. The absence of amorphous carbon is revealed by the TEM image in Figure 1.16(c). TEM images also suggest that the bundles grow and have diameters in the range of 20–50 nm after the hydrogen treatment. The metal nanoparticles agglomerate in the 750–850 °C range (Figure 1.16c) and are removed in the second, shorter duration, acid treatment. The second acid treatment is followed by the high-temperature hydrogen treatment at 1000 °C to obtain pure SWNTs. A TEM image of such purified SWNTs is shown in Figure 1.16(d). We do not see hollow onion-like structures often found in SWNTs purified by other methods.

Following purification of the SWNTs to remove foreign materials such as catalyst, amorphous carbon, and carbon-coated nanoparticles the sorting of the SWNTs according to their length becomes particularly important in light of their potential applications. Since SWNTs occur as large bundles with micron lengths, it is desirable to break them from the bundles, for purposes of further manipulations. Liu *et al.*[144] subjected SWNTs to ultrasound treatment in an acidic medium to break-up the bundles into open-ended small fragments 100–300 nm long. The smaller fragments were functionalized. Various techniques have been employed for obtaining SWNTs sorted by length.[145–149] Chromatographic techniques are useful for the length fractionation of shortened SWNTs with sizes of less than 300 nm.[145a,146,147,149] For longer SWNTs, techniques such as capillary electrophoresis[146] and field-flow fractionation appear to be more applicable.[148]

The stochastic nature of SWNT growth generates mixtures of metallic and semiconducting SWNTs in a 1:2 ratio, packed together in bundles, and their separation is difficult.[109] Several recent reports, however, provide evidence for such separation. By exploring the complex physicochemical interactions of surfactant amines with carbon nanotubes, the bulk separation of metallic from the semiconducting SWNTs was achieved by using the enhanced chemical affinity of the semiconducting SWNTs to octadecylamine [ODA, $CH_3(CH_2)_{17}NH_2$].[150] Spectroscopic and thermogravimetric studies indicate that the amine interacts more strongly with the semiconducting than with the metallic SWNTs. This reduces the tendency of semiconducting SWNTs to aggregate as the concentration is increased. Based on these observations, as well as by utilizing resonance Raman spectroscopy, Samsonidze *et al.*[151] has developed octadecylamine-assisted analytical method to provide quantitative evaluation of the separation efficiency between semiconducting and metallic SWNTs. The measured relative changes in the integrated intensities of the radial-breathing mode region in the Raman spectrum provide a quantitative evaluation of the separation efficiency between semiconducting and metallic SWNTs along with diameter separation. They found this separation technique provides an enrichment of semiconducting nanotubes by a factor of 5 for SWNTs prepared by high pressure CO decomposition and higher enrichment of semiconducting SWNTs with diameters below 1 nm.

Metallic/semiconducting SWNT separation by exposing an aqueous suspension of SWNTs in Triton X-100 surfactant to bromine solution, followed by centrifugation at 24000 *g* for 12 h, leads to enrichment of the supernatant in semiconducting nanotubes and the sediment in metallic nanotubes.[152] This separation relies on the important finding that *individual* nanotubes in aqueous surfactant suspension remain suspended during high-speed centrifugation, despite the fact that the specific gravities of the SWNTs are 20–40% greater than that of water.[153] As indicated by modelling, the micelles have a specific gravity that is lower than unity (ascribed to the exclusion of water from the region of the hydrophobic tails contacting the nanotubes), so that the combined object (SWNT plus micelle) acquires a specific gravity of close to 1.

Krupke *et al.*[154] have developed a method to separate metallic from semiconducting SWNTs from suspensions using alternating current dielectrophoresis. For the ac dielectrophoresis, the generator was operated at a frequency (*f*) of 10 MHz and a peak-to-peak voltage (*V*p-p) of 10 V range. The method utilizes the difference in relative dielectric constants of the two species with respect to the solvent, resulting in an opposite movement of metallic and semiconducting tubes along the electric field gradient. Metallic tubes are attracted toward a microelectrode array, leaving semiconducting tubes in the solvent. Proof of the effectiveness of separation is given by a comparative Raman spectroscopy study on the dielectrophoretically deposited tubes and on a reference sample.

Strano *et al.*[155] have reported a selective reaction pathway of carbon nanotubes in which covalent chemical functionalization is controlled by differences in the nanotube electronic structure. They demonstrate the utility of these chemical pathways for manipulation of nanotubes of distinct electronic types by the selective functionalization of metallic nanotubes. They show that diazonium reagents functionalize single-walled carbon nanotubes suspended in aqueous solution with high selectivity and enable manipulation according to electronic structure. For example, metallic nanotubes react to the exclusion of semiconducting nanotubes under controlled conditions. Selectivity is dictated by the availability of electrons near the Fermi level to stabilize a charge-transfer transition state preceding bond formation. The chemistry is reversed by thermal treatment that restores the pristine electronic structure of the nanotubes.

Zheng *et al.*[156] report that single-stranded DNA (ssDNA) interacts strongly with CNTs to form a stable DNA-CNT hybrid that effectively disperses CNTs in aqueous solution. They also demonstrate that DNA-coated carbon nanotubes can be separated into fractions with different electronic structures by ion-exchange chromatography. They have not discussed whether DNA wrapping on a CNT depends on the specific sequence of the DNA strand. Zheng *et al.*[157] later showed that an oligonucleotide sequence self-assembles into a highly ordered structure on CNTs, allowing not only markedly improved metal from semiconducting tube separation but also diameter-dependent separation. They find that anion exchange chromatography provides a macroscopic means to assay for electrostatic properties of nanoscale DNA-CNT hybrids. More specifically, the outcome of anion exchange-based DNA-CNT separation, as measured by optical absorption spectral changes from fraction to fraction, depends on the DNA sequence. A search of the ssDNA library selected a sequence d(GT)n, $n = 10$ to 45 that self-assembles into a helical structure around individual nanotubes in such a manner that the electrostatics of the DNA-CNT hybrid depends on the tube diameter and the electronic properties, thereby enabling nanotube separation by anion exchange chromatography. Optical absorption and Raman spectroscopy showed that early fractions are enriched in the smaller diameter and metallic tubes, whereas late fractions were enriched in the larger diameter and semiconducting tubes.

Haddon *et al.*[158] have reviewed the methods available for measuring the purity of SWNTs and the current status of processes designed to purify them. They emphasize the need for a hierarchy of purification steps that must be developed to obtain high-quality material suitable for advanced applications. The electronic (near IR) absorption spectra seem to be the best means of establishing purity.

3 Structure, Spectra and Characterization

General Structural Features

Transmission electron microscope (TEM) observations clearly reveal that the carbon nanotubes prepared by the arcing process generally consist of multi-layered, concentric cylinders of single graphitic (graphene) sheets. The diameter of the inner tubes is generally of the order of a few nanometres. The outer diameter could be as large as 10–30 nm (Figure 1.1b). During the curling of a graphene sheet into a cylinder, helicity is introduced. Electron diffraction patterns show the presence of helicity, suggesting that the growth of nanotubes occurs as in the spiral growth of crystals. The concentric cylinders in MWNTs are about 3.45 Å apart, which is close to the separation between the (002) planes of graphite. These are the lowest energy surfaces of graphite with no dangling bonds, so that the nanotubes are in fact the expected structures. In the electron microscope images, one typically observes nanotubes along their lengths, with the electron beam falling perpendicular to the axis of the nanotube. In high-resolution images, it is possible to see spots due to the lattice planes running along the length of the nanotubes. Iijima[159] has published such an image for the (110) planes separated by 2.1 Å. Ring-like patterns are found due to individual tubes that consist cylindrical graphitic sheets that are independently oriented (with no registry between the sheets) with helical symmetry for the arrangement of the hexagons.

Zhang *et al.*[20c] interpreted the electron diffraction patterns of MWNTs in terms of the reciprocal space. They found that streaks perpendicular to the axis of the tubule occur at most spots in the diffraction patterns. The streaks often exhibit a fine structure. Tilting experiments about an axis perpendicular to the needle axis, allowing the exploration of reciprocal space and a quantitative analysis of the geometry of the electron diffraction patterns in terms of the tilting angle and reciprocal vectors, are shown by them to confirm the model. A typical diffraction pattern of MWNTs (Figure 1.1d) exhibits spots from both chiral and non-chiral tubes. This pattern was recorded from a MWNT with 18 individual layers and an innermost diameter 1.3 nm; the incident electron beam is normal to the tube axis (20c). The $\{000l\}$ spots that result from the parallel graphene layers perpendicular to the beam can be seen running horizontally on either side of the central spot. The other arrowed spots correspond to reflections from achiral (*i.e.* zigzag or armchair) tubes, and are of the $\{10\bar{1}0\}$ or $\{11\bar{2}0\}$ type. The other weak reflections are due to chiral tubes. An analysis of the spots from the chiral nanotubes enables the chiral angle to be determined. In such experiments, the tube must be aligned exactly perpendicular to the electron beam.

Graphitic cylinders would have dangling bonds at the tips, but the carbon nanotubes are capped by dome-shaped hemispherical fullerene-type units. The capping units consist of pentagons to provide the curvature necessary for closure. Ajayan *et al.*[8a] studied the distribution of pentagons at the caps of carbon nanotubes and found that the caps need not be perfectly conical or hemispherical, but can form skewed structures. The simplest possible single-walled carbon nanotube can be visualized by cutting the C_{60} structure across the middle and adding a cylinder of graphite of the same diameter. If C_{60} is bisected normal to a five-fold axis, an armchair tube is formed, and if it is bisected normal to a three-fold axis, a zigzag tube is formed. Armchair and zigzag tubes are non-chiral. In addition to these, various chiral tubes can be formed with the screw axis along the axis of the tube. Figure 1.17 shows the models of the three types of nanotubes formed by bisecting the C_{60} molecule and adding a cylinder of graphite. Nanotubes can be defined by a chiral angle θ and a chiral vector C_h, given by

$$C_h = na_1 + ma_2 \qquad (1)$$

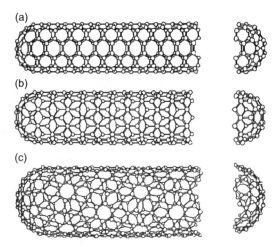

Figure 1.17 *Models of* (a) *armchair,* (b) *zigzag, and* (c) *chiral nanotubes*
(Reproduced from ref. 17)

The vector C_h connects two crystallographically equivalent sites on a 2D graphene sheet and the chiral angle is the angle it makes with respect to the zigzag direction, Figure 1.18. A tube is formed by rolling up the graphene sheet in such a way that the two points connected by the chiral vector coincide. Several possible chiral vectors can be specified by Equation (1) in terms of pairs of integers (n,m). Many such pairs are shown in Figure 1.18 and each pair (n,m) defines a different way of rolling up the graphene sheet to form a carbon nanotube of certain chirality. The limiting cases are $n \neq 0$, $m = 0$ (zigzag tube), and $n = m \neq 0$, (armchair tube). For a carbon nanotube defined by the index (n,m), the diameter, d, and the

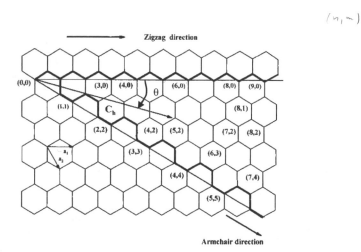

Figure 1.18 *A 2D graphene sheet showing chiral vector C_h and chiral angle (θ)*

chiral angle, θ, are given by Equations (2) and (3), where $a = 1.42\ (3)^{1/2}$ and $0 \leq \theta \leq 30°$.

$$d = a(m^2 + mn + n^2)^{\frac{1}{2}}/\pi \tag{2}$$

$$\theta = \arctan\left[-(3)^{\frac{1}{2}}m\right]/2n + m \tag{3}$$

The (n,m) indices determine the nanotube electronic structure. Armchair SWNTs are metals with $n - m = 3q$, where q is a nonzero integer, are semiconductors with a tiny band gap; and all others are semiconductors with a band gap that inversely depends on the nanotube diameter.[8b,17,18] SWNTs for which $|n-m| = 3q$ are metallic, and those for which $|n-m| = 3q \pm 1$ are semiconducting. Here, q is an integer. Thus, we see that 2/3 of the nanotubes correspond to semiconducting nanotubes and 1/3 to metallic nanotubes. This remarkable property about carbon nanotubes is a phenomenon based on the special symmetry of 2D graphite. It is not, however, straightforward to measure (n,m) experimentally and then to carry out property measurements on the same SWNT.

MWNTs consist of capped concentric cylinders separated by 3.45 Å (which is slightly larger than interlayer spacing in graphite) because the number of carbon atoms increases as we go from an inner cylinder to an outer cylinder and it is not possible to maintain perfect ABAB. ... stacking as in graphite. Thus, an interlayer spacing close to that in turbostratic graphite is observed in MWNTs. In addition to pentagons and hexagons, carbon nanotubes can also have heptagons. Pentagons impart a positive curvature whereas heptagons give rise to a negative curvature, to the otherwise flat graphene sheet made of hexagons. Thus, nanotubes with pentagons and heptagons will have unusual curvatures and shapes. Bent nanotubes arising from the presence of pentagons and heptagons on opposite sides of the tube have been observed.[159] Based on force field calculations, Tersoff and Ruoff[160] suggest that the nanotubes will form cylindrical bundles in a crystal and that large tubes will be hexagonal to maximize the van der Waals contact between the tubes. Simulation studies indicate radical p-orbital character and large pyramidalization angles at the sites of local deformation.[161]

An interesting observation with SWNTs is that of rings in the electron microscopic images.[162] The rings are formed during the ultrasound treatment in acidic medium, followed by the settling of nanotube dispersions on the substrate. Similar ring morphologies are observed in atomic force microscope (AFM) and SEM images of catalytically produced multi-walled nanotubes.[163] Huang et al.[164] have observed "crop circles" of aligned nanotubes in a direction normal to the substrate surface, on pyrolysing Fe(II)phthalocyanine. Yet another important nanostructure is the presence of encapsulated fullerenes inside SWNTs, as observed by high resolution TEM.[118] Fullerenes were observed inside the nanotubes after annealing laser-synthesized SWNTs. A TEM image showing the presence of C_{60} in SWNTs is given in the inset of Figure 1.12(a). An analysis of the size distribution of the encapsulated fullerenes inside arc-produced SWNTs revealed significant quantities of fullerenes in the size range of C_{36}–C_{120} in the nanotube capillaries.[165] Another noteworthy discovery concerns the double-walled nanotubes discussed earlier (Figure 1.12b and 1.12c).

X-ray diffraction (XRD) measurements are routinely employed to characterize carbon nanotubes.[166,167] XRD patterns of nanotubes show only the (*hk*0) and (00*l*) reflections but no general (*hkl*) reflections. This is the case in turbostratically modified graphites.[168] Warren[169] has suggested special methods for the analysis of the (*hk*0) reflections, which support electron microscopy data in showing that structural correlations exist along the direction perpendicular to the carbon nanotube axis as well as within each individual tube, but not in any combination of these. The correlation lengths obtained from the analysis of the XRD patterns are in the same regime as that from microscopy.

Scanning electron microscopy (SEM) is extensively used to study carbon nanotubes and the technique constitutes a good way to check bulk yields as well as the alignment of nanotubes. Nanotubes seem to form in bundles at the cathode, held together by van der Waals interactions. Alignment of the tubes in these bundles seems to depend on the stability of the arc.[170–172] Wang *et al.*[172] have observed tightly packed buckybundles in crosssectional SEM images of the cathodic deposit. The bundles are formed when the arc is maintained at very small separations between the electrodes. Ebbesen *et al.*[170] consider the nanotube bundles to be self-similar in the sense that large cylindrical bundles consist of smaller ones and the smaller ones are made up of nanotubes and so on. STM has been used to probe the electronic structure of carbon nanotubes deposited on various substrates.[173–175] STM has also been used to probe sp^3 defect structures, closure of the tips and pentagon-induced changes in the electronic structure in carbon nanotubes.[176] Venema *et al.*[177] have obtained atomically resolved STM images of SWNTs, wherein the chirality of the nanotubes is unambiguously determined, which in turn influences the electronic property of nanotubes.

Atomic-resolution imaging and site-specific force measuremens on a single-walled nanotube have been achieved by dynamic force microscopy and 3-D force field spectroscopy. The imaged topography reveals the trigonal arrangement of the hollow sites.[178] Electron diffraction intensities have been employed to obtain an atomic-resolution image of a double-walled nanotube.[179] Figure 1.19 shows a section of the reconstructed image at 1 Å resolution. Zamkov *et al.*[180] have presented experimental evidence for the existence of image-potential states in carbon nanotubes. The observed features constitute a new class of surface image states due to their quantized centrifugal motion. Measurements of binding energies and the temporal evolution of image state electrons were performed using femtosecond time-resolved photoemission. The associated lifetimes are longer than those of $n = 1$ image state on graphite, indicating a substantial difference in electron decay dynamics between tubular and planar graphene sheets. These states provide a novel means for the investigation of nanotube surface phenomena, structural and optical properties at interfaces as well as electron transport in nanotube heterojunctions.

Raman and Other Spectroscopies

Raman spectroscopy provides invaluable insights into the structure of the nanotubes. Jishi *et al.*[181] calculated the Raman-active phonon modes using a

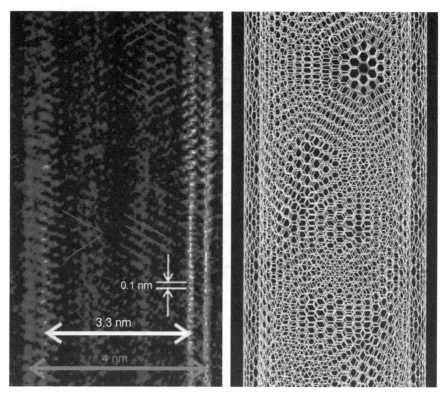

0.1 nm

3.3 nm

4 nm

Figure 1.19 *A section of the reconstructed DWNT image at 1 Å resolution and (right) a structural model constructed with the use of the chiral vectors determined from the image and diffraction pattern*
(Reproduced from ref. 179)

zone-folding method for a 2D graphene sheet, and demonstrated that there are 15 allowed Raman modes for each diameter of the tube. The frequency of the allowed mode depends on the tube diameter and the chiral angle, the number of modes being independent of the diameter. Hiura *et al.*[182] found that the linewidth of phonon peaks in the Raman spectrum was narrow, of the order of 20 cm^{-1}. The Raman phonon frequency of nanotubes is softer than that of highly oriented pyrolytic graphite (HOPG), probably due to the curvature of the nanotubes. Softening of phonon modes can be related to the larger *c*-axis lattice parameter in the nanotube as compared to graphite. Holden *et al.*[183] examined the spectra of SWNTs produced by using Co catalysts and have compared them with the predictions of Jishi *et al.*[181] Rao *et al.*[184] have described the characteristic normal modes of armchair (*n,n*) carbon nanotubes and demonstrated a diameter-selective resonance behaviour. The resonance results from the 1D quantum confinement of the electrons in the nanotubes. Kasuya *et al.*[185] provided the first evidence for a diameter-dependent dispersion arising from the cylindrical symmetry of the nanotubes. They carried out Raman scattering studies on SWNTs with mean diameters of 1.1, 1.3 and 2 nm and found size-dependent multiple splitting of the

Table 1.1 *Vibrational modes observed for Raman scattering in SWNTs*
(Reproduced from ref. 188)

Notation	Frequency (cm^{-1})	Symmetry	Types of modes
RBM[a]	$248/d_t$	A	In phase radial displacements
D-band	~1350	–	Defect-induced dispersive
G-Band	1550–1605	A, E_1, E_2	Graphite-related optical mode[b]
G′-band	~2700	–	Overtone of D-band, highly dispersive

[a]RBM denotes radial breathing mode. [b]The related 2D graphite mode has E_{2g} symmetry. In 3D graphite, the corresponding mode is denoted E_{2g2}.

optical phonon peak corresponding to the E_{2g} mode of graphite. In Table 1.1 we list characteristic Raman features of SWNTs. Figure 1.20, shows typical Raman spectra of laser-synthesized SWNTs. Assignment of bands due to nanotubes of different diameters is indicated in the figure. Polarized Raman studies on aligned

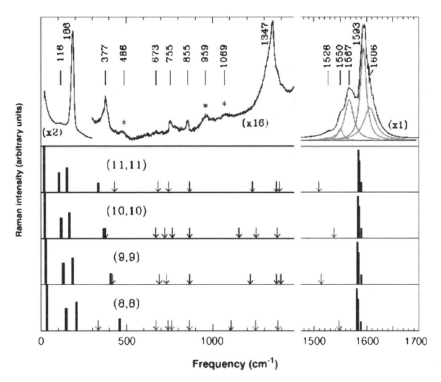

Figure 1.20 *Raman spectra showing diameter-dependent scattering in SWNTs. An asterisk in the spectrum represents a feature assigned tentatively to second-order Raman scattering. The four bottom panels show calculated Raman spectra for armchair (n,n) nanotubes (n = 8–11)*
(Reproduced from ref. 184)

multi-walled carbon nanotubes show a strong dependence of the graphite-like G-band and disorder induced D-band on the polarization geometry.[186]

Resonance Raman spectra of isolated nanotubes enable the determination of (n,m). The ability to use resonance Raman spectroscopy to determine the geometrical structure is a remarkable property of carbon nanotubes.[187,188] Generally, Raman spectroscopy measures phonon frequencies. Under resonance conditions, the technique provides information about the electronic structure. Since the electronic structure of a nanotube is uniquely determined by its (n,m) indices, it becomes possible to determine the geometrical structure of a SWNT from the resonance Raman spectrum. Observation of the Raman spectrum from an isolated SWNT is an important development that has become possible because of the large density of electron states close to the van Hove singularities (see section on Electronic Structure). When the incident or scattered photons are in resonance with an electronic transition between the singularities in the valance and conduction bands, the Raman cross section become large due to the strong coupling between the electrons and phonons. This has enabled the study of the dependence of the various features in the Raman spectrum on the nanotube diameter and chiral angle, and allowed one to obtain spectroscopic details of each Raman feature. Figure 1.21 shows the radial mode Raman bands of three isolated single-walled nanotubes of different diameters. Thus, the electronic transition energies and the radial mode frequencies of different types of metallic and semiconducting SWNT's can be discussed on the basis of geometrical parameters.[189]

Raman spectra of double-walled nanotubes have been investigated and the features associated with the interior nanotube identified.[119a,190] *In situ* Raman spectroscopy has been employed to study the changes in the radial mode with

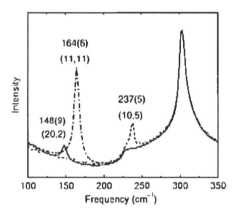

Figure 1.21 *Superposition of the three Raman spectra (solid, dashed, and dash–dotted curves) come from three different spots on the Si substrate, showing the presence of only one resonant nanotube and one RBM frequency for each of the three laser spots. The RBM frequencies (line-widths) and the (n, m) assignments for each resonant SWNT are displayed. The 303 cm^{-1} feature comes from the Si substrate and is used for calibration purposes* (Reproduced from refs. 187 and 188)

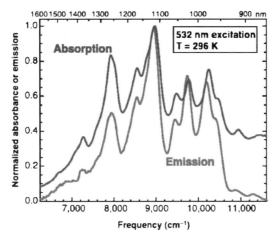

Figure 1.22 *Emission spectrum of individual fullerene nanotubes suspended in SDS micelles in D_2O excited by 8 ns, 532 nm laser pulses, overlaid with the absorption spectrum of the sample in this region of first van Hove band gap transitions. The correspondence between the absorption and the emission features indicates that the emission is band gap photoluminescence from various semiconducting nanotube structures*
(Reproduced from refs. 153)

electrochemical biasing, to demonstrate how the energy gaps between the van Hove singularities are affected by changes in the π overlap integrals.[191]

Spectrofluorometric measurements show how the absorption and emission transitions are distinct for different SWNTs depending on (n,m).[192] Fluorescence from individual SWNTs has been observed directly across the band gap.[153] Typical absorption and emission spectra are given in Figure 1.22. The electronic properties of the individual nanotubes can be derived from the fluorescence spectra.[193] Photoluminescence and excitation spectra of individual nanotubes have been investigated.[194] Measurement of optical spectra in magnetic fields has revealed the dependence of band structure on the magnetic flux threading – consistent with the Aharonov–Bohm effect.[195]

Pressure-induced Transformations

Pressure-induced phase transformations under static and dynamic loading between the many allotropes of carbon, like diamond, graphite, C_{60} and C_{70} and their polymeric and amorphous forms, are of academic and practical importance. Pressure-effects on SWNT bundles have been probed by Raman spectroscopy upto a maximum pressure of 25.9 GPa (1 GPa $= 10^9$ N m^{-2}) in a diamond anvil cell.[196,197] Spectra arising from the radial and tangential modes at 0.1 GPa are similar to those reported earlier at atmospheric pressure.[184] The two dominant radial bands in the spectrum of the sample recorded at 0.1 GPa were at 172 and 182 cm^{-1}. For an isolated SWNT, the calculated frequencies of the radial mode ω_R(cm^{-1}) for a tube of diameter d(nm) fit to $\omega_R = 223.75/d$, irrespective of the nature of the tube.[198] This gives $\omega_R = 164$ cm^{-1} for the (10,10) tube and 183 cm^{-1} for the (9,9) tube. Inclusion of van der Waals interaction between the (9,9) tubes shifts the radial mode frequency from 171.8 cm^{-1} (for an isolated tube) to 186.2 cm^{-1}. This blue-shift of 14.4 cm^{-1} is due to intertube interaction, and is independent of the tube diameter.[199] Accordingly, the empirical relation for the diameter dependence of the radial mode frequency in a SWNT bundle is given by $\omega_R = 14.4 + 209.9/d$, which retains the $1/d$ dependence of ω_R and reproduces $\omega_R = 186.2$ cm^{-1} for the (9,9) tube.[200] The tangential modes are assigned in terms of the irreducible representations of D_{nh} (D_{nd}) for even n (odd n), with 1531 cm^{-1} as E_{1g}, 1553 and 1568 cm^{-1} as E_{2g}, 1594 cm^{-1} with unresolved doublet $A_{1g} + E_{1g}$ and 1606 cm^{-1} with E_{2g} symmetry.[201,202] The intensities of the radial modes fall rapidly with increasing pressure, and were not discernible beyond 2.6 GPa, but the features are reversible. Intensities of the tangential modes also decrease with pressure. The intensities of the radial modes fall rapidly with increasing pressure, and were not discernible beyond 2.6 GPa, but the features are reversible. Intensities of the tangential modes also decrease with pressure. Figure 1.23 shows the pressure dependence of the 172 cm^{-1} radial mode in the increasing and decreasing pressure runs, along with the calculated curves for three models.[199] The tangential mode frequencies are plotted as a function of pressure in Figure 1.24(a–c). The modes at $\omega_T = 1568$ and 1594 cm^{-1} show softening between ~10 and 16 GPa beyond 16 GPa, the band position increases with pressures, Figure 1.24(a–c). Remarkably, when the pressure is reduced from the highest pressure of 25.9 GPa, the peak

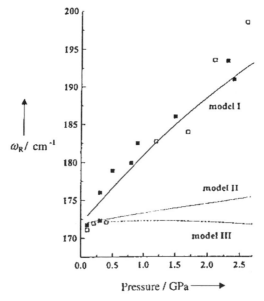

Figure 1.23 *Pressure variation of the radial mode Raman bands ω_R. Squares ■ (□) indicate a run with increasing (decreasing) pressure. Theoretical curves for models I, II, and III are also shown*
(Reproduced from ref. 196)

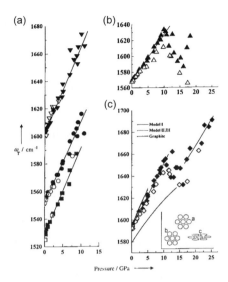

Figure 1.24 *Pressure variation of* (a) *1531, 1553, and 1606,* (b) *1568, and* (c) *1594* cm^{-1}
*Raman frequencies. Data points for increasing (decreasing) pressure are
filled (open) symbols. Inset: Faceting in SWNT bundles leads to the
conversion of* (a) *a circular cross-sectional bundle to that of* (b) *hexagonal
and* (c) *elliptical cross sections*
(Reproduced from ref. 196)

positions (open symbols) follow the same trend as in the increasing pressure run.
The slope dω/dP obtained by fitting $\omega(P) = \omega(0) + aP$ to the tangential modes and
a second-order polynomial to the radial mode have been used to calculate a mode
Gruneissen parameter, $\gamma = Bd(\ln\omega)/dP$, where B (136 Pa) is the bulk modulus of
the SWNT bundle. There is a large difference between the values of γ for tangential
and radial modes, just as in graphite and other layered as well as molecular crystals.
Pressure-induced changes in the energy difference between the singularities in the
valence and conduction bands in the 1D density of states would reduce the
resonance enhancement in the Raman cross section, causing a decrease in the intensities under pressure. Enhanced intertube interactions at high pressures, resulting in increased density of electronic states near the Fermi level, can contribute
to reduced Raman intensities and increased line-widths. Venkateswaran *et al.*[199]
have used generalized tight binding molecular dynamics simulations to calculate
the pressure dependence of the radial and tangential modes upto 5 GPa. They
consider three situations for pressure transmittance to the SWNT bundle. Model I
takes uniform external radial compression of the entire triangular lattice, which
corresponds to the situation that the pressure-transmitting medium is not present in
the interstitial channels in the tube lattice. This is not the case in Models II and III,
where all the individual nanotubes in a triangular lattice feel the pressure medium.
However, in Model II, intertube interactions are ignored and, therefore, the
compression is symmetric for all the tubes, whereas these interactions are included
in Model III. The pressure dependence of the radial mode is consistent with only

Model I (Figure 1.23). This is surprising since the assumption of non-penetration of the pressure-transmitting fluid inside the bundle is inconsistent with the small blue-shift of the radial mode frequency ($\sim$2–4 cm^{-1}), when the SWNT bundles are soaked in alcohol mixture.[197] For the tangential modes, the prediction of Models I–III are similar (Figure 1.24c). The calculations, done only upto 5 GPa, had included intertube van der Waals interactions in the usual Lennard-Jones form. At high pressures, electron hopping between the tubes has to be taken into account and hence the results of Venkateswaran *et al.*[199] cannot be extrapolated to higher pressures. Studies of SWNTs under high pressure confirm the potential of these materials as the strongest ever carbon nanofibres, and also their remarkable resilience.[200,203]

Valence-force model calculations[199] predict that the tubes flatten against each other, forming a honeycomb structure. Smaller diameter tubes can also facet under pressure[202] and the tube cross-section can change from circular to hexagonal or elliptical (inset of Figure 1.24c). This will considerably affect the Raman intensities of the modes, more so for the radial one, as seen in our experiments. The faceting will result in considerable overlap of the surfaces of the neighbouring tubes, similar to stacking of two AB planes in graphite. The tangential mode frequencies in the SWNT bundle then will be close to the E_{2g} in-plane vibration of graphite at 1579 cm^{-1}. The pressure dependence of the E_{2g} 1579 cm^{-1} mode of graphite[204] measured to 14 GPa is shown by the thick solid line in Figure 1.24(c). The frequency of the 1594 cm^{-1} mode of SWNT bundles starts approaching that of graphite at $\sim$10 GPa and this process is completed at $\sim$16 GPa. Beyond 16 GPa, the pressure derivative $d\omega/dP$ is similar to that of graphite.[204] At 23 GPa Graphite transforms into amorphous carbon, which reverts to the crystalline form on pressure release.[205] Apparently therefore, the SWNT bundle does not transform into bulk graphite under hydrostatic pressure, as evident from the reversibility of the Raman spectra under pressure. X-ray diffraction studies under pressure reveal that, around 10.4 GPa, the SWNT bundles lose translational order.[206] This may be the cause of the anomaly found in the tangential Raman modes around this pressure. High-resolution electron microscope observations and atomistic simulations of the bending of SWNT under mechanical stress also reveal the remarkable flexibility of the hexagonal network, which resists bond breaking and stretching upto high strain values.[207]

Electronic Structure

As with fullerenes, the curvature of the graphitic sheets in the nanotubes would be expected to influence the electronic structure. The electronic properties of perfect MWNTs are somewhat similar to those of perfect SWNTs, because the coupling between the cylinders is weak in MWNTs. Theoretical calculations show that nanotubes may be as good conductors as copper, although a combination of the degree of helicity and the number of six-membered rings per turn around the tube can tune the electronic properties in the metal–semiconductor range.[208–210] Thus, the electronic properties of CNTs are sensitive to their geometric structure. Although graphene is a zero-gap semiconductor, theory predicts that CNTs can be metals or semiconductors with different sized energy gaps, depending sensitively on the diameter and helicity of the tubes, *i.e.* on the indices (n,m). Such a sensitivity of the electronic properties of the CNTs to their structure can be understood within a band-folding picture. This is because of the unique band structure of the graphene sheet, with states crossing the Fermi level at only two inequivalent points in k-space, and the quantization of the electron wave vector along the circumferential direction.

An isolated sheet of graphite is a zero-gap semiconductor whose electronic structure near the Fermi energy is given by an occupied π band and an empty π^* band. These two bands have linear dispersion and meet at the Fermi level at the K point in the Brillouin zone (Figure 1.25). The Fermi surface of an ideal graphite sheet consists of the six corner K points. When forming a tube, owing to the periodic boundary conditions imposed in the circumferential direction, only a certain set of k states of the planar graphite sheet is allowed. The allowed set of k (indicated by the lines in Figure 1.25) depends on the diameter and helicity of the tube. Whenever the allowed k include the point K, the system is a metal with a non-zero density of states at the Fermi level, resulting in a 1D metal with two linear dispersing bands. When the point K is not included, the system is a semiconductor with different sized energy gaps. The states near the Fermi energy in both the metallic and the semiconducting tubes are all from states near the K point, and hence their transport and other electronic properties are related to the properties of the states on the allowed lines. For example, the conduction band and valence bands of a semiconducting tube come from states along the line closest to the K point.

The general rules related to the metallicity of SWNTs are as follows. The (n,n) tubes (armchair) are metals; (n,m) tubes with $n-m = 3q$, where q is a non-zero integer, are small gap semiconductors while all others are large-gap semiconductors.

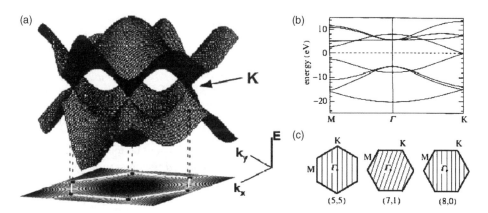

Figure 1.25 (a) *Three-dimensional view of the graphene π/π^* bands and its 2D projection.* (b) *Tight-binding band structure of graphene (a single basal plane of graphite), showing the main high symmetry points.* (c) *Allowed k-vectors of the (5,5), (7,1) and (8,0) tubes (solid lines) mapped onto the graphite Brillouin zone* (Reproduced from refs. 22 and 208–210)

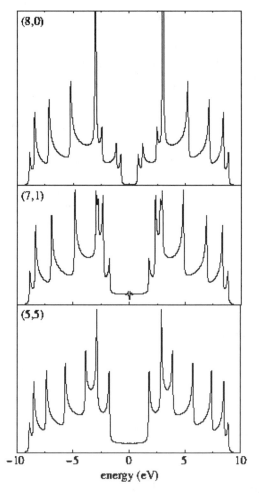

Figure 1.26   *Electronic densities of states for the (5,5), (7,1) and (8,0) tubes, showing*
singularities characteristic of 1D systems. The (5,5) armchair nanotube is
metallic for symmetry reasons. The (7,1) chiral tube displays a tiny gap due to
curvature effects, but will display a metallic behaviour at room temperature.
The (8,0) zigzag tube is a large-gap semiconductor
(Reproduced from refs. 22 and 355a)

Within the band-folding scheme, the $n-m = 3q$ tubes would be metals, but a small
gap opens because of tube curvature for the case that q is non-zero (Figure 1.26).
Hence, CNTs can be of three varieties: large gap, small gap and zero gap. The (n,n)
tubes are always metallic within the single-electron picture, independent of
curvature because of their symmetry (Figure 1.26). As the tube radius, R, increases,
the band gap decreases with a $1/R$ or $1/R^2$ dependence. Thus, for most
experimentally observed carbon nanotube sizes, the gap in the small-gap variety
arising from curvature effects is small and, for all purposes, the $n-m = 3q$ tubes
can be considered as metallic at room temperature. Accordingly, in Figure 1.25,

a (7,1) tube would be metallic, whereas a (8,0) tube would be semiconducting; the (5,5) armchair tube would always be metallic. Such a band-folding picture, based on the tight-binding approach, seems to be valid for larger-diameter tubes.

The continuous electronic density of states (DOS) in graphite divides into a series of spikes in SWNTs because of the radial confinement of the wave function which are referred to as van Hove singularities (Figure 1.26). Electronic transitions between these singularities give rise to prominent features in scanning tunnelling spectroscopy (STS) and electronic spectroscopy. As can be seen in Figure 1.27, most SWNTs exhibit the first and second electronic transitions in the semiconducting nanotubes (S_{11} and S_{22}) and the first transition in the metallic nanotubes (M_{11}).[211] The transitions at the Fermi level of the metallic SWNTs (M_{00}) are observable in the far-IR region of the electromagnetic spectrum.[212] This is an informative region of the spectrum, and the low-energy feature in acid-purified SWNTs arises from the doping of the SWNTs, presumably by nitric acid. A peak occurs around 0.01 eV, probably due to the curvature-induced pseudogap in chiral SWNTs.

The electronic structure and chemical reactivity of carbon nanotubes can also be understood based on the traditional chemical concepts.[213] Thus, the frontier orbital picture of Hoffmann can be used to understand the electronic structure. Interactions between occupied orbitals as well as that between empty orbitals (zero electron) would be different in metallic and semiconducting nanotubes. In Figure 1.28, we show the frontier orbital representation of zero electron interactions between a molecule and (a) a metallic and (b) semiconducting nanotube. The molecular orbitals are designated as HOMO and LUMO. The nanotube orbitals are given by the density of state plots. Interaction with the metallic nanotube becomes attractive

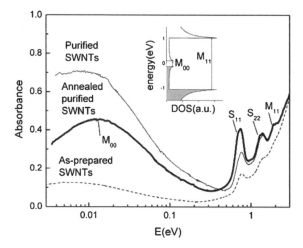

Figure 1.27 *Fermi level electronic structure of common forms of SWNTs. The low energy (~0.01 eV) features arise from a combination of transitions that are intrinsic to the metallic SWNTs, transitions due to the curvature-induced gap (M00) in the chiral metallic SWNTs, and transitions due to purification-induced acid doping of the semiconducting SWNTs*
(Reproduced from ref. 212)

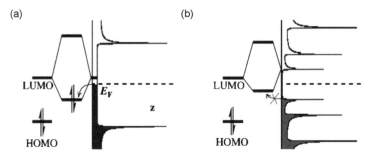

Figure 1.28 *Frontier-orbital picture representation of zero-electron interactions between a molecule and a metallic (a) or a semiconducting (b) single-wall carbon nanotube. The molecule frontier orbitals are designated as HOMO and LUMO, and the nanotube orbitals are represented by the density of states plots. Interaction with the metallic nanotube becomes attractive due to charge transfer, but is not effective with the semiconducting nanotube* (Reproduced from ref. 213)

due to charge transfer. The frontier orbitals of a semiconducting nanotube can be related to the first van Hove singularities in the valance and conduction bands.

The electronic structure of bulk SWNTs has been investigated by resonant inelastic soft X-ray scattering. The electronic states are derived from the K point and located close to the Fermi level, showing the existence of metallic nanotubes in the bundles, apart from a substantial semiconducting fraction.[214] [13]C nuclear magnetic resonance (NMR) spectroscopy of SWNTs shows two types of nuclear spins with different spin–lattice relaxation rates. The fast relaxing component follows the relaxation behaviour expected of metals.[215]. Femtosecond time-resolved photoemission studies of SWNTs indicate a decrease in the lifetime of the electrons excited to the π^* bands, with increase in energy relative to the Fermi level. This in turn leads to lifetime-induced broadening of van Hove singularities in the nanotube density of states.[216] Employing ultraviolet photoelectron spectroscopy, Ago *et al.*[217] have obtained the work function and DOS of multi-walled nanotubes. While pure MWNTs show a lower work function, the acid-oxidized nanotubes show a higher work function due to the disruption of π-conjugation and the introduction of surface dipole moments.

4 Chemically Modified Nanotubes

Doping with Boron and Nitrogen

Since the discovery of the carbon nanotubes, there has been interest in substituting carbon with other elements. Accordingly, boron–carbon (B-C), boron–carbon–nitrogen (B-C-N) and carbon–nitrogen (C-N) nanotubes have been prepared and characterized. Boron-substitution in the carbon nanotubes gives rise to p-type doping and nitrogen-doped carbon nanotubes correspond to n-type doping. Novel electron transport properties are expected of such doped nanotubes.[218] Boron-doped carbon nanotubes have been synthesized by pyrolysing mixtures of acetylene and diborane, and characterized by employing microscopic and spectroscopic techniques.[219] The average composition of these nanotubes is $C_{35}B$. B-C-N nanotubes have been prepared by striking an arc between a graphite anode filled with B-N and a pure graphite cathode in helium.[220] B-C-N nanotubes have also been obtained by laser ablation of a composite target containing B-N, carbon, Ni and Co at 1000 °C under flowing nitrogen.[221] Terrones *et al.*[222] pyrolysed the addition compound $CH_3CN:BCl_3$ over Co powder at 1000 °C to obtain B-C-N nanotubes. B-C-N as well as C-N nanotubes have been prepared by Sen *et al.*[223] by the pyrolysis of appropriate precursors. Pyrolysis of aza-aromatics such as pyridine over Co catalysts gives C-N nanotubes ($C_{33}N$ on average). Pyrolysis of the 1:1 addition compound of BH_3 with $(CH_3)_3N$ produces B-C-N nanotubes. Typical TEM images of a few nanotubes are shown in Figure 1.29, exhibiting bamboo-shape, nested cone-shaped cross section as well as unusual morphology, including coiled nanotubes. The composition of the B-C-N nanotubes varies with the

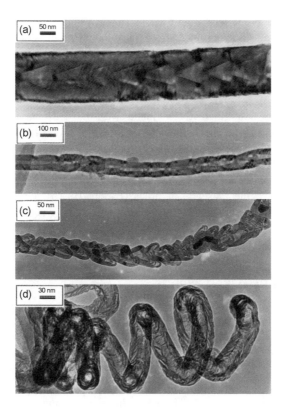

Figure 1.29 *TEM images of some of the unusual carbon nanotube structures obtained by the pyrolysis of pyridine (flow rate = 30 cm^3 min^{-1}) over Fe/SiO$_2$ substrates at 900 °C for 1.5 h under Ar (120 cm^3 min^{-1}) flow. The nanotubes show (a) bamboo shape, (b) nested cone, (c) and (d) other unusual morphologies. (d) TEM image of a coiled nanotube (Reproduced from ref. 225a)*

preparation. Furthermore, considerable variability exists in the composition in any given batch of B-C, B-C-N or C-N nanotubes obtained by the pyrolysis of precursors. Zhi *et al.*[224] discuss the design, synthesis and properties of boron carbonitride in some detail.

Nath *et al.*[225a] have obtained aligned carbon–nitrogen nanotube bundles by the pyrolysis of pyridine over sol–gel-derived iron/silica or cobalt/silica substrates. Employing anodic alumina, Sung *et al.*[225b] synthesized the C-N nanotubes by electron cyclotron resonance CVD, using C_2H_2 and N_2. Suenaga *et al.*[226] carried out CVD of Ni-phthalocyanine to obtain aligned C-N nanotubes.

Goldberg *et al.*[227] have employed a method wherein SWNTs were thermally treated with boron trioxide in a nitrogen atmosphere to obtain boron- or boron and nitrogen-doped SWNTs. EELS analysis showed the boron content to be (10 at% in B-C nanotubes. An interesting aspect of the B-C-N nanostructures is that phase separation occurs in which the BC_3 islands segregate in the graphene sheets. Tunnelling conductance measurements of doped nanotubes demonstrate acceptor-like states near the Fermi level, arising out of the BC_3 islands.[228] Efsarjani *et al.*[229] propose that a nanotube with donor atoms at one side and acceptor atoms on the other can function as a nano-diode. An experimental situation near to this effect is the observation of rectification in a SWNT.[230] The presence of an impurity in one of the segments of a SWNT influences its nonlinear transport behaviour.

Terrones *et al.*[231] have reviewed the literature on the production and characterization of B- and N-doped carbon nanotubes (CNTs) and nanofibres. They also demonstrate how CNTs doped with B or N should exhibit novel electronic, chemical, and mechanical properties that are not found in their pure carbon counterparts. To exploit the novel properties, low concentrations of dopants (*e.g.* <0.5%) should be incorporated within these tubes, so that the electronic conductance is significantly enhanced and the mechanical properties are not be altered. Because of the presence of holes (B-doped) or donors (N-doped), their surface would become more reactive. Such reactivity would be useful in the development of field-emission sources, nanoelectronics, sensors, and strong composite materials.

Intercalation by Alkali Metals

MWNTs can be intercalated either by electron donors or electron acceptors in intershell spaces, and in between the individual tubes or inside the tubes in the case of SWNTs. Intercalation can be carried out in the vapour or liquid phase, and electrochemically. The first doping reactions by K and Rb were performed on MWNTs prepared by the electrical arc-discharge method.[232] The samples were intercalated in the vapour phase using the two-bulb method, leading to the saturation MC_8 composition (M = K or Rb). Interlayer distances were similar to the first stages graphite intercalation compounds ($I_c = 0.568$ nm in RbC_8), indicating that the host was fully intercalated with all the intershell spaces occupied by the alkali cations. HREM studies after successive intercalation with K and deintercalation showed the presence of individual nanoparticles, indicating that intercalation in the graphitic shell of nanotubes had induced the formation of defects and a partial disintegration of the tubes, in agreement with a Russian doll morphology. Mordkovitch *et al.*[233] have observed a characteristic beadline pattern with swollen (intercalated) regions alternating with nonintercalated necks on intercalation with K, rather than the MC_8 saturated composition, the XRD pattern showing no evidence of residual pristine buckybundles. K intercalation is accompanied by a suppression of the conductivity and modification of inelastic scattering processes. Alkali metal intercalation gives rise to metallic character in SWNTs.

To perform intercalation of a perfect MWNT (Russian doll tube closed at the extremities), the diffusion of the dopant within the host nanotube along the defects (the edge planes accessible at the surface of the cylindrical layers of the tube) is needed. This means that intercalation is limited by the dopant diffusion along the defects. Dopant intercalation, therefore, depends on the structural order and the morphology (basal planes parallel with the tube axis or at an angle to the tube axis) of the tubes. Thus, the reactivity of the MWNTs depends on whether they are scroll nanotubes, Russian dolls, or catalytic nanotubes (straight or conical). For example, the easy diffusion of dopant in the scroll nanotubes (buckybundles helical nanotubes) allows an intercalation with the preservation of the tube even if some disordered domains only are intercalated, forming intercalated beads.[233,234] Furthermore, large-sized molecules, such as metal chlorides, intercalate to a lower extent in the scroll tubes than in graphite. The Russian doll tubes are intercalated by alkali metals in disordered graphite-like domains and are not reactive with Br_2 and metal chlorides.[235] Catalytic tubes react with alkali metals to give compositions and metallic properties similar to graphite intercalation compounds.[232,236] Intercalation in MWNTs induces a decrease in the cohesive energy of the tubes, yielding to their partial or complete destruction. Russian doll tubes are destroyed, as observed by TEM, due to a mechanical distortion of the cylindrical planes needed for the accommodation of the dopant species in the intershell spacing.[232]

By Li doping of SWNTs in THF solution[237] or by electrochemical means,[238] the composition of LiC_6 and $Li_{1.2}C_6$ reversible capacity are obtained respectively. Ball-milling of SWNTs induces fracture of the tubes and the electrochemical doping is increased ($Li_{2.7}C_6$ reversible capacity) due to the diffusion of Li atoms into the inner cores of the fractured tubes.[239]

Opening and Filling of Nanotubes

Multi-walled nanotubes are generally closed at either end, the closure being made possible by the presence of five-membered rings. MWNTs can be uncapped by oxidation with carbon dioxide or oxygen at elevated temperatures.[11,12,166] High yields of uncapped MWNTs are, however, obtained by boiling them in concentrated HNO_3. Treatment with HNO_3 gives rise to carbonyl and other functionalities in the nanotubes. Opened nanotubes have been filled with metals. The established method is to treat the nanotubes with boiling HNO_3 in the presence of metal salts such as $Ni(NO_3)_2$.[240] The nanotubes are opened by HNO_3 and filled by the metal salt. On drying and calcination, the metal salt transforms into the metal oxide and reduction of the encapsulated oxide in hydrogen at around 400 °C gives rise to the metal inside the nanotubes. MWNTs have been opened using various oxidants,[241–243] and the opened nanotubes have been filled with Ag, Au, Pd, or Pt by different chemical means, rather than by reduction with hydrogen at high temperatures.[243] By employing *in situ* techniques in a TEM, Bower *et al.*[244] observed alkali metal incorporation into the SWNTs. Sealed tube reactions of SWNTs and metal salts also yield metal intercalated SWNTs.[245] In an effort to realize the conversion of the sp^2 carbon of the nanotubes into sp^3, Hsu *et al.*[246] treated potassium-intercalated MWNTs with CCl_4 hydrothermally, and obtained crystallization of KCl inside the nanotubes and within the tube walls. Possible ways of closing the nanotubes, opened by oxidants, have been examined.[243] Besides opening, filling and closing nanotubes, highly functionalized MWNTs have been prepared by treatment with acids.[243,247] SWNTs are readily opened by mild treatment with acids and filled with metals.[63,248] Acid-treated nanotube surfaces can be decorated by nanoparticles of metals such as Au, Ag or Pt.[63,249]

The enhanced reactivity of SWNTs by acceptor dopants such as HNO_3,[250] H_2SO_4,[251] bromine[252] and iodine[253] can be explained on the basis of intercalation both at the outside and inside of the nanotubes. By oxidation with iodine, the tubes are opened, prior to intercalation. HREM images show that the iodine is accommodated not only in between the tubes, but also inside the tubes, forming helical chains.

Green and co-workers[254] have investigated molecules and 1D crystals inside SWNTs by HREM and have obtained atomic scale images. Usually, direct lattice imaging by conventional HREM can be used to image molecular species and aligned 1D crystals formed within SWNTs. Indeed, the technique has proved invaluable, since both the microstructure and the helical nature of multilayer nanotubes were correctly demonstrated by a combination of HREM and electron diffraction in early 1993. More recently, considerable progress has been made with regard to the imaging of SWNT-incorporated molecules and 1D crystals by direct, super-resolved, and spectroscopic methods. Figure 1.30(a) shows the conventional HREM image obtained at close-to optimum Scherzer defocus conditions of a 2×2 potassium iodide (KI) crystal formed within a SWNT ~1.4 nm in diameter. In this image, the I–K or K–I columns are clearly resolved (see the structure model in Figure 1.30b), whereas the graphene walls appear as continuous and featureless parallel black lines. In the example in Figure 1.30(a), only the strongly scattering I atoms ($Z = 53$) contribute significantly to the image contrast, while the K atoms ($Z = 19$) make a negligible contribution. In Figure 1.30(c), we see a super-resolved image of 3×3 KI formed within a ~1.6 nm diameter SWNT and, in Figure 1.30(d), the corresponding derived structure model. As the encapsulated crystal is visible in a [110] projection relative to bulk KI, all the atom columns are visible as pure atom columns. In a conventional image, the heavy I atoms are visible, whereas the K atoms are invisible. In the reverse-contrast restored image (Figure 1.30c), the I columns are much more prominent, but are now interspersed with the weaker K columns. The intensities of the bright spots corresponding to the I and K columns are modulated according to atom column thickness. Along the SWNT, we therefore identify two alternating layers, corresponding to I-2K-3I-2K-I and K-2I-3K-2I-K,

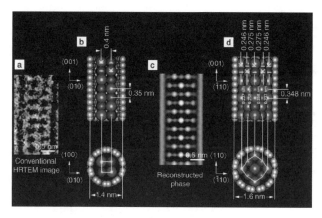

Figure 1.30   (a) *High-resolution transmission electron microscope image of a 2×2 KI crystal formed within a ~1.4 nm diameter single-walled carbon nanotube (SWNT). (b) Structure model derived from (a). (c) Super-resolved HRTEM image of a 3×3 KI crystal formed in a ~1.6 nm diameter SWNT. (d) Corresponding structure model derived from (c)*
(Reproduced from ref. 254)

which is consistent with the $3 \times 3 \times \infty$ 1D KI crystal. Notably, both 1D crystals display considerable lattice distortions compared with their bulk structures. In the 2×2 case, a lattice expansion of ~17% occurs across the SWNT capillary, whereas in the 3×3 case a differential expansion is observed, with the I columns being more compressed than the K columns (Figure 1.30d).

Reactivity, Solubilization and Functionalization

The reactivity of fullerenes is driven by the enormous strain arising from their spherical geometry, as reflected in the pyramidalization angles of the carbon atoms. For an sp^2-hybridized (trigonal) carbon atom, planarity is strongly preferred, and this implies a pyramidalization angle of $\theta_P = 0°$, whereas an sp^3-hybridized (tetrahedral) carbon atom requires $\theta_P = 19.5°$ (Figure 1.31). All the carbon atoms in C_{60} have $\theta_P = 11.6°$, and their geometry is more appropriate for tetrahedral than trigonal hybridization. Chemical conversion of a trivalent carbon in C_{60} into a tetravalent carbon relieves the strain at the point of attachment and mitigates the strain at the 59 remaining carbon atoms. Reactions that saturate the carbon atoms are facilitated by strain relief, and this strongly favours the addition chemistry of fullerenes. As with a fullerene, a perfect SWNT is without functional groups and these cylindrical aromatic macromolecules are, therefore, chemically inert. The curvature-induced pyramidalization and misalignment of the π-orbitals of the carbon atoms induces a local strain (Figure 1.31), and the nanotubes are expected to be more reactive than a flat graphene sheet.[255] It is conceptually useful to divide the carbon nanotubes into two regions: the end caps and the side wall. The end caps of the carbon nanotubes resemble a hemispherical fullerene. Since it is impossible to reduce the maximum pyramidalization angle of a fullerene below

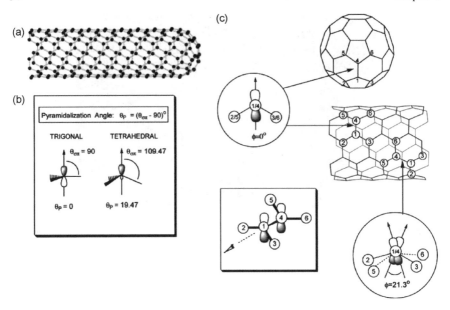

Figure 1.31 *Diagrams of (a) metallic (5,5) SWNT, (b) pyramidalization angle (θ_P), and (c) the θ_P – orbital misalignment angles (ϕ) along C_1–C_4 in the (5,5) SWNT and its capping fullerene, C_{60} (Reproduced from ref. 255)*

about $\theta_P^{max} = 9.7°$, the end caps will always be quite reactive, irrespective of the diameter of the carbon nanotube. In the (5,5) SWNT shown in Figure 1.31(a), which is capped by a hemisphere of C_{60}, the pyramidalization angles are $\theta_P \approx 11.6°$ (end cap) and $\theta_P \approx 6.0°$ (side wall). The strain in the fullerenes is primarily from pyramidalization, and the large strain imposed by the spheroidal structure accounts for their well-developed (addition) chemistry. As with fullerenes, the reactivity of carbon nanotubes arises out of their topology, but for different reasons. Furthermore, since the pyramidalization angles and the π-orbital misalignment angles of SWNTs scale inversely with the diameter of the tubes, a differentiation is expected between the reactivity of carbon nanotubes of different diameters.

Although Chen *et al.*[256] and others attempted chemistry with the as-prepared SWNTs, the difficulties in characterizing the products made it necessary to develop a dissolution process for the SWNTs by chemical attachment of organic functional groups by other means. SWNTs are highly resistant to wetting.[257] They typically exist as ropes or bundles, 10–25 nm in diameter, that are a few micrometres long; the SWNT ropes are entangled in the solid state to form a highly dense, complex network structure. Furthermore, these pseudo-1D graphitic cylinders do not have any surface functional groups, making them difficult to disperse in organic media.

SWNT raw soot can be wetted in refluxing nitric acid, whereby the end caps of the tubes are oxidized to carboxylic acid and other weakly acidic functionalities.[144,258] The acid-treated SWNTs can be dispersed in various amide-type organic solvents under the influence of an ultrasonic force field. The nitric acid purifies the carbon nanotubes, but also introduces defects on the nanotube surface, oxidizes (hole dopes) the carbon nanotubes, and produces impurity states (Figure 1.27) at the Fermi level of the nanotubes.[212] This latter effect may be viewed as an intercalation of the nanotube lattice by oxidizing agents such as nitric acid with partial expoliation, with concomitant effects on the electronic properties of the nanotubes, as revealed by Raman scattering.[250,259] Defect sites that are introduced into the carbon nanotubes can be used to shorten and even destroy the nanotubes under oxidizing conditions.[135,139,144,260,261] The shortened tubes (s-SWNTs) are better solvated by amide solvents than are the full-length SWNTs.

Acid–base titration methods have been used to determine the acidic and basic characters of the surfaces of a wide variety of carbon materials, including carbon MWNTs.[13,240,243,247] The density of surface acidic groups in such MWNTs treated with nitric acid is in the range 0.2–0.5 at%. The total acidic sites (including carboxylic acids, lactones and phenols) can be determined by titration with NaOH, while the carboxylic acid groups can be individually determined by titration with $NaHCO_3$. Boiling these MWNTs in a HNO_3/H_2SO_4 mixture dissolves them and the solid obtained from the solution is highly functionalized by acid groups and the density of acidic groups present exceeds 1 at%.[243,247] Hu *et al.*[258] reported the measurement of the acidic sites in full-length nitric acid purified SWNTs. Titration with NaOH and $NaHCO_3$ gives the total percentage of acidic sites in full-length, purified SWNTs to be: SWNT–COOH functionality, 1–2%; total SWNT acid functionality, 1–3%.

Chen *et al.*[262] have derivatized SWNT fragments with halogen and amine moieties in order to dissolve them in organic solvents. Mickelson *et al.*[263a] studied the solvation of fluorinated SWNTs in alcohol solvents. Fluorinated single-walled carbon nanotubes (F-SWNTs) are obtained by the reaction with elemental fluorine at elevated temperatures. They have a highest degree of functionalization (up to F/C = 1/2) of any reported derivatized carbon-nanotube material. Among the interesting properties of F-SWNTs are their solubility in alcohols and their further derivatization with nucleophiles. F-SWNTs can be reduced with hydrazine to yield SWNTs, while the reaction with Grignard reagents or peroxides allows further functionalization. Bettinger[263b] has covered the experimental and computational investigations of F-SWNTs with a focus on the nature and the strength of the C–F linkage. STM studies of fluorinated SWNTs reveal an interesting banded structure followed by atomically resolved regions, indicating sidewall functionalization.[264] Starting from fluorinated SWNTs, Boul *et al.*[265] have carried out alkylation by reaction with alkyl-magnesium bromides or alkyllithium. Individual SWNTs have been deposited controllably on chemically functionalized nanolithographic templates.[70]

Liu *et al.*[70] envisioned that the addition of a long-chain hydrocarbon at the ends of the shortened (100–300 nm) carbon nanotubes might render the functionalized SWNTs soluble in organic solvents. They convert the acid functionality in the short SWNTs into the amide of octadecylamine (Scheme 1.1), thereby obtaining shortened soluble SWNTs. The formation of the amide bond is readily monitored using mid-IR spectroscopy. The SWNTs exhibit several clear spectroscopic signatures to show that the material is in solution. Direct reaction of the acid-purified

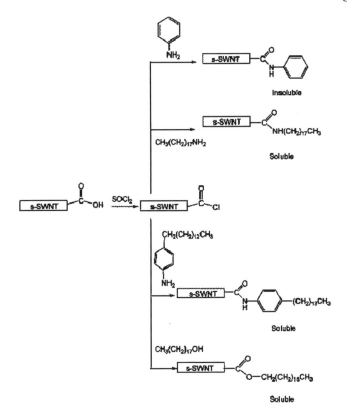

Scheme 1.1   *Covalent chemistry at the open ends of short-SWNTs*
             (Reproduced from refs. 255, 262 and 266)

short SWNTs with long-chain amines led to soluble materials by the formation
of zwitterions (Scheme 1.2).[266] Some insight into the mechanism of dissolution
was obtained by functionalization with 4-tetradecylaniline and with aniline
(Scheme 1.1). While the tetradecylaniline functionalization yielded short, soluble
SWNTs, soluble in THF, CS_2, and aromatic solvents, the aniline derivatives were
soluble only in aniline. The hydrocarbon long-chain appears to play a important role in disrupting and compensating for the loss of van der Waals attraction
between the carbon nanotubes.

Full-length SWNTs are ideal for use in composites and nanoscale conductors.
Thus, a dissolution process of unshortened, full-length SWNTs was accomplished

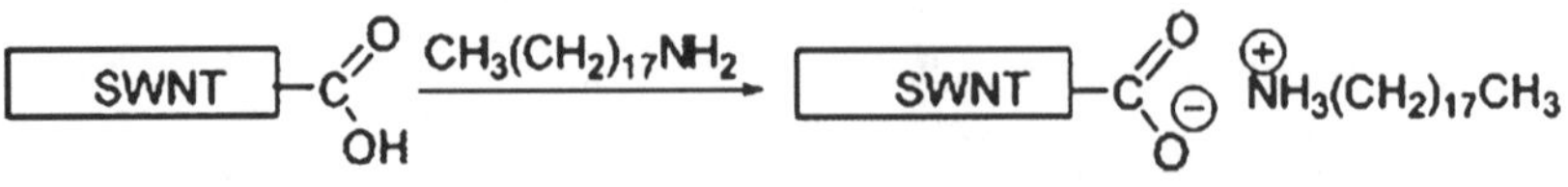

Scheme 1.2   *Zwitterionic functionalization of SWNT (short-length-SWNT)*
             (Reproduced from ref. 267)

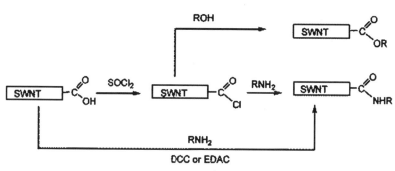

Scheme 1.3 *Carboxylic acid derivatization protocol*
(Reproduced from refs. 266 and 267)

by the direct reaction of the SWNT–COOH with octadecylamine (Scheme 1.2).[267] Ionic functionalization has certain advantages. Firstly, the acid–base reaction represents the simplest possible route to soluble SWNTs and can be readily scaled-up at low cost. Secondly, unlike the covalent amide bond, the cation, $^+NH_3(CH_2)_{17}CH_3$, in the ionic bond of $SWNT-COO^{-+}NH_3(CH_2)_{17}CH_3$ can be readily exchanged with organic and inorganic cations. This feature allows electrostatic interactions between the SWNTs and biological molecules, which can serve as the basis for developing biocompatible SWNTs.

The protocol of adding a bulky side chain to the SWNT ends *via* the amide or the ester bond formation (Scheme 1.3) has been applied by different groups. Riggs *et al.*[268] have studied the strong luminescence of solubilized carbon nanotubes. Sun *et al.*[269] have studied the preparation, characterization and properties of soluble dendron-functionalized carbon nanotubes. Banerjee and Wong[270] have reported the synthesis and characterization of carbon nanotube–nanocrystal heterostructures. Huang *et al.*[271] attached proteins to carbon nanotubes *via* diimide-activated amidation. Self-organization of PEO-graft-single-walled carbon nanotubes in solutions and Langmuir–Blodgett films have been generated by Shinkai and co-workers.[272] Water solubilization of single-walled carbon nanotubes by functionalization with glucosamine was achieved by Pompeo and Resasco,[273] while Huang *et al.*[274] achieved sonication-assisted functionalization and solubilization of carbon nanotubes. Basiuk *et al.*[275] have studied the interaction of oxidized SWNTs with vaporous aliphatic amines.

An alternative method (Scheme 1.4) of functionalization makes use of side wall chemistry by employing electrochemical reduction of aryl diazonium salts using a Bucky paper electrode.[276] Bahr and Tour[277] have synthesized highly functionalized carbon nanotubes using *in situ* generated diazonium compounds. Side-wall functionalization of carbon nanotubes was carried out by Holzinger and co-workers[278] with concommitant loss of the conjugated electronic structure of the nanotubes in order to render the SWNTs soluble. The dissolution methods generally make use of ionic or covalent functionalization of the carbon nanotubes, but noncovalent functionalization is also effective in the dissolution of the small diameter carbon nanotubes.[279]

arylation

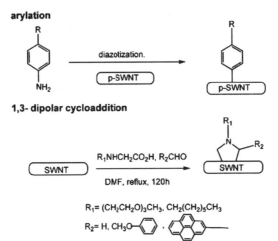

1,3- dipolar cycloaddition

R₁NHCH₂CO₂H, R₂CHO
────────────────────────→
DMF, reflux, 120h

SWNT → SWNT

$R_1 = (CH_2CH_2O)_3CH_3, CH_2(CH_2)_5CH_3$

$R_2 = H, CH_3O-\bigcirc-$, pyrenyl

nitrene cycloaddition

$R-O-C(=O)-N^{\ominus}-N^{\oplus}\equiv N$
────────────160°C────────────→
SWNT

R= *tert*-butyl or ethyl

nucleophilic addition

KO*t*Bu, THF, -60°C
────────────────────→
SWNT

radical addition

$CF_3(CF_2)_6CF_2I$
────────hv , 4h────────→
SWNT

nucleophilic substitution

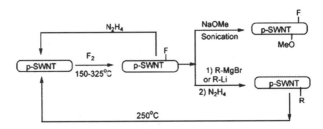

Scheme 1.4 *Side wall chemistry on SWNTs*
(Reproduced from ref. 255)

Covalent chemistry on the walls of the SWNTs is a viable route to soluble materials. The ability to carry out controlled (covalent) chemistry on the side-walls of the SWNTs may become useful for several applications. Covalent side-wall chemistry of SWNTs was first achieved by carbene reactions with the short, soluble SWNTs using phenyl(bromodichloromethyl)mercury in toluene to obtain modifications of the band electronic structure of the SWNTs.[262] Characterization of the functionalized species is a major difficulty in carbon nanotube chemistry, but for wall chemistry, where the band electronic structure is disrupted, solution spectroscopy is useful. Raman spectroscopy can also be used to study the effects of side-wall functionalization. Dichlorocarbene is an electrophilic reagent that will add to deactivated double bonds and to fullerenes. Electronic spectroscopy has shown that the band-gap transitions in the semiconducting tubes are completely disrupted at a functionalization level of 2% of the available SWNT carbon atoms. Apart from carbene chemistry, other chemical processes such as nitrene addition, hydrogenation *via* the Birch reduction, fluorination, alkylation, arylation, and 1,3-dipolar cycloaddition can be carried out with nanotubes. The chemistry of SWNTs has been reviewed comprehensively by Niyogi *et al.*[255]

Unlike the perpendicular alignment of carbon nanotubes discussed earlier, the *post-synthesis* ordering of chemically and/or physically modified carbon nanotubes has not been examined in detail. Perpendicularly aligned carbon nanotubes have, however, been constructed from end-functionalized carbon nanotubes on certain substrates. For example, Liu *et al.*[280] have prepared aligned SWNTs by self-assembling nanotubes end-functionalized with thiol groups on a gold substrate. The carboxy-terminated short SWNTs prepared by acid oxidation were used as the starting material for further functionalization with thiol-containing alkyl amines through the amide linkage. The self-assembly of aligned SWNTs was obtained by dipping a gold(111) ball into the thiol-functionalized SWNT suspension in ethanol, followed by ultrasonication and drying in nitrogen. The self-assembled aligned nanotube film thus obtained was stable and ultrasonication could not remove it from the gold substrate. The packing density of the self-assembled, aligned carbon nanotubes depended strongly on the incubation time.

Shimoda *et al.*[281] have demonstrated the preparation of ordered/micropatterned nanotubes through the self-assembly of preformed nanotubes on glass, and other substrates, by vertically immersing the substrate into an aqueous solution of acid-oxidized short SWNTs. Carbon nanotube deposition did not occur on a hydrophobic substrate (*e.g.* a polystyrene spin-coated glass slide) under the conditions. Such a selectivity enabled the fabrication of carbon nanotube patterns by using glass substrates with prepatterned hydrophobic and hydrophilic regions. We must recall here that Chen and Dai[71] prepared micropatterns of nanotubes by patterned growth on surfaces prepatterned with plasma polymer or region-specific adsorption of certain chemically modified nanotubes. As an alternative to solution phase functionalization,[248,264] carbon nanotube (MWNT) probes in an AFM have been functionalized by a discharge process, in the presence of different gases.[282] Such functionalized carbon nanotube probes can be employed for chemically sensitive imaging of materials, especially biomolecules.

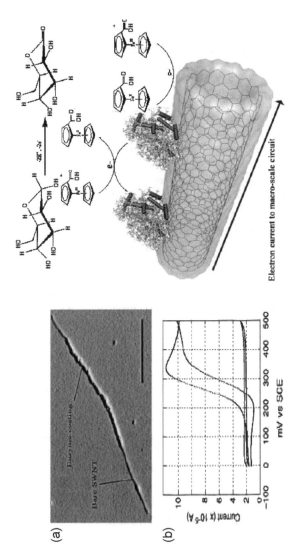

Figure 1.32 (a) *TMAFM amplitude micrograph of a GOX modified SWNT in which a high degree of enzyme loading is apparent. Scale bar 200 nm.*
(b) *Voltammetric response of such nanotubes in the absence (lower curves) and presence (upper curves) of substrate, β-D-glucose.*
(c) *Schematic representation of the SWNT glucose biosensor. Solution-phase D-glucopyranose is turned over by oxidase enzymes immobilized on the nanotubes. This redox process at the enzyme flavin moieties is communicated to the nanotube system through the diffusive mediator ferrocene monocarboxylic acid. The redox action of the ferrocene at the nanotube surface ultimately generates a quantifiable catalytic current characteristic of substrate detection and turnover* (Reproduced from ref. 283)

Glucose sensing with SWNTs has been possible by functionalizing the side-walls or ends of oxidized nanotubes with colloidal particles or polyamine dendrimers through carboxylate chemistry.[283] Proteins adsorb individually along the nanotube length. The nanotube–protein conjugates are characterized at the molecular level by AFM. Several metalloproteins and enzymes are bound on both the sidewalls and termini of SWNTs. Though coupling can be controlled to some degree, AFM studies suggest that the immobilization is strong and does not require covalent bonding. Protein attachment appears to occur with retention of native biological structure. Nanotube electrodes exhibit useful voltammetric properties with direct electrical communication possible between a redox-active biomolecule and the delocalized system of its carbon nanotube support. A biomodified SWNT acts as a good glucose biosensor as (Figure 1.32).

Richard *et al.*[284] have assembled of surfactants and synthetic lipids on the surface of carbon nanotubes. Above the critical micellar concentration c.m.c., sodium dodecyl sulphate (SDS) forms supramolecular structures made of rolled-up half-cylinders on the nanotube surface. Depending on the symmetry and the diameter of the carbon nanotube, it forms rings, helices, or double helices. Such self-assemblies could also be obtained with several synthetic single-chain lipids designed for the immobilization of histidine-tagged proteins. At the nanotube–water interface, permanent assemblies were obtained from mixed micelles of SDS and different water-insoluble double-chain lipids after the dialysis of the surfactant. Such arrangements can be exploited for developing biosensors and bioelectronic nanomaterials

Kartz and Willner[285] have reviewed biomolecule-functionalized carbon nanotubes and their applications in nanobioelectronics. These authors review the various synthetic strategies to chemically modify the side-walls or tube ends by molecular or biomolecular components. The tailoring of hybrid systems consisting of CNTs and biomolecules (proteins and DNA) has rapidly expanded and attracted substantial research effort. The integration of biomaterials with CNTs enables the use of the hybrid systems as active field-effect transistors or biosensor devices (enzyme electrodes, immunosensors, or DNA sensors). Their integration also helps to generate complex nanostructures and nanocircuitry of controlled properties and functions.

Song *et al.*[286] have reported evidence for the formation of a lyotropic liquid crystalline phase of surface-treated MWNTs (with HNO_3–H_2SO_4 mixture) in aqueous dispersions. By examination of a series of aqueous dispersions of the surface-treated/functionalized MWNTs of different concentrations, they showed a phase transition from isotropic to a Schlieren texture typical of lyotropic nematic liquid crystals above a critical concentration of 4.3% by volume.

5 Electronic Properties

Carbon nanotubes have many fascinating properties, the significant ones being the varied electronic properties which depend on the structure, high thermal conductivity and superior mechanical strength. There are also properties related to the tubular one-dimensional nature. They show liquid crystalline characteristics and good gas adsorption properties with internal surface areas of ~ 300 m^2 g^{-1}.

Some of these properties are potentially useful for technological applications. The electronic properties are likely to be exploited in many devices such as field-effect transistors. Nanotubes may be used as AFM or STM tips. We shall examine the salient electronic properties in this section. Some of the other properties are covered in the next two sections. STM–STS studies, along with tight–binding calculations, reveal interesting electron transport characteristics of as-prepared, shortened as well as bent SWNTs,[20a,287] and demonstrate how a range of electronic transport characteristics can be achieved with SWNTs. Fundamentals of electronic properties of carbon nanotubes have been reviewed by Lieber and co-workers[288] and Dresselhaus *et al.*[22] Electron transport properties of SWNTs have been reviewed by McEuen and Park.[289]

Ultrahigh-vacuum scanning tunnelling microscopy (STM) has allowed atomic resolution imaging of the surface of SWNTs, and *I–V* spectroscopy has provided the direct measurement of the electronic band structure. As noted earlier, SWNTs can behave as metals, semiconductors, or small band-gap semiconductors, depending upon their diameter and chirality. Electronic transitions between the energy bands of SWNTs (Figure 1.33) can be observed by standard spectroscopic techniques. Besides the heterogeneity of the samples with respect to the tube diameters and helicities, impurity doping contributes to the breadth of the absorption features. Since the band gaps are inversely proportional to the tube diameters, structural information can be derived from the band transition energies.[211,290] Figures 1.34 and 1.35 show typical results obtained by employing

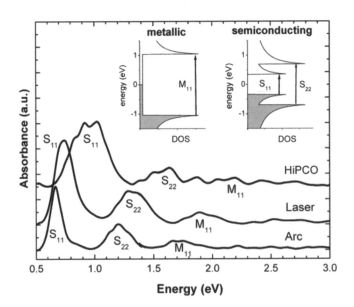

Figure 1.33   *Electronic transitions between the energy bands of SWNTs observed by electronic spectroscopy of films (purified HiPCO, purified Laser, and soluble arc produced SWNTs) after baseline correction, together with a schematic of the nomenclature used to designate the interband transitions* (Reproduced from ref. 211)

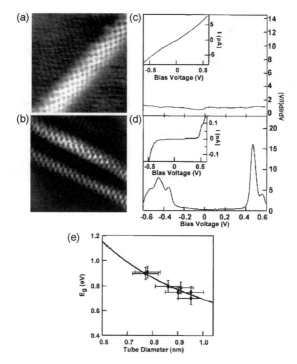

Figure 1.34 (a), (b) *STM images of atomically resolved SWNTs.* (c), (d) *The current–potential (I–V) characteristics of metallic and semiconducting SWNTs, respectively.* (e) *Inverse scaling of tunnelling gap with the nanotube diameter* (Reproduced from refs. 20a and 287)

STM. The STM images in the figures show atomically resolved images of SWNTs.[20a,287] The *I–V* characteristics show both metallic and semiconducting behaviour with the tunnelling gap varying inversely with nanotube diameter. Calculated and observed density of states showing van Hove singularities can be seen from the figure.

Electrical transport properties of bundles of single-walled carbon nanotubes have been measured by Bockrath *et al.*[291] Tans *et al.*[292] reported electron transport measurements on individual SWNTs and found conduction to occur through well separated, discrete electron states that are quantum-mechanically coherent over long distances (~140 nm). Venema *et al.*[293] imaged the electronic wave functions on the short metallic nanotubes by employing STM and STS. These wave functions correspond to the quantized energy levels in single-walled nanotubes. The implications of topological defects in carbon nanotubes has been investigated by Chico *et al.*[294] The inclusion of pentagon–heptagon defects in a carbon nanotube leads to confined electronic states, reminiscent of a quantum dot. By varying the distance between the pentagon–heptagon pairs, modifications in the energy difference, spatial confinement and the number of discrete levels can be achieved.

SWNTs show a linear dependence of resistivity on temperature.[295] The role of twistons (long-wavelength torsional shape fluctuations) on the conduction in

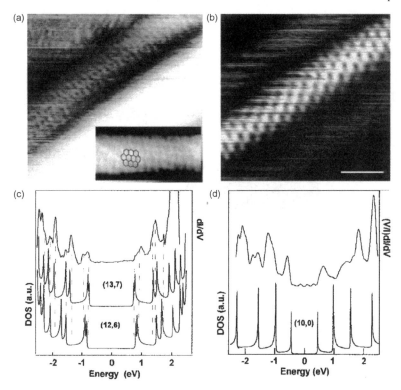

Figure 1.35   (a), (b) *STM images of SWNTs.* (c), (d) *Corresponding I–V characteristics of the tubes shown in (a, b). The calculated and observed density of states curves are also included (The scale bar is 1 nm)*
(Reproduced from refs. 20a and 287)

SWNTs has been analysed. Electrical contacts to carbon nanotubes influence the contact resistance, and hence their ultimate use in devices. In measurements of low-temperature transport properties, SWNTs exhibit high contact resistance, probably due to the weak electronic coupling of nanotubes with the Fermi level of the metal contacts.[296] The resistance of SWNTs decreases at high pressures, due to van der Waals compaction of the intertube spacing.[297] By employing CVD, Kong *et al.*[298] synthesized SWNTs on patterned catalyst islands and characterized the low-temperature transport properties. The observed two-terminal resistance (at 4.2 K) of a metallic SWNT was around 16.5 kΩ. In a modified technique towards gating the individual SWNTs, AFM tips were used to contact the nanotubes with Ti electrodes and the transport characteristics measured.[299] Avouris *et al.*[300] could manipulate multi-walled nanotubes using AFM, to observe radial and axial structural deformations arising from the interaction between the nanotubes and the substrate. The strong substrate–nanotube interaction was utilized to position the individual nanotubes on electrical contacts and their electrical characteristics have been evaluated, apart from their use as tips for AFM lithography. A semiconducting SWNT-based field-effect transistor action was observed by Tans *et al.*[301] at room temperature, making the nanotube-based devices a distinct possibility. A semi-conducting MWNT-based field-effect transistor has also been reported.[302] The metallic nanotubes generally show gate-voltage-independent transport characteristics.

Single-walled carbon nanotubes with highly reproducible Pd ohmic contacts and lengths ranging from several microns down to 10 nm, have been investigated.[303] The mean-free path (MFP) for acoustic phonon scattering is estimated to be l_{ap} ~300 nm, and that for optical phonon scattering is l_{op} ~15 nm. Transport through very short (~10 nm) nanotubes is free of significant acoustic and optical phonon scattering and thus ballistic and quasiballistic at the low- and high-bias voltage limits, respectively. High currents of up to 70 μA can flow through a short nanotube. Possible mechanisms for the eventual electrical breakdown of short nanotubes at high fields are discussed. These results have implications in high-performance nanotube transistors and interconnects. Coskun *et al.*[304] report experiments on quantum dot single-electron-tunnelling (SET) transistors made from short multi-wall nanotubes and threaded by magnetic flux. The system allowed the authors to probe the electronic energy spectrum of the nanotube and its dependence on the magnetic field. Evidence was obtained for the interconversion between gapped (semiconducting) and ungapped (metallic) states. The tubes exhibited *h/e*-period magnetic flux dependence, in agreement with simple tight-binding calculations.

The current-induced electromigration causes conventional metal wire inter-connects to fail when the wire diameter becomes too small. The covalently bonded structure of carbon nanotubes helps to avoid such a breakdown of nanotube wires, and, because of ballistic transport, the intrinsic resistance of the nanotube nearly vanishes. Experimental results show that metallic SWNTs can carry up to 10^9 A cm^{-2}, whereas the maximum current densities for normal metals are 10^5 A cm^{-2}.[305,306] Unfortunately, the ballistic current carrying capability is less useful for presently envisioned applications because of necessarily large contact resistances. An electronic circuit involving electrical leads to and from a SWNT will have a resistance of at least $h/4e^2$ or 6.5 kΩ, where h is Planck's constant and e is the charge of an electron.[307] Contacting all layers in a MWNT could reduce this contact resistance, but it cannot be totally eliminated.

SWNTs exhibit large thermoelectric power (TEP) with hole-like behaviour at high temperatures.[308] Calculations on a single metallic SWNT predict values sig-nificantly lower than the measured ones, signifying the breaking-up of electron–hole symmetry due to the self-assembly of nanotubes into crystalline ropes. The presence of transition metals along with the SWNTs strongly influences the electrical resistance and TEP. This transport behaviour has been assigned to Kondo effect[309] (*i.e.* interaction between magnetic moment of the metal and the spin of the conduction electrons of nanotubes). Magnetoresistance behaviour of SWNTs in the 4–300 K range is consistent with 2D variable-range-hopping and weak localiza-tion.[310] Transport through doped nanotube junctions has been investigated by Farajian *et al.*[311] who find that regions of negative differential resistance exist in the *I–V* characteristics of metallic nanotubes, while an asymmetric transport char-acteristic with respect to the applied bias is expected for semiconducting nanotubes.

An interesting manifestation of charge transport in carbon nanotubes in the presence of an applied magnetic field is that the magnetoresistance shows oscillations as a function of field at low temperatures. This effect, referred to as Aharanov–Bohm effect, has been observed in MWNTs.[312] Short period oscillations found in some nanotubes are attributed to defect-induced anisotropic electron currents.

Doping is expected to alter the electrical properties of pristine carbon nanotubes. After doping semiconducting SWNT ropes with potassium, through reversible intercalation, Bockrath *et al.*[313] measured their electrical properties. Doping changes the carriers in the ropes from holes to the electrons. The majority carrier (electrons) density is similar to that (for holes) in the pristine nanotubes.[302] Transport as well as optical conductivity measurements on potassium-doped SWNT films show that doping influences the on-tube and the intertube transport differently.[314] Four-probe transport measurements on SWNT ropes demonstrate that doping involves charge transfer and the SWNTs are inherently p-type materials, because of defects or inadvertent doping by exposure to air.[315] Transport properties of SWNTs can also be modified by employing the redox reactions.[237a] Van Hove singularities characteristic of 1D systems disappear due to the loss of long-range ordering on intercalation.

Because of the nearly 1D electronic structure, electron transport in metallic SWNTs and MWNTs occurs ballistically over long nanotube lengths, enabling them to carry high currents with essentially no heating.[305,316] Phonons also propagate easily along the nanotube. The measured room temperature thermal conductivity for an individual MWNT (>3000 W m^{-1} K^{-1}) is greater than that of natural diamond and the basal plane of graphite (both 2000 W m^{-1} K^{-1}).[317] Superconductivity has been observed in SWNTs at low temperatures, with transition temperatures of ~0.55 K for 1.4 nm diameter SWNTs[318] and ~5 K for 0.5 nm diameter nanotubes grown in zeolites.[319]

6 Carbon Nanotube Composites and their Properties

Nanotube–polymer composites have been investigated by several workers in the last few years. Of interest are the electrical and mechanical properties of the composites, especially the latter. Determining the mechanical properties of SWNTs is difficult. Some measurements have been made using a nanostressing stage.[320] These have yielded values for the Young's modulus in the range 320–1470 GPa (mean 1002 GPa) and an average breaking strength of 30 GPa. For comparison, the stiffest conventional carbon fibres have Young's modulus of ~800 GPa, while glass fibres typically have a modulus of ~70 GPa. Direct measurements on individual nanotubes using atomic force microscopy show that they can accommodate extreme deformations without fracturing.[321] They also have the extraordinary capability of returning to their original structure after deformation. Carbon nanotubes are also excellent electrical conductors, with current densities of up to 10^{11} A m^{-2}, and have high thermal conductivities.[36,317] These properties can be exploited by incorporating the nanotubes into a matrix. Preparation and characterization of nanotube composites are therefore of value.[322] The composites generally employ polymer matrices, but there is interest in other matrix materials such as ceramics and metals as well.

A commonly used method for preparing nanotube–polymer composites involves mixing nanotube dispersions with solutions of the polymer and then evaporating the solvents in a controlled manner. Functionalization has been used for the formation of surface carboxyl and other groups to facilitate the dispersion of nanotubes in solvents and help to improve bonding between tubes and matrix. As noted in previous sections, carbon nanotubes are often pretreated chemically to facilitate solubilization. Shaffer *et al.*,[323] have shown that acid treatment enables stable aqueous solutions of catalytically produced MWNTs. A nanotube–PVA [poly(vinyl alcohol)] composite could be prepared simply by mixing one of these aqueous nanotube dispersions with an aqueous solution of the polymer and then casting the mixtures as films and evaporating the water.[324] Solution-based methods have also been used to produce nanotube–polystyrene composites.

Hill *et al.*[325] have solubilized both SWNTs and MWNTs by functionalizing with a polystyrene copolymer. This was done by acid treating the nanotubes, followed by esterification of the surface bound carboxyl groups. The polymer-modified carbon nanotubes were soluble in common organic solvents. To prepare composites, polystyrene was dissolved in the nanotube solution, and nanotube–polystyrene thin films prepared using wet casting. Functionalization of the nanotubes is not always necessary to prepare a polymer composite. Qian *et al.*[326] used a high energy ultrasonic probe to disperse MWNTs in toluene and then mixed the dispersed suspension with a dilute solution of polystyrene in toluene under ultrasonic agitation. The low viscosity of the polymer solution allowed the nanotubes to move freely through the matrix. The mixture was cast on glass and the solvent removed to yield composite films.

Nanotube–polystyrene composites have been prepared by *in situ* polymerization. Using thermoplastic polymers, shear mixing of the melt with the nanotubes can be employed to produce homogeneous dispersions. Extrusion produces aligned nanotubes; artefacts are produced by injection moulding. Andrews *et al.*[327] employed shear mixing to disperse catalytically produced nanotubes in several polymers such as high-impact polystyrene, acrylonitrile and polypropylene. Haggenmueller *et al.*[328] used a combination of solvent casting and melt mixing methods to disperse SWNTs in poly(methyl methacrylate) (PMMA).

Composite fibres with a high degree of nanotube orientation are obtained by melt spinning. Polycarbonate, which is a thermoplastic polymer, can be used as a matrix for nanotube composites.[329] Sennett *et al.*[330] have used melt processing techniques to disperse and align carbon nanotubes in polycarbonate. Dispersion was achieved by mixing catalytically produced MWNTs and SWNTs with the polycarbonate resin in a conical twin-screw extruder, and alignment was carried out using a fibre spinning apparatus. By optimizing mixing time and fibre draw rates, excellent dispersion and alignment were accomplished. Nanotube–epoxy composites have also been prepared.[331] The earliest work in this area was carried out by Ajayan *et al.*[68] who embedded the nanotubes in an epoxy resin, which was then cut into thin slices with a diamond knife. The nanotubes were found to be aligned in the direction of the knife movement (Figure 1.36). The alignment is

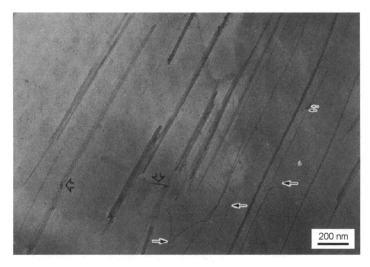

Figure 1.36 *Alignment of nanotubes in a polymer matrix following cutting with microtome: arrows indicate buckled nanotubes*
(Reproduced from ref. 68)

considered to be primarily a consequence of extensional or shear flow of the matrix produced by the cutting. Alignment is more readily achieved when the epoxy is in a liquid state.

Using spin coating methods, Xu *et al.*[331d] produced some alignment of nanotubes in nanotube–epoxy composites. Nanotube–epoxy thin films showed excellent mechanical properties. Compared to neat resin thin films, for which the elastic modulus was 4.2 GPa, a 20% increase in modulus was seen when 0.1 wt% MWNTs were added. This was attributed to partial alignment of the MWNTs induced by spin coating. Fracture behaviour of the films, investigated by SEM, showed that pulled-out tubes were often covered with polymer, suggesting strong interfacial adhesion.

Several other nanotube–polymer composites have been prepared using *in situ* polymerization, including MWNT–polystyrene[332] and SWNT-polyimide.[333] Electrochemical polymerization has been employed to grow porous composite films of MWNT and polypyrrole for use as supercapacitors.[334]

In most composite preparations, the aim is to produce samples in which the nanotubes are distributed evenly throughout the polymer (either randomly oriented or aligned). For some applications, however, a layered arrangement is advantageous. Photovoltaic devices containing layers of nanotubes or nanotube–polymer composites have been prepared. The preparation of multilayer SWNT–polymer composites with excellent mechanical properties has also been described.[335] The method involved layer-by-layer deposition of SWNTs and polymer onto a substrate, followed by crosslinking. Composites with SWNT loadings as high as 50 wt% could thus be obtained. Coating was carried out by alternate dipping of a glass slide or silicon wafer into dispersions of SWNTs (stabilized by acid treatment) and polymer solutions. The layers were held together by van der Waals forces and by electrostatic attraction between the negatively charged SWNTs and a positively charged polyelectrolyte, such as branched polyethyleneimine (PEI). Coatings containing up to 50 SWNT-PEI bilayers could thus be built up. When the procedure was complete, the multilayer films were heated to 120 °C to promote crosslinking. The films could then be lifted off the substrate to obtain uniform free-standing membranes. A similar method has been used to prepare multilayer composites films of the polyelectrolyte polydiallyldimethylammonium chloride and SWNTs.[336]

To-date, most studies have been directed towards exploiting the mechanical properties of carbon nanotubes rather than their electronic or optical properties, although interest in these properties is growing. The mechanical performance of carbon nanotube composites depends on whether the nanotubes are from the arc or catalytically produced. Qian *et al.*[326] found catalytically produced nanotubes to be well dispersed throughout the polymer matrix with no clustering. During composite fracture, the nanotubes often broke, suggesting that they had relatively poor strength. Watts and Hsu,[337] however, employed more perfect nanotubes, produced by arc evaporation, and found superior mechanical properties, with no breaking. The nanotubes tended to cluster together so that during crack growth the tubes could quite easily slide past one another. The best mechanical performance is achieved with arc grown nanotubes.

Interfacial bonding between the nanotubes and the polymer matrix can be enhanced by functionalizing the nanotubes before embedding them in the matrix. Gojny *et al.*[338] have shown that functionalizing arc-produced MWNTs with amine groups greatly enhances bonding with the epoxy matrix. As a result, the outer shells of the tubes often remained embedded in the matrix following pullout. A possible disadvantage of functionalization is that it could weaken the nanotubes by disrupting the perfect cylindrical structure. Functionalization of the nanotubes may not, however, be necessary to produce bonding with a polymer matrix. Liao and Li[339] have modelled the interfacial characteristics of a nanotube–polystyrene composite, assuming no covalent bonding between the tubes and the matrix. They find that electrostatic and van der Waals interactions contributed significantly to the interfacial stress transfer. A nanotube pull-out simulation suggested that the interfacial shear stress of the nanotube–polystyrene system is about 160 MPa, significantly higher than for most carbon fibre reinforced polymer composite systems. Star and co-workers[340] have demonstrated that poly{(5-alkoxymphenylenevinylene)-*co*-[(2,5-dioctyloxy-*p*-phenylene)vinylene]} (PAmPV) derivatives could be induced to wrap around SWNTs, forming strong non-covalent bonds. Single-walled nanotube–epoxy composites have been prepared by Biercuk *et al.*[341] The Vickers hardness of the polymer increased monotonically with addition of SWNTs, up to a factor of 3.5 at 2 wt% nanotube loading. Greatly enhanced thermal conductivities were also observed. Thus, samples loaded with 1 wt% unpurified SWNT material showed a 70% increase in thermal conductivity at 40 K, rising to 125% at room temperature.

Pristine SWNTs when mixed and ground with imidazolium ion-based roomtemperature ionic liquid form a physical gel.[342] The heavily entangled nanotube bundles untangle within the gel to form much finer bundles. The gels appear to be formed by physical crosslinking of the nanotube bundles, mediated by local molecular ordering of the ionic liquids rather than by entanglement of the nanotubes. The gels were thermally stable and did not shrivel, even under reduced pressure resulting from the nonvolatility of the ionic liquids, but they readily underwent a gel-to-solid transition on absorbent materials. The use of a polymerizable ionic liquid as the gelling medium allows the fabrication of a highly electroconductive polymer–nanotube composite material, showing a considerable enhancement in dynamic hardness.

Mamedov *et al.*[335] have prepared SWNT–polyelectrolyte composites by the layer-by-layer method. Films produced in this way have exceptional mechanical properties. The average ultimate tensile strength of the films was 220 MPa, with some measurements as high as 325 MPa. This is several orders of magnitude greater than the tensile strength of strong industrial plastics, and indicates that the layer-by-layer method has great promise for the production of strong nanotube-containing composites.

Li *et al.*[343] spun fibres and ribbons of carbon nanotubes directly from the CVD synthesis zone of a furnace using a liquid source of carbon and an iron nanocatalyst. MWNT yarns with much higher strengths could be directly spun from aerogels during nanotube synthesis by CVD. This process was realized through the appropriate choice of reactants, control of the reaction conditions, and continuous withdrawal of the product with a rotating spindle used in various geometries. This direct spinning from a CVD reaction zone is extendable to other types of fibre and to the spin coating of rotating objects in general. The resulting fibre can then be postimpregnated with epoxy to make a composite. These fibres presently have failure strengths of up to 1 GPa, compared with an expected ultimate strength of 30–50 GPa for a single nanotube. In fact, fibres made by each method creep before failure, suggesting that a great deal of straightening and aligning occurs before failure. It also shows that much optimization is needed to share the load better across the many individual CNTs. By introducing twist during spinning of multi-walled carbon nanotubes from nanotube forests to make multiply, torque-stabilized yarns, Zhang *et al.*[344] achieved yarn strengths greater than 460 MPa. These yarns deform hysteretically over large strain ranges, reversibly providing up to 48% energy damping, and are nearly as tough as fibres used for bulletproof vests. Unlike ordinary fibres and yarns, these nanotube yarns are not degraded in strength by overhand knotting. They also retain their strength and flexibility after heating in air at 450 °C for an hour or when immersed in liquid nitrogen. High creep resistance and high electrical conductivity are observed and are retained after polymer infiltration, which substantially increases yarn strength. PVA infiltration decreased yarn electrical conductivity by only ∼30%, leading to MWNT-PVA composite yarns whose electrical conductivity is more than 150× that of coagulation spun nanotube composite fibres containing this insulating polymer.

One way of preparing nanotube–polymer composites is to use the monomer rather than the polymer as a starting material, and then carry out *in situ* polymerization. Cochet *et al.*[345] used this method to prepare a MWNT–polyaniline (PANI) composite. Nanotubes prepared by arc evaporation were used, and were sonicated in a solution of HCl to achieve a dispersion. The aniline monomer, again in HCl, was added to the suspension and a solution of an oxidant was slowly added with constant sonication and cooling. Sonication was continued in an ice bath for 2 h, and the composite was then obtained by filtering, rinsing, and drying. In this way, it was possible to prepare composites with high MWNT loadings (up to 50 wt%). Transport measurements on the composite revealed major changes in the electronic behaviour, confirming strong interaction between nanotubes and polymer. The room temperature resistivity of the composite was an order of magnitude lower than that of pure polyaniline, while the low temperature resistivity was much smaller than that of either polyaniline or of MWNTs. The temperature dependence of resistivity was also weaker than that of polyaniline alone. These observations were explained by assuming that *in situ* polymerization favours charge transfer between polyaniline and MWNTs, resulting in an overall material that is more conducting than the starting components. Vivekchand *et al.*[346] prepared composites of PANI by the *in situ* polymerization of aniline by ammonium persulphate with pristine multi-walled and single-walled nanotubes as well as with nanotubes subjected to acid treatment and subsequent reaction with thionyl chloride. The well-characterized PANI-nanotube composites exhibit electrical resistivities different from those of the parent nanotubes and of PANI. The electrical resistivity of the PANI-nanotube composites could be manipulated through the variation of the composition as well as by the prior treatment of the nanotubes. Figure 1.37(a) shows the electrical resistivities of the parent nanotubes, PANI and the 2:1, 1:2 composites of PANI with multi-walled nanotubes. The resistivities of the MWNTs are low and nearly temperature independent (curve a–c). PANI shows an insulating type behaviour with a strong temperature dependence of resistivity (curve d). The resistivity of the MWNTs decreases slightly on acid treatment but the resistivity of the chlorinated MWNTs is similar to that of the pristine MWNTs. The resistivities of the 2:1 PANI-MWNT composites fall between those of PANI and MWNTs (curves e–g). The resistivities of the 1:2 composites show similar behaviour (h–j). Figure 1.37(b) shows the electrical resistivities of the 1:1 PANI-SWNT composites. The PANI-SWNT (1:1) composite shows a slightly higher resistivity than PANI while the PANI-acid-treated-SWNT composite shows a resistivity close to that of PANI. The resistivity of the PANI-chlorinated SWNT composite is close to that of the SWNTs.[346]

Sandler *et al.*[331b] have investigated the electrical properties of nanotube–epoxy composites. Matrix resistivities of about 100 Ω m with filler volume fractions as low as 0.1 vol.% were achieved. These figures represented an advance on the best conductivities previously obtained with carbon black in the same epoxy matrix. These and other published studies suggest that carbon nanotubes have great promise in reducing electrostatic charging of bulk polymers. Nanotubes also have other advantages over conventional fillers, such as carbon black and carbon fibres, in that they are more amenable to processing and can be more easily dispersed throughout the matrix. In addition to the studies available in the open literature, there is, undoubtedly, much commercial work being carried out on the use of nanotubes to reduce the electrostatic charging of plastics. In fact, nanotube-containing plastics are already being used in commercial products.[347] These

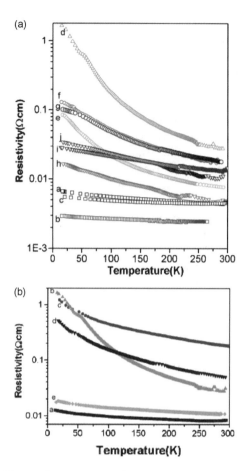

Figure 1.37 (A) *Electrical resistivities of PANI-MWNT composites*: (a) *as prepared MWNTs*, (b) *acid treated MWNTs*, (c) *chlorinated MWNTs*, (d) *PANI, 2:1 composite with* (e) *as-prepared MWNTs*, (f) *with acid – treated MWNTs and* (g) *with chlorinated MWNTs and 1:2 composite with* (h) *as prepared MWNTs,* (i) *with acid-treated MWNTs* (j) *with chlorinated MWNTs*. (B) *Electrical restivities of PANI-SWNT composites*: (a) *as prepared SWNTs*, (b) *PANI, 1:1 composite with* (c) *as-prepared SWNTs*, (d) *with acid-treated SWNTs and* (e) *with chlorinated SWNTs*
(Reproduced from ref. 346)

include fuel lines in automobiles, where the nanotubes help to dissipate any dangerous charge that may build up. Thermoplastic polymers containing nanotubes are also used in some exterior automobile parts, so that they can be earthed during electrostatic painting.

Several groups have prepared nanotube composites using the conjugated polymer poly(*p*-phenylene vinylene) (PPV) and its derivatives. Polymers of this class have electroluminescent properties, and are widely used in light-emitting diodes. They are also used in photovoltaic devices. In an early study, workers from Ireland and the USA described the synthesis of a composite material containing arc-produced MWNTs in a PPV derivative, PmPV.[348] The composite was prepared using a solution mixing method, and good wetting of the tubes by the polymer was achieved. One of the aims of adding nanotubes to the polymer was to increase its electrical conductivity. Previous attempts to achieve this by doping had resulted in degradation of the optical properties of the polymer. For nanotubes, the electrical conductivity of the polymer was increased by up to eight orders of magnitude, with no concomitant degradation of optical properties. This was, apparently, because the tubes acted as nanometric heat sinks, preventing the build up of large thermal effects, which damage these conjugated systems. This work also found that the composite could be used as the emissive layer in an organic light-emitting diode (LED). Several groups have been interested in using nanotube–polymer composites in photovoltaic devices.

Kymakis and Amaratunga[349] used a different approach to prepare a nanotube-conjugated polymer photovoltaic device. In this work, composite SWNT-poly(3-octylthiophene) films were drop or spin cast from a solution onto an indium–tin oxide (ITO) coated quartz substrate. A clear diode characteristic was present in the dark, and, upon illumination through the aluminium electrode, a photocurrent was observed. A quantum efficiency approximately twice that of the standard indium–tin oxide device was reported. Devices were also prepared using the pure P3OT polymer. The photovoltaic effect was enhanced considerably with the P3OT-SWNTs blend device compared with the pure polymer device.

Suspensions of both SWNTs and MWNTs have nonlinear optical properties. Nanotube–polymer composites may, therefore, have applications in optical devices. To date, however, there has been little work on the optical limiting properties of nanotube-containing composites. In one of the first such studies, Chen *et al.*[350] demonstrated third-order optical nonlinearity in a SWNT-polyimide composite. O'Flaherty *et al.*[351] prepared a composite in which MWNTs were dispersed in the polymer poly(9,9-di-n-octylfluorenyl-2,7′-diyl) (PFO). Maximum optical limiting was observed at carbon nanotube loadings in excess of 3.8 wt%, relative to the polymer.

Han *et al.*[352a] have described a novel nanotube–silica composite material with potentially useful optical properties. They synthesized films consisting of silica spheres with diameters ranging from 200 to 650 nm. Molybdenum/cobalt catalyst particles were then deposited onto the silica spheres to catalyse the growth of SWNTs. Since SWNTs are highly nonlinear and fast-switching materials, they can be incorporated into an optically confining environment to achieve these characteristics at relatively low levels of laser intensities. Curtin and Sheldon[352b] have briefly reviewed the incorporation of CNTs into ceramic and metal matrices to form composite structures with an emphasis on processing methods, mechanical performance, and prospects for successful applications.

7 Applications, Potential and Otherwise

The wide range of fascinating properties of carbon nanotubes provide attractive opportunities for technological applications.[353,354] Some are realistic and likely to become commercial in the near future, while others are in the development stage. Notably, however, numerous patents and proto-type devices have been reported in the last few years.

Electronic Applications

Applications based on electronic properties have been reviewed[288] and some of them, such as transistor action, were indicated in earlier sections. We shall survey some of the possible device applications and the properties related to them in this section. Device miniaturization in semiconductor technology is expected to reach its limits due to the inherent quantum effects as one goes towards smaller size. In such a scenario, a feasible alternative is nanoelectronics based on molecules. The possible use of carbon nanotubes in nanoelectronics has aroused considerable interest. For such applications it becomes important to be able to connect nanotubes of different diameters and chirality.[355] Nanostructures with well-defined crystalline interfaces are of possible use in electronic devices. The *I–V* (current–voltage) curves of a SWNT reported by Tans *et al.*[301] illustrate field-effect transistor action (Figure 1.38). Because of their unique electronic properties, SWNTs can be interfaced with other materials to form novel heterostructures. Thus, SWNT-carbide nanorod heterostructures have been prepared by Zhang *et al.*[356] The interfaces of SWNTs with carbides such as SiC, TiC or NbC may possess interesting electronic properties.

The simplest device one can imagine with carbon nanotubes is that involving a bend or a kink, arising from the presence of a diametrically opposite pentagon–heptagon pair. The resultant junction connects two nanotubes of different chirality

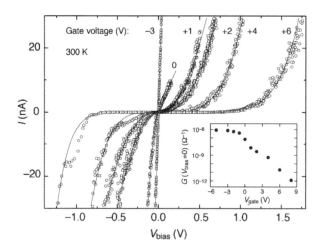

Figure 1.38   *Current – potential characteristics at various gate voltages (V_{gate}) over a -3 to $+6$ V range for a SWNT-based field-effect transistor (300 K). Inset: Conductance S versus gate voltage at bias $V_{bias} = 0$ V* (Reproduced from ref. 301)

and hence of different electronic structure, leading to the realization of an intramolecular device. Such a device in SWNTs behaves like a diode rectifier.[357] Complex three-point nanotube junctions have been proposed as the building blocks of nanoelectronics and, in this regard, Y- and T-junctions have been considered as prototypes.[358,359] The Y- and T-junctions appear to defy the traditional models involving an equal number of five- and seven-membered rings to create nanotube junctions. Instead, such junctions can be created with an equal number of five- and eight-membered rings.[359] To date, there have been no practical devices made of real three-point nanotube junctions. However, junctions consisting of crossed nanotubes have been fabricated to study their transport characteristics.[360] Y-junction nanotubes have been produced by using Y-shaped nanochannel alumina as templates.[361] By carrying out simple pyrolysis of a mixture of nickelocene with thiophene, Y-junction carbon nanotubes have been synthesized in good quantities.[362] Figure 1.39(a) shows a TEM image of such a Y-junction nanotube. A TEM image revealing the presence of several Y-junction carbon nanotubes is shown in

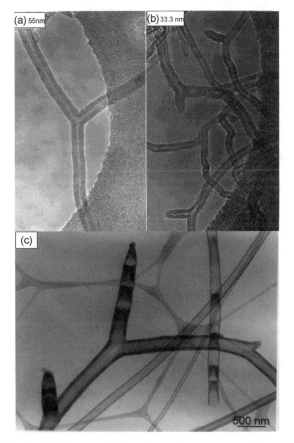

Figure 1.39 (a)–(c) *TEM images of Y-junction carbon nanotubes obtained by the pyrolysis of nickelocene and thiophene at 1000 °C* (Reproduced from ref. 362)

Figure 1.39(b) and 1.39(c). STM and STS studies of Y-junction carbon nanotubes show interesting characteristics at the junctions. Using tight-binding molecular dynamics, Menon *et al.*[363] simulated the formation of single-wall carbon nanotube T junctions *via* the fusing of two nanotubes. They have presented energetically efficient all-sp^2 pathways for the formation of SWNT junctions. Silicon nanowire–carbon nanotube heterojunctions also exhibit rectification behaviour.[364]

To realize multi-functional devices made of carbon nanotubes, it is desirable to develop microprocesses capable of identification and manipulation of the nanotubes. Nishijima *et al.*[365] have developed such a novel microprocess to attach individual nanotubes to scanning probe microscope tips (in a SEM), which are later used as probes to image biological and industrial specimens. Nanotube device properties are also observed. Employing tapping-mode AFM, Lafebvre *et al.*[366] have developed a method to controllably transport, rotate, cut and place the individual SWNTs, by varying the tip–sample force and tip speed. Complex SWNT circuits so assembled were contacted to metal electrodes by electron beam lithography and their transport characteristics studied.

Frank *et al.*[316] have studied quantum transport in the MWNTs by attaching the nanotubes with a nanotube fibre and contacting them into liquid metal to establish electrical contact. A conductance of 77.5 µS ($=2e^2/h$), which is nothing but one unit of conductance quantum, was obtained for the MWNTs. The charge transport is ballistic, even at room temperature. Using SWNTs attached to an AFM cantilever as the probe and atomically flat titanium surfaces on an α-Al$_2$O$_3$ substrate, Cooper *et al.*[367] have demonstrated an areal data storage density of the order of terabits per square inch. This method employs SWNT-based lithography which offers sub-10 nm nanofabrication capabilities. In contrast to conventional electronic devices operating on the basis of charge transport, spin-electronic devices operate upon the concept of spin transport. Tsukagoshi *et al.*[368] have studied the transport of spin-polarized electrons, injected from a ferromagnetic metal contact into MWNTs, and observed coherent transport of electron spins. The phase coherence length is 250 nm. It would be of interest to examine spin transport behaviour through SWNTs.

Mason *et al.*[369] have made carbon nanotube quantum dots with multiple electrostatic gates and used the resulting enhanced control to investigate a nanotube double quantum dot. Transport measurements reveal honeycomb charge stability diagrams as a function of two nearly independent gate voltages. The device can be tuned from weak to strong interdot tunnel-coupling regimes, and the transparency of the leads can be controlled independently. They extracted energy-level spacings, capacitances, and interaction energies for this system. This ability to control electron interactions in the quantum regime in a molecular conductor is important for applications such as quantum computation.

Dierking and co-workers[370] exploited the ability of nematic liquid crystals to self-organize to induce alignment of dispersed nanotubes. In addition, the director reorientation on application of electric or magnetic fields to liquid crystals, commonly known as the Freedericksz transition, was employed to manipulate the direction of the nanotubes. Thus the reorientation of the nanotube follows the reorientation of the liquid crystal through elastic interactions *via* the liquid crystal director field. Such dynamically changing nanotube orientation is due to elastic interactions with the anisotropic host, which leads towards the realization of a nanosized on-off (i.e., conducting to nonconducting) electrical switches.

The spectrum of findings mentioned above opens up the possibility of assembling carbon nanotubes that possess such novel device-like properties[301,302,357,362,371] into multi-functional circuits, and ultimately towards the realization of a carbon nanotube based computer chip. Thus, Rueckes *et al.*[371] have described the concept of carbon-nanotube based nonvolatile random access memory for molecular computing. The viability of the concept has also been demonstrated.

Field-effect Transistors and Related Devices

We indicated the field-effect transistor action of nanotubes in Figure 1.38. In nanotube field-effect transistors (NT-FETs), gating has been achieved by applying a voltage to a submerged gate beneath a SWNT (Figure 1.40a and 1.40b), which is contacted at opposite nanotube ends by a metal source and drain leads.[301] A typical nanoelectronic device of NT-FET consists of a semiconducting nanotube, which is

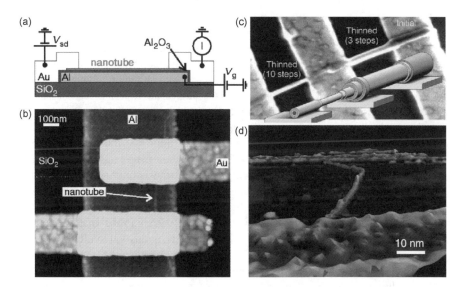

Figure 1.40 *Nanoelectronic devices. (a) Schematic diagram for a carbon NT-FET. V_{sd}, source-drain voltage; V_g, gate voltage. (b) Scanning tunneling microscope (STM) picture of a SWNT field-effect transistor made using the design of (a). The aluminium strip is overcoated with aluminium oxide. (c) Image and overlaying schematic representation for the effect of electrical pulses in removing successive layers of a MWNT, so that layers having desired transport properties for devices can be revealed. (d) STM image of a nanotube having regions of different helicity on opposite sides of a kink, which functions as a diode; one side of the kink is metallic, and the opposite side is semiconducting. The indicated scale bar is approximate*
(Part (a) and (b) reproduced from ref. 372; Part (c) reproduced from ref. 373; and Part (d) from ref. 357)

on top of an insulating aluminium oxide layer, connected at both ends to gold electrode. The nanotube is switched by applying a potential to the aluminium gate under the nanotube and aluminium oxide. Transistors were fabricated by lithographically applying electrodes to nanotubes that were either randomly distributed on a silicon substrate or positioned on the substrate with an atomic force microscope.[302,372] A transistor assembled in this way may or may not work, depending on whether the chosen nanotube is semiconducting or metallic, over which the operator generally has no control. Outer layers can be selectively peeled from a MWNT (Figure 1.40c) until a nanotube cylinder with the desired electronic properties is obtained,[373] but this process is not yet very reliable and is probably unsuitable for mass production. Nanoscopic NT-FETs aim to replace the source-drain channel structure with a nanotube. A more radical approach is to construct entire electronic circuits from interconnected nanotubes. Because the electronic properties depend on helicity, it should be possible to produce a diode, for example, by grafting a metallic nanotube to a semiconducting nanotube. Such a device has been demonstrated. The bihelical nanotube was not, however, rationally produced; rather, it was fortuitously recognized in a normal nanotube sample by its kinked structure (Figure 1.40d), which was caused by the helicity change.[354]

Yaish *et al.*[374] used an atomic force microscope (AFM) tip to locally probe the electronic properties of semiconducting carbon nanotube transistors. A gold-coated AFM tip serves as a voltage or current probe in three-probe measurement setup. Using the tip as a movable current probe, he investigated the scaling of the device properties with channel length. Using the same tip as a voltage probe, he studied the properties of the contacts. He found that Au makes an excellent contact in the *p* region, with no Schottky barrier. In the *n* region, large contact resistances were found, which dominate the transport properties. Wind *et al.*[375] have fabricated carbon-nanotube (CN) field-effect transistors with multiple, individually addressable gate segments. They have been able to probe their properties as a function of gate length. The devices exhibit markedly different transistor characteristics when switched using gate segments controlling the device interior *versus* those near the source and drain. They ascribe this difference to a change from Schottky-barrier modulation at the contacts to bulk switching. They also found that the current through the bulk portion is independent of gate length for any gate voltage, offering direct evidence for ballistic transport in semiconducting carbon nanotubes over at least a few hundred nanometres, even for relatively small carrier velocities. In addition, they can clearly distinguish between Schottky-barrier switching and bulk switching in CNFETs, which suggests paths, which may be pursued for future nanoelectronic applications.

Appenzeller *et al.*[376] have presented a detailed study on the impact of multimode transport in carbon nanotube field-effect transistors. Under certain field conditions electrical characteristics of tube devices are a result of the contributions of more than one 1D subband. They discuss the importance of scattering for a stepwise change of current as a function of gate voltage and explain the implications of their observations for the performance of nanotube transistors. While no stepwise change in current as a function of V_{gs} may be observed due to the small or vanishing scattering probability inside the tube, more than one subband can be involved in current transport for high V_{gs}. Experimentally, subbands have been made visible through extensive doping with potassium. A model has been presented to explain the results, and the implications of these findings for the electrical characteristics of nanotube-based FETs have been discussed.

The combination of their electronic properties and dimensions makes carbon nanotubes ideal building blocks for molecular electronics. However, the advancement of carbon nanotube-based electronics requires assembly strategies that allow their precise localization and interconnection. Using a scheme based on recognition between molecular building blocks, Keren *et al.*[377] have reported the realization of a self-assembled carbon nanotube field-effect transistor operating at room temperature. A DNA scaffold molecule provides the address for precise localization of a semiconducting SWNT as well as the template for the extended metallic wires contacting it. The realization of a FET in a test tube promotes self-assembly as a realistic strategy for the construction of carbon nanotube-based electronics. This approach can be generalized to form a functional circuit on a scaffold DNA network. Numerous molecular devices could be localized at different addresses on the network and interconnected by DNA-templated wires. The scheme appears to be robust and sufficiently general to allow flexibility in the integration of other active electronic components into circuits. Realization of a functional circuit would, however, require improving the electronic properties of the transistor and individual gating to each device. The latter can be achieved by means of a three-armed DNA junction as a template with the SWNT localized at the junction and by developing a method for turning one of the arms into a gate.

The CNT-based FETs have been combined with biomaterials to design nano-devices for sensing biorecognition events and biocatalytic processes. Such a FET, sensitive to streptavidin was fabricated using a biotinfunctionalized carbon nanotube bridging two microelectrodes (source and drain); Figure 1.41(a).[378] The SWNT operating as a gate in the CNTFET device was coated with a mixture of two polymers: polyethyleneimine (PEI) and polyethylene glycol (PEG). The former provided amino groups for further coupling of biotin-*N*-hydroxysuccinimidyl ester, the latter prevented nonspecific adsorption of proteins on the functionalized carbon nanotube (Figure 1.41b). Figure 1.41c shows an AFM image of the device after its exposure to streptavidin labelled with gold nanoparticles (10 nm). Light dots represent gold nanoparticles, and thus indicate the presence of streptavidin bound to the biotinylated carbon nanotube. The source-drain current, I_{sd}, dependence on the gate voltage of the FET showed significant change upon the streptavidin binding to the biotin-functionalized carbon nanotube (Figure 1.41d). Control experiments revealed specific binding of the streptavidin, which occurs only at the biotinylated interface. The mechanism of the effect was discussed in terms of the charge redistribution at the nanogate upon binding of charged streptavidin molecules.[379]

Field Emission

As mentioned earlier, carbon nanotubes can be used as electron sources for field-emission (FE) displays and these aspects have been reviewed by deJonge and

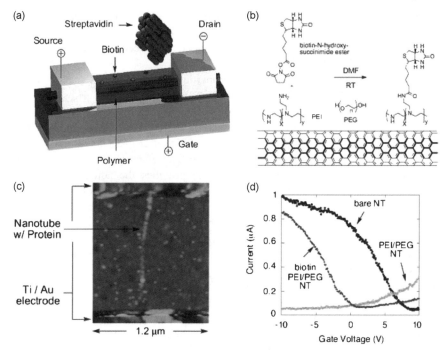

Figure 1.41   (a) *A carbon nanotube field-effect transistor with a biotin-functionalized SWNT operating as a gate sensitive to streptavidin. (b) Biotinylation reaction of the polymer layer (PEI and PEG) on a sidewall of the SWNT. (c) AFM image of the polymer-coated and biotinylated CNTFET after exposure to streptavidin labeled with gold nanoparticles (10 nm diameter). (d) Source-drain current (I_{sd}) dependence on the gate voltage of the FET device based on carbon nanotube functionalized with biotin in the absence (a) and presence (b) of streptavidin*
(Reproduced from ref. 378)

Bonard.[380] FE is the emission of electrons from a solid under an intense electric field. The simplest way to create such a field is by field enhancement at the tip of a sharp object. Si or W tips were initially used, made by anisotropic etching or deposition. CNTs have an advantage over Si or W tips in that their strong, covalent bonding means they are physically inert to sputtering, chemically inert to poisoning, and can carry a huge current density of 10^9 A cm^{-2} before electromigration. In addition, when driven to high currents, their resistivity decreases, so that they do not tend to electric-field-induced sharpening, which causes instabilities in metal tip field emitters.[381] CNTs have better FE performance than other forms of carbon such as diamond and diamond-like carbon.

There has been much research on the possible uses of SWNTs and MWNTs as field emission electron sources[382,383] for flat panel displays,[384] lamps,[385] gas discharge tubes providing surge protection,[386] and X-rays.[387] A potential applied between a carbon nanotube-coated surface and an anode produces high local fields, as a result of the small radius of the nanofibre tip and the length of the nanofibre.

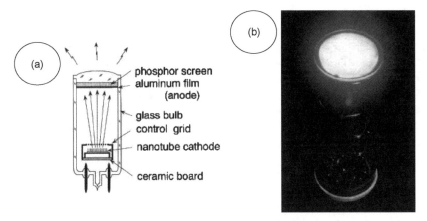

Figure 1.42 (a) *Schematic drawing of a longitudinal cross section of a field-emission fluorescent display with a field-emission cathode made of carbon nanotubes. The front glass forms a convex lens to condense emitted light in the forward direction. (b) A field-emission fluorescent display based on carbon nanotubes emitting visible light. The anode current and voltage are 200* mA *and 10* kV, *respectively*
(Reproduced from ref. 391)

These local fields cause electrons to tunnel from the nanotube tip into the vacuum. Electric fields direct the emitted electrons toward the anode, where a phosphor produces light for the flat panel display (Figure 1.42). The picture is not, however, as simple. Unlike in bulk metals, electron emission from nanotube tips arises from discrete energy states, rather than continuous electronic bands.[388] The emission behaviour also depends on the nanotube tip structure. For example, enhanced emission results from opening SWNT[383] or MWNT tips.[385] Nanotube field-emitting surfaces are easy to manufacture by screen-printing nanotube pastes and do not deteriorate in moderate vacuum (10^{-8} Torr). Tungsten and molybdenum tip arrays generally require a vacuum of 10^{-10} Torr and are more difficult to fabricate.[389] Nanotubes provide stable emission, long lifetimes, and low emission threshold potentials.[382,385] Current densities as high as 4 A cm^{-2} have been obtained, compared with the 10 mA cm^{-2} needed for flat panel field emission displays and the >0.5 A cm^{-2} required for microwave power amplifier tubes.[390] Some of the advantages of nanotube displays over liquid crystal displays are low power consumption, high brightness, wide viewing angle, fast response rate, and wide operating temperature range.

MWNT-based field emission lighting devices have been built and their luminescence characteristics studied.[391]. One such field emission lighting device is shown in Figure 1.42. Choi *et al.*[392a] have assembled a sealed 4.5 inch^2 field-emission display device using vertically aligned SWNTs along with organic binders. The display in three primary colours has an emission current of 1.5 mA at 3 V μm^{-1}, with a brightness of 1800 Cd m^{-2}. Lee *et al.*[392b] have shown that aligned nanotube bundles exhibit a high emission current density of around 2.9 mA cm^{-2} at 3.7 V μm^{-1}. Lovall *et al.*[393] have investigated the emission properties of SWNT ropes by employing a field ion microscopy. The field-emitted electron energy distribution (FEED) of SWNT field emitters shows a large density of states near the Fermi energy. Emission characteristics of CVD produced MWNTs as well as SWNTs have been examined by Groning *et al.*,[394] who obtained an emission site density of 10000 emitters cm^{-2} at fields of around 4 V μm^{-1}. A work function of 5 eV was obtained with MWNTs, and a smaller value with the SWNTs. The use of dense, quasi-aligned carbon nanotubes produced by the pyrolysis of ferrocene on a pointed tungsten tip exhibit high emission current densities with good performance characteristics.[395] Figure 1.43(a) shows a typical *I–V* plot for the carbon nanotube covered tungsten tip for currents ranging from 0.1 nA to 1 mA. The applied voltage was 4.3 kV for a total current of 1 μA and 16.5 kV for 1000 μA. The Fowler–Nordheim (F-N) plot shown in Figure 1.43(b) has two distinct regions. The behaviour is metal-like in the low-field region, while it saturates at higher fields as the voltage is increased. We have obtained a field emission current density of 1.5 A cm^{-2} at a field of 290 V mm^{-1}, a value considerably higher than that found with planar cathodes. Accordingly, the field enhancement factor calculated from the slope of the F-N plot in the low-field region is also large. Field emission micrographs reveal the lobe structure symmetries typical of carbon nanotube bundles. The emission current is remarkably stable over an operating period of more than 3 h for various currents in the 10–500 mA range. The relative fluctuations decrease with increasing current level, and the emitter can be operated continuously at the high current levels for at least 3 h without any

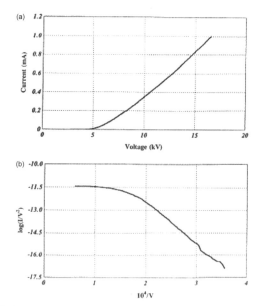

Figure 1.43 *I–V characteristics showing field emission currents in the range of 0.1 nA to 1 mA. (b) Fowler-Nordheim plot corresponding to the data in (a)* (Reproduced from ref. 395)

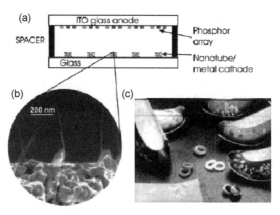

Figure 1.44 (a) *Schematic of a flat panel display based on carbon nanotubes. ITO, indium tin oxide.* (b) *SEM image of an electron emitter for a display, showing well-separated SWNT bundles protruding from the supporting metal base* (c) *Photograph of a 5* inch *(13* cm) *nanotube field emission display made by Samsung*
(Reproduced from ref. 32, 384)

degradation in the current. Samsung has produced several prototypes (Figure 1.44), including a 9 inch (23 cm) red–blue–green colour display that can reproduce moving images.[384].

Charlier *et al.*[396] have reported that B-doped MWNTs exhibit enhanced field emission (turn on voltages of 1.4 V μm^{-1}) compared with carbon MWNTs (turn on voltages of 3 V μm^{-1}). This is because of the presence of B atoms at the nanotube tips, which results in an increased DOS close to the Fermi level. Tight binding and *ab initio* calculations demonstrate that the work function of B-doped CNTs is much lower (1.7 eV) than that of pure carbon MWNTs. Similarly, Golberg *et al.*[397] have demonstrated that N-doped MWNTs can emit electrons at relatively low turn-on voltages (2 V μm^{-1}) and high current densities (0.2–0.4 A cm^{-2}). Thus, both B- and N-doped CNTs may have potential for use as stable and intense field-emission sources. Further work should be carried out on the field-emission properties of B- and N-doped SWNTs.

Supercapacitors and Actuators

Because of the high electrochemically accessible surface area of porous nanotube arrays, combined with their high electronic conductivity and useful mechanical properties, these materials are attractive as electrodes for devices that use electrochemical double-layer charge injection. Examples include supercapacitors, which have giant capacitances in comparison with those of ordinary dielectric-based capacitors, and electromechanical actuators that may eventually be used in robots. Like ordinary capacitors, carbon nanotube supercapacitors[398–400] and electromechanical actuators[401a] typically consist of two electrodes separated by an electronically insulating material, which is ionically conducting in electrochemical devices. Because this separation is about a nanometre for nanotubes, as compared with the micrometre or larger separations in ordinary dielectric capacitors, very large capacitances result from the high nanotube surface area accessible to the electrolyte. These capacitances (typically between ~15 and ~200 F/g^{-1}, depending on the surface area of the nanotube array) result in large amounts of charge injection when only a few volts are applied.[398–401a] The charge injection can be used for energy storage in nanotube supercapacitors and to provide electrode expansions and contractions that can do mechanical work in electromechanical actuators. The capacitances (180 and 102 F g^{-1} for SWNT and MWNT electrodes, respectively) and power densities (20 kW kg^{-1} at energy densities of ~7 W h kg^{-1} for SWNT electrodes)[398,399] are attractive, especially because performance can, likely, be improved by replacing SWNT bundles and MWNTs with unbundled SWNTs. An extraordinarily short discharge time of 7 ms was reported[400] for 10 MWNT capacitors connected in series, which operated at up to 10 V. Nanotube electromechanical actuators function at a few volts, compared with the ~100 V used for piezoelectric stacks and the ≥1000 V used for electrostrictive actuators.

Nanotube actuators have been operated at up to 350 °C. Operation above 1000 °C should be possible, on the basis of SWNT thermal stability and industrial carbon electrode electrochemical application above this temperature.[20b] From observed nanotube actuator strains that can exceed 1%, order-of-magnitude advantages over commercial actuators in work per cycle and stress generation capabilities are predicted if the mechanical properties of nanotube sheets can be increased to close to the inherent mechanical properties of the individual nanotubes.[20b] The maximum observed isometric actuator stress of SWNT actuators is presently 26 MPa.[20b] This is >10× the stress initially reported for these actuators and ~100× that of the stress generation capability of natural muscle, and it approaches the stress generation capability of high-modulus commercial ferroelectrics (~40 MPa). However, the ability to generate stress is still >100× lower than that predicted for nanotube fibres with the modulus of the individual SWNTs. The success of actuator technology based on carbon nanotubes will depend on improvements in the mechanical properties of nanotube sheets and fibres with a high surface area by increasing nanotube alignment and the binding between nanotubes. Fennimore et al.[402] constructed and operated the first rotational nanoscale electrochemical actuator based on CNTs by incorporating a rotatable metal plate, with a MWNT serving as the key motion-enabling element.

Sensors and Probes

Chemical sensor applications of nonmetallic nanotubes are interesting, because nanotube electronic transport and thermopower (voltages between junctions caused by interjunction temperature differences) are sensitive to substances that affect the amount of injected charge.[403–405] Advantages are the minute size of the nanotube sensing element and the small amount of material required for the response. However, major challenges remain in making devices that can differentiate absorbed species in complex mixtures and provide rapid responses. Some sensor applications were indicated earlier and we shall present a brief survey of the efforts in this direction.

Various groups[403,404,406] have demonstrated that pure carbon SWNTs and MWNTs can be used to detect toxic gases and other species, because small concentrations can produce large changes in the nanotube conductance, shifting the Fermi level to the valence band, and generating hole-enhanced conductance. N-doped MWNTs have proved to be efficient in this context. CN_x MWNTs display a fast response, on the order of milliseconds, when exposed to toxic gases and organic solvents (Figure 1.45) and reach saturation within 2–3 s.[407] The increase in electrical resistance is caused by the presence of molecules bound to the pyridine-like sites present within the CN_x nanotubes (Figure 1.45). For ethanol, acetone, and NH_3, a permanent change to higher resistance has been observed. In particular, a clear decrease in the DOS at the Fermi level is observed – indicative of lower conduction and chemisorption. Thus, CN_x nanotubes appear to be more efficient in detecting hazardous gaseous species. Clearly, B-doped CNTs should also be tested for such applications. Theoretical *ab initio* calculations show that CO and HCN_2O do not react with the surface of pure carbon SWNTs.[408] If the surface of the tube is doped with a donor or an acceptor, changes in the electronic properties should occur as a result of the binding of the molecules to the doped locations[407,408] suggesting that either CN_x nanotubes or B-doped MWNTs may be useful to detect low concentrations of ethanol.

Carbon nanotubes are used as electron conductors in enzyme-based electrochemical sensors. Surface functionalization of nanotubes renders them more biocompatible. Amperometric biosensors are based on the ability of an enzyme adlayer to transduce the turnover of substrate into a detectable, reliably quantifiable current. This transduction is carried out by monitoring the direct voltammetric response of the enzyme or the catalytic enhancement in diffusive voltammetry of a

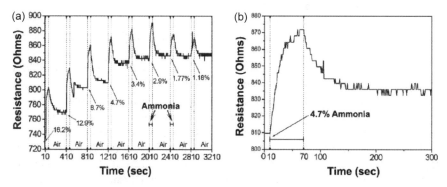

Figure 1.45 *Resistance versus time plots for NH_3 on N-doped MWNT sensors. (a) Clearly the sensor is sensitive to 1% NH_3. Chemisorption is readily observed, which can be attributed to strong interactions between the pyridinic sites of the tube surface with the NH_3; (b) graph indicating the response time for NH_3 gas (4.7%). The results demonstrate that N-doped MWNTs could be used in the fabrication of novel and fast responsive gas sensors*
(Reproduced from ref. 407)

suitable mediator (redox iron complexes, such as ferrocene). Glucose oxidase is a flavin enzyme used commercially on a massive scale to monitor the blood glucose levels in diabetics. Prolonged incubation of SWNTs (oxidized or otherwise) with glucose oxidase gives rise to an effective coating of the nanotube with enzyme[409] (Figure 1.32a). Immobilization can take place without gross loss of enzyme activity. The treatment of such Bio-SWNT-electrodes with both a diffusive mediator and equilibriated glucose substrate leads to a catalytic anodic wave (Figures 1.32b and 1.32c). The magnitude of the catalytic response is greater than one order of magnitude (for the same enzyme, substrate and mediator concentrations) than that found with an activated macro-carbon electrode. Since metalloproteins in solution can communicate electrochemically with oxidized SWNTs at an electrode surface and bioimmobilization at high loading occurs with retention of activity, metalloproteins immobilized on a nanotube surface can, seemingly, communicate directly with the nanotube π system. Direct electro-chemical communication between the flavin active site of glucose oxidase and the nanotube itself is not possible since the tunnelling distance between the (redox) active site and the underlying support/electrode is much too large.

Since CNTs possess a high conductivity as well as high surface area-to-weight ratio, they can be used as supports for the immobilization of biomolecules that are further used for electrochemical and quartz crystal microbalance sensing of biorecognition processes, including immunosensing and DNA sensing. Thus, immunosensing systems with an electrochemiluminescence read-out signal have been designed using CNTs as supports. Liposomes have been employed as carriers for biocatalysts and the recognition sites in biosensor systems. Functionalized CNTs have also been employed as carriers of multiple enzyme labels for electrochemical DNA sensing and immunosensing.[410] CNTs were loaded with about 9600 alkaline phosphatase (AlkPh) molecules per CNT by covalently coupling the protein to the carboxylic groups of the oxidized CNTs. Modification of the enzyme-functionalized CNTs with a biorecognition unit (such as an oligonucleotide and an antibody, respectively for DNA sensing and immunosen-sing) enables the amplified electrochemical detection of the biorecognition events. For example, magnetic particles functionalized with the DNA primer were reacted with the complementary analyte DNA, resulting in the double-stranded DNA complex – further reaction is shown in Figure 1.46(a and b). Figure 1.46(c) corresponds to the magnetic beads-DNA-CNT assembly produced during 20 min hybridization with 10 pg mL^{-1} DNA target sample. Chronopotentiometric responses obtained upon analysis of the different concentrations of the DNA analyte using a CNT-modified electrode and a nucleic acid-functionalized CNT–enzyme hybrid as the amplifying label are shown in Figure 1.46(d). The calibration plot (bottom) shows that the detection limit was about 1 fg mL^{-1} for the analyte DNA. Enzyme-loaded CNTs, functionalized with an antibody, have similarly been applied to amplified electrochemical immunosensing. Katz and Willner[285] have reviewed nanobioelectronic applications of carbon nanotubes.

Cao *et al.*[411] carried out a systematic investigation of the electromechanical properties of SWNTs under tensile strain. In the small strain range, small bandgap semiconducting (or quasi-metallic) nanotubes exhibit the largest resistance changes

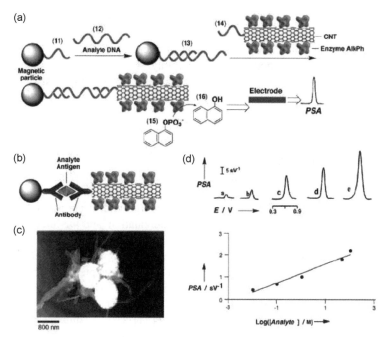

Figure 1.46 (a) *Electrochemical DNA sensing and* (b) *immunosensing using the alkaline phosphatase-functionalized CNTs as the biocatalytic amplifying tags.* (c) *TEM image of the magnetic beads-DNA-CNT assembly produced following a 20 min hybridization with the 10 pg mL^{-1} DNA target sample.* (d) *Chronopotentiometric signals for various concentrations of the DNA target:* (a) *0.01;* (b) *0.1;* (c) *1;* (d) *50;* (e) *100 pg mL^{-1}. (PSA = potentio metric stripping analysis). Bottom: The derived calibration plot* (Reproduced from ref. 410)

or highest sensitivity to tensile stretching (β_{GF} up to 600–1000), while metallic nanotubes are the least sensitive, in qualitative agreement with existing theoretical expectations. The result suggests that, at room temperature, quasi-metallic SWNTs are potentially useful for highly sensitive eletromechanical sensors and could present a new type of strain gauge material (conventional doped-Si strain gauges have $\beta_{GF} \sim 200$).

Nishijima *et al.*[365] have attached individual nanotubes to scanning probe microscope tips, which are then used as probes to image biological and industrial specimens. Carbon nanotube scanning probe tips for atomic probe microscopes are commercially available. The cylindrical shape and small tube diameter enable imaging in narrow, deep crevices and improve resolution in comparison with conventional nanoprobes, especially for high sample feature heights.[412,413] Covalently modifying the nanotube tips, such as by adding biologically responsive ligands, enables the mapping of chemical and biological functions.[406] Nanoscopic tweezers have been made by Kim and Lieber[414] that are driven by the electrostatic interaction between two nanotubes on a probe tip. They attached carbon nanotubes to electrodes fabricated on the pulled glass micropipettes. Voltages are applied to the electrodes to achieve closing and opening of the free ends of the nanotubes, to facilitate the grabbing and manipulation of submicron clusters and nanowires. They may be used as nanoprobes for assembly. These uses may not have real applications, but they are likely to increase the value of measurement systems for characterization and manipulation on the nanometric scale.

Lithium Batteries

When using graphite-like materials in Li^+ batteries, the ions are intercalated between the graphite layers, so that Li^+ migrates from a graphitic anode to the cathode (usually $LiCoO_2$, $LiNiO_2$, and $LiMn_2O_4$). The theoretical Li storage capacity in graphite is 372 mA h g^{-1} (LiC_6), and the charge and discharge phenomenon in these batteries is based upon the Li^+ intercalation and de-intercalation.[415] Such batteries are widely used in portable computers, mobile telephones, digital cameras *etc.* Endo and co-workers have shown that B-doped vapour grown carbon fibres and nanofibres are superior to any other carbon source present in the graphitic anode. This may be because of Li^+ has a stronger affinity to B-doped sites, resulting in a higher energy storage for the battery.[415,416] The use of nanotubes as electrodes in lithium batteries is a possibility because of the high reversible component of storage capacity at high discharge rates. The maximum reported reversible capacity is 1000 mA h g^{-1} for SWNTs that are mechanically milled in order to enable the filling of nanotube cores, as compared to 372 mA h g^{-1} for graphite[417a] and 708 mA h g^{-1} for ball-milled graphite.[417b] N-doped CNTs and nanofibres also exhibit efficient reversible Li storage (480 mA h g^{-1}), much higher than commercial carbon materials used for Li^+ batteries (330 mA h g^{-1}).[417c]

Gas Adsorption and Hydrogen Storage

SWNTs have nano-sized channels that can facilitate adsorption of liquids or gases. Eswaramoorthy *et al.*[418] studied the adsorption properties of SWNTs with respect to methane, benzene and nitrogen. The studies indicate that SWNTs are good microporous materials with a total surface area above 400 m^2 g^{-1}. Figure 1.47 shows typical adsorption isotherms for SWNTs. The unique hexagonal packing of the SWNTs in the bundles offers ideal channels, thus allowing the realization of 1D adsorbates. Helium adsorption in SWNTs has been studied and the high binding energy observed is considered to be due to 1D adsorption in the interstitial sites of SWNT bundles.[419] Opening the SWNTs by thermal activation increases both the kinetic rate and saturation adsorption of nanotubes for Xe atoms at 95 K.[420] Notably, CNTs have been used as catalysts themselves, but they have been used as metal supports.

Carbon nanotubes were considered to be good hosts for hydrogen storage (*e.g.* for fuel cells that power electric vehicles) but there is considerable doubt about the magnitude of the hydrogen uptake and, hence, the use of CNTs for storage.[20a,421–425] Reversible adsorption of molecular hydrogen in carbon nanotubes was first reported by Dillon *et al.*[426] These workers measured the hydrogen

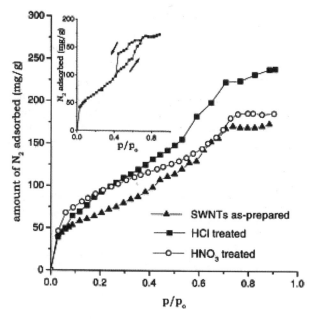

Figure 1.47 *N₂ adsorption isotherms of SWNTs at 77 K; as-prepared (▲), HCl treated (■), and HNO₃ treated (○). Inset: Hysteresis in the adsorption–desorption isotherms for SWNTs*
(Reproduced from ref. 418)

adsorption capacity of the as-prepared SWNT bundles (0.1–0.2 wt%) containing unidentified carbonaceous materials as well as large fractions of cobalt catalyst particles (20 wt%). Composition (H/C) *versus* pressure isotherms at 80 K (−193 °C) of as-prepared SWNTs, sonicated SWNTs and a high surface area carbon (Saran) reported by Ye *et al.*[425] These workers find the hydrogen storage capacity in arc-derived SWNTs to be 8.25 wt% at 80 K and ~4 MPa. A hydrogen storage capacity of 4.2 wt% for SWNTs was reported by Liu *et al.*[427] at 27 °C and 10.1 MPa. The SWNTs used in this study had a large mean diameter of 1.85 nm. Moreover, 78.3% of the adsorbed hydrogen (3.3 wt%) could be released under ambient pressure at room temperature, while the release of the residual hydrogen (0.9%) required heating of the sample. Gundiah *et al.*[428] performed a comparative study of high-pressure hydrogen uptake and electrochemical hydrogen storage with different types of carbon nanotube samples. For hydrogen storage studies they used the following samples: SWNTs synthesized by the arc-discharge method (as-synthesized), I; SWNTs synthesized by the arc-discharge method (treated with conc. HNO₃), II; MWNTs synthesized by the pyrolysis of acetylene (as-synthesized), III; MWNTs synthesized by the pyrolysis of acetylene (treated with conc. HNO₃), IV; MWNTs synthesized by the arc-discharge method, V; aligned MWNT bundles synthesized by the pyrolysis of ferrocene (as-synthesized), VI; aligned MWNT bundles synthesized by the pyrolysis of ferrocene (treated with acid), VII; aligned MWNT bundles synthesized by the pyrolysis of ferrocene and acetylene (as-synthesized), VIII; and aligned MWNT bundles synthesized by the pyrolysis of ferrocene and acetylene (treated with acid), IX. Figure 1.48(a) shows plots of hydrogen adsorption *versus* time for the various carbon nanostructured samples. By eliminating many of the common errors encountered in such experiments, these workers achieved a maximum storage capacity of 3.75 wt% (143 bar, 27 °C) with densely aligned nanotubes, prepared by the pyrolysis of ferrocene–hydrocarbon mixtures. SWNTs and MWNTs (arc-generated) showed a high-pressure hydrogen storage capacity, which is much less than 3 wt%. Figure 1.48(b) shows plots of electrochemical charging capacity of various types of carbon nanotubes. Electrodes

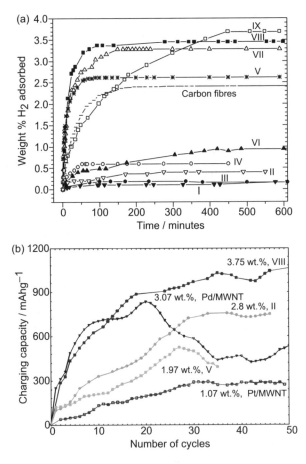

Figure 1.48 (a) *Amount of hydrogen adsorbed in wt% as a function of time for various carbon nanostructures (I–IX see text for details). The broken curve represents the blank data obtained in the absence of a carbon sample. (b) Plot of the charging capacity against the number of cycles for different carbon nanostructures. Also shown are the corresponding weight percentages of* H_2 *stored*
(Reproduced from ref. 428)

made out of aligned MWNTs demonstrate higher electrochemical charging capacities, up to 1100 mA h g^{-1}, corresponding to a hydrogen storage capacity of 3.75 wt%. SWNTs and MWNTs (arc-generated), however, show capacities in the range 2–3 wt%.

Other Useful Properties

Optical limiting properties of the carbon nanotubes are of importance for applications involving high-power lasers. Optical limiting behaviour of visible nanosecond laser pulses in SWNT suspensions occurs mainly due to nonlinear scattering.[429] Yoshino *et al.*[430] observed increased conductivity at relatively low nanotube concentrations in polymer composites and an enhancement of photoconductivity, implying possible use in optoelectronic devices.

Electromechanical actuators based on SWNT-sheets appear to generate stresses higher than natural muscles and higher strains than the ferroelectric counterparts.[401]. This behaviour of nanotubes may be useful in the direct conversion of electrical energy into mechanical energy, relevant to applications in robotics. Wood and Wagner[431] observed a significant shift in the Raman peaks of SWNTs upon immersion in liquids. This allows the use of nanotubes as molecular sensors. Flow of liquids and gases over nanotubes generate a voltage in the tubes.[405,432] The voltage so produced fits a logarithmic velocity dependence over nearly six decades of velocity. The magnitude of the voltage/current depends on the ionic conductivity and the polar nature of the liquid. The dominant mechanism responsible for this highly nonlinear response appears to involve a direct forcing of the free charge carriers in the nanotubes by the fluctuating Coulombic field of the liquid flowing past it. This work suggests the potential of nanotubes as flow sensors. While this is an interesting property, it is unlikely that electricity can be generated by this means. Karl and Tomanek[433a] propose a molecular pump based on carbon nanotubes for the transport of atoms. The concept is based upon the excitation of a nanotube by two laser beams in a coherent control scheme, thus generating an electron current, which in turn drives the intercalated atoms. Regan *et al.*[433b] reported the achievement of controllable, reversible atomic scale mass transport (electromigration) along CNTs, using indium metal as the prototype species.

References

1. H. W. Kroto, J. R. Heath, S. C. O'Brien, R. F. Curl and R. E. Smalley, *Nature*, 1985, **318**, 162.
2. W. Krätschmer, L. D. Lamb, K. Fostiropoulos and D. R. Huffman, *Nature*, 1990, **347**, 354.
3. S. Iijima, *Nature*, 1991, **354**, 56.
4. S. Iijima and T. Ichihashi, *Nature*, 1993, **363**, 603.
5. D. S. Bethune, C. H. Kiang, M. S. de Vries, G. Gorman, R. Savoy, J. Vazquez and R. Bayers, *Nature*, 1993, **363**, 605.

6. W. K. Hsu, J. P. Hare, M. Terrones, P. J. F. Harris, H. W. Kroto and D. R. M. Walton, *Nature*, 1995, **377**, 687.

7. M. Endo, K. Takeuchi, S. Igarashi, K. Kobori, M. Shiraishi and H. W. Kroto, *J. Phys. Chem. Solids*, 1993, **54**, 1841.

8. (a) P. M. Ajayan, T. Ichihashi and S. Iijima, *Chem. Phys. Lett.*, 1993, **202**, 384; (b) S. G. Louie, *Top. Appl. Phys.*, 2001, **80**, 113.

9. S. Iijima, P. M. Ajayan and T. Ichihashi, *Phys. Rev. Lett.*, 1992, **69**, 3100.

10. V. P. Dravid, X. Lin, Y. Wang, X. K. Wang, A. Yee, J. B. Ketterson and R. P. H. Chang, *Science*, 1993, **259**, 1601.

11. P. M. Ajayan, T. W. Ebbesen, T. Ichihashi, S. Iijima, K. Tanigaki and H. Hiura, *Nature*, 1993, **362**, 522.

12. S. C. Tsang, P. J. F. Harris and M. L. H. Green, *Nature*, 1993, **362**, 520.

13. S. C. Tsang, Y. K. Chen, P. J. F. Harris and M. L. H. Green, *Nature*, 1994, **372**, 159.

14. C. N. R. Rao, R. Seshadri, A. Govindaraj and R. Sen, *Mat. Sci. Eng. R.*, 1995, **15**, 209; P. M. Ajayan, *Chem. Rev.*, 1999, **99**, 1787.

15. Special issue on Carbon Nanotubes, *Appl. Phys. A*, 1998, **67**, 1; *Appl. Phys. A*, 1999, **69**, 245; "Carbon Nanotubes" *J. Mater. Res.*, 1998, **13**, 2355; articles in *J. Phys. Chem.*, 2000, **104** and *Physics World*, June 2000.

16. *Electronic Properties of Novel Materials; Science and Technology of Molecular Nanostructures*, ed. H. S. Kuzmany, J. Fink, M. Mehring and S. Roth,"AIP Conference Proceedings" vol. 442, 1999, pp.237–506; American Institute of Physics (AIP), USA.

17. M. S. Dresselhaus, G. Dresselhaus and P. C. Eklund in *Science of Fullerenes and Carbon Nanotubes*, Academic Press, San Diego, 1996, p.756.

18. Carbon Nanotubes: (special issue), *Acc. Chem. Res.*, 2002, **35**, 998.

19. M. Terrones, W. K. Hsu, H. W. Kroto and D. R. M. Walton, *Top. Curr. Chem.* 1999, **199**, 190.

20. (a) C.N.R. Rao, B.C. Satishkumar, A. Govindaraj and M. Nath, *ChemPhysChem*, 2001, **2**, 78; (b) R. H. Baughman, A. A. Zakhidov and W. A. de Heer, *Science*, 2002, **297**, 787; (c) X. B. Zhang, X. F. Zhang, S. Amelinckx, G. Van Tendeloo and J. Van Landuyt, *Ultramicroscopy*, 1994, **54**, 237.

21. S. Reich, C. Thomsen and J. Maultzsch, *Carbon Nanotubes, Basic Concepts and Physical Properties*, Wiley-VCH Verlag GmbH & Co., Weinheim, 2004.

22. M. S. Dresselhaus, G. Dresselhaus, J. C. Charlier and E. Hernandez, Electronic, thermal and mechanical properties of carbon nanotubes, in *Phil. Trans. R. Soc. Lond. A* ('Nanotechnology of carbon and related materials', a special issue), 2004, **362**, 2065 and other articles therein.

23. A. Govindaraj and C.N.R. Rao, in *The Chemistry of Nanomaterials*, ed. C. N. R. Rao, A. Müller and A. K. Cheetham, Wiley-VCH Verlag GmbH & Co, Weinheim, 2004, Vol. 1.

24. Advances in Carbon Nanotubes, ed. M. S. Dresselhaus and H. Dai, *MRS Bull.*, **29**, 4 April 2004.

25. R. Saito, G. Dresselhaus and M. S. Dresselhaus, *Physical Properties of Carbon nanotubes*, Imperial College Press, London, 1998.

26. P. J. F. Harris, *Carbon Nanotubes and Related Structures (New Materials for the 21st Century)*, Cambridge University Press, UK 1999.
27. Special issue on carbon nanotubes, *Carbon*, 2002, **40**, issue 10.
28. P. Yang, Y. Wu and R. Fan, *Int. J. Nanosci.*, 2002, 1.
29. C. N. R. Rao and M. Nath, *J. Chem. Soc., Dalton Trans.* 2003, 1.
30. R. Tenne and C. N. R. Rao, *Phil. Trans. R. Soc. Lond. A*, ('Nanotechnology of carbon and related materials', a special issue), 2004, **362**, 2099; R. Tenne, *Chem. Eur. J.* 2002, **8**, 5296.
31. G. R. Patzke, F. Krumeich and R. Nesper, *Angew. Chem. Int. Ed.*, 2002, **41**, 2446.
32. (a) T. W. Ebbesen and P. M. Ajayan, *Nature*, 1992, **358**, 220; (b) Y. Ando, X. Zhao, T. Sugai and M. Kumar, *Mater. Today*, 2004, **7**(10) 22 and the references therein.
33. N. Hatta and K. Murata, *Chem. Phys. Lett.*, 1994, **217**, 398.
34. D. T. Colbert, J. Zhang, S. M. McClure, P. Nikolaev, Z. Chen, J. H. Hafner, D. W. Owens, P. G. Kotula, C. B. Carter, J. H. Weaver and R. E. Smalley, *Science*, 1994, **266**, 1218.
35. S.-H. Jung, M.-R. Kim, S.-H. Jeong, S.-U. Kim, O.-J. Lee, K.-H. Lee, J.-H. Suh and C.-K. Park, *Appl. Phys. A-Mater. Sci & Process.*, 2003, **76**, 285.
36. K. Anazawa, K. Shimotani, C. Manabe, H. Watanabe and M. Shimizu, *Appl. Phys. Lett.*, 2002, **81**, 739.
37. S. J. Lee, H. K. Baik, J. Yoo and J. H. Han, *Diamond Relat. Mater.*, 2002, **11**, 914.
38. M. Ge and K. Sattler, *Science*, 1993, **260**, 515.
39. (a) W. K. Hsu, M. Terrones, J. P. Hare, H. Terrones, H. W. Kroto and D. R. M. Walton, *Chem. Phys. Lett.*, 1994, **262**, 161; (b) Y. Gogotsi, J. A. Libera and M. Yoshimura, *J. Mater. Res.*, 2000, **15**, 2591.
40. M. Jose-Yacaman, M. Miki-Yoshida, L. Rendon and T. G. Santiesteban, *Appl. Phys. Lett.*, 1993, **62**, 202.
41. V. Ivanov, J. B. Nagy, Ph. Lambin, A. Lucas, X. B. Zhang, X. F. Zhang, D. Bernaerts, G. Van Tendeloo, S. Amelinckx and J. Van Landuyt, *Chem. Phys. Lett.*, 1994, **223**, 329.
42. K. Hernadi, A. Fonseca, J. B. Nagy, D. Bernaerts, J. Riga and A. Lucas, *Synth. Metals*, 1996, **77**, 31.
43. N. M. Rodriguez, *J. Mater. Res.*, 1993, **8**, 3233.
44. Z. F. Ren, Z.P. Huang, J. W. Xu, J. H. Wang, P. Bush, M. P. Siegel and P. N. Provencio, *Science*, 1998, **282**, 1105.
45. Z. F. Ren, Z. P. Huang, D. Z. Wang, J. G. Wen, J. W. Xu, J. H. Wang, L. E. Calvet, J. Chen, J. F. Klemic and M. A. Reed, *Appl. Phys. Lett.*, 1999, **75**, 1086.
46. S. B. Sinnott, R. Andrews, D. Qian, A. M. Rao, Z. Mao, E. C. Dickey and F. Derbyshire, *Chem. Phys. Lett.*, 1999, **315**, 25.
47. M. Yudasaka, R. Kikuchi, T. Matsui, Y. Ohki, S. Yoshimura and E. Ota, *Appl. Phys. Lett.*, 1995, **67**, 2477; M. Yudasaka, R. Kikuchi, Y. Ohki, E. Ota and S. Yoshimura, *Appl. Phys. Lett.*, 1997, **70**, 1817.
48. M. Chen, C. M. Chen and C. F. Chen, *J. Mater. Sci.*, 2002, **37**, 3561; www.iljinnanotech.co.kr/en/material/r-4-3htm.

49. Z.P. Huang, D. Z. Wang, J. G. Wen, M. Sennett, H. Gibson and Z. F. Ren, *Appl. Phys. A-Mater. Sci. Process.*, 2002, **74**, 387.

50. J.-B. Park, G.-S. Choi, Y.-S. Cho, S.-Y. Hong, D. Kim, S.-Y. Choi, J.-H. Lee and K.-I. Cho, *J. Cryst. Growth*, 2002, **244**, 211.

51. R. Sen, A. Govindaraj and C. N. R. Rao, *Chem. Phys. Lett.*, 1997, 267, 276.

52. R. Sen, A.Govindaraj and C. N. R. Rao, *Chem. Mater.*, 1997, **9**, 2078.

53. C. J. Lee, S. C. Lyu, H.-W. Kim, C.-Y. Park and C.-W. Yang, *Chem. Phys. Lett.*, 2002, **359**, 109.

54. S. R. C. Vivekchand, L. M. Cele, F. L. Deepak, A. R. Raju and A. Govindaraj, *Chem. Phys. Lett.*, 2003, **386**, 313.

55. W. A. de Heer, J. M. Bonard, K. Fauth, A. Chatelain, L. Forro and D. Ugarte, *Adv. Mater.*, 1997, **9**, 87, and references therein.

56. W. Z. Li, S. S. Xie, L. X. Qian, B. H. Chang, B. S. Zou, W. Y. Zhou, R. A. Zhao and G. Wang, *Science*, 1996, **274**, 1701.

57. G. Che, B. B. Laxmi, C. R. Martin, E. R. Fisher and R. S. Ruoff, *Chem. Mater.*, 1998, **10**, 260.

58. M. Terrones, N. Grobert, J. Olivares, J. P. Zhang, H. Terrones, K. Kordatos, H. K. Hsu, J. P. Hare, P. D. Townsend, K. Prassides, A. K. Cheetham, H. W. Kroto and D. R. M. Walton, *Nature*, 1997, **388**, 52.

59. M. Terrones, N. Grobert, J. P. Zhang, H. Terrones, J. Olivares, H. K. Hsu, J. P. Hare, A. K. Cheetham, H. W. Kroto and D. R. M. Walton, *Chem. Phys. Lett.*, 1998, **285**, 299.

60. S. Fan, M. C. Chapline, N. R. Franklin, T. W. Tombler, A. M. Cassel and H. Dai, *Science*, 1999, **283**, 512.

61. Z. W. Pan, S. S. Xie, B. H. Chang, L. F. Sun, W. Y. Zhou and G. Wang, *Chem. Phys. Lett.*, 1999, **299**, 97.

62. (a) B. C. Satishkumar, A. Govindaraj and C. N. R. Rao, *Chem. Phys. Lett.*, 1999, **307**, 158; (b) C. N. R. Rao, R. Sen, B. C. Satishkumar and A Govindaraj, *Chem. Commun.*, 1998, 1525.

63. C. N. R. Rao, A. Govindaraj, R. Sen and B. C. Satishkumar, *Mater. Res. Innov.*, 1998, **2**, 128.

64. (a) R. Andrews, D. Jacques, A. M. Rao, F. Derbyshire, D. Qian, X. Fan, E. C. Dickey and J. Chen, *Chem. Phys. Lett.*, 1999, **303**, 467; (b) S. Huang, A. W. H. Mau, T. W. Turney, P. A. White and L. Dai, *J. Phys. Chem. B*, 2000, **104**, 2193.

65. (a) J. Li, C. Papadopoulos, J. M. Xu and M. Moskovits, *Appl. Phys. Lett.*, 1999, **75**, 367; (b) K. Mukhopadhyay, A. Koshio, T. Sugai, N. Tanaka, H. Shinohara, Z. Konya and J. B. Nagy, *Chem. Phys. Lett.*, 1999, **303**, 117.

66. (a) C. N. R. Rao and A. Govindaraj, *Acc. Chem. Res.*, 2002, **35**, 998; (b) B. C. Satishkumar, A. Govindaraj, P.V. Vanitha, A. K. Raychaudhuri and C. N. R. Rao, *Chem. Phys. Lett.*, 2002, **362**, 301.

67. L. Dai, A. Patil, X. Gong, Z. Guo, L. Liu, Y. Liu and D. Zhu, *ChemPhysChem*, 2003, **4**, 1150.

68. P. M. Ajayan, *Adv. Mater.*, 1995, **7**, 489.

69. M. Burghard, G. Duesberg, G. Philipp, J. Muster and S. Roth, *Adv. Mater.*, 1998, **10**, 584.

70. J. Liu, M. J. Casavant, M. Cox, D. A. Walters, P. Boul, W. Lu, A. J. Rimberg, K. A. Smith, D. T. Colbert and R. E. Smalley, *Chem. Phys. Lett.*, 1999, **303**, 125.

71. Q. Chen and L. Dai, *Appl. Phys. Lett.*, 2000, **76**, 2719.

72. S. Huang, L. Dai and A. W. H. Mau, *J. Phys. Chem. B*, 1999, **103**, 4223.

73. Y. Yang, S. Huang, H. He, A. W. H. Mau and L. Dai, *J. Am. Chem. Soc.*, 1999, **121**, 10832.

74. Q. Chen and L. Dai, *J. Nanosci. Nanotechnol.*, 2001, **1**, 43.

75. R. J. Jackman and G. M. Whitesides, *Chemtech*, 1999, **29**, (May) 18.

76. G. Zheng, H. Zhu, Q. Luo, Y. Zhou and D. Zhao, *Chem. Mater.*, 2001, **13**, 2240.

77. Do -H. Kim, D -S. Cho, H. -S. Jang, C. -D. Kim and H.-R. Lee, *Nanotechnology*, 2003, **14**, 1269.

78. H. Dai, *Acc. Chem. Res.*, 2002, **35**, 1035.

79. C. Journet and P. Bernier, *Appl. Phys. A-Mater. Sci. Process.*, 1998, **67**, 1.

80. H. Dai, A. Z. Rinzler, P. Nikolaev, A. Thess, D. T. Colbert and R. E. Smalley, *Chem. Phys. Lett.*, 1996, **260**, 471.

81. Y. Saito, M. Okuda and T. Koyama, *Surf. Rev. Lett.*, 1996, **3**, 863.

82. Y. Saito, K. Kawabata and M. Okuda, *J. Phys. Chem.*, 1995, **99**, 16076.

83. C. Journet, W. K. Maser, P. Bernier, A. Loiseau, M. Lamy de la Chapelle, S. Lefrant, P, Deniard, R. Lee and J. E. Fischer, *Nature*, 1997, **388**, 756.

84. S. Farhat, M. L. de La Chapelle, A. Loiseau, C. D. Scott, S. Lefrant, C. Journet and P. Bernier, *J. Chem. Phys.*, 2001, **115**, 6752.

85. I. Hinkov, S. Farhat, M. Lamy De La Chapelle, S. S. Fan, H. X. Han, G. H. Li and C. D. Scott, *Proc. VI Appl. Diamond Conference/Sec. Frontier Carbon Technology Joint Conference*, Auburn, AL, Ohio, USA, August 6–10, 2001 and *NASA Conference Publication*, 2001.

86. P. Costa, C. Xu, K. Coleman, J. Sloan and M. L. H. Green, *Trends in Nanotechnology (Conference Proceedings)* (TNT2002), Abstracts published by CMP/Cientifica Santiago de Compostela, Spain, September 9–13, 2002.

87. H. J. Huang, J. Marie, H. Kajiura and M. Ata, *Nano Lett.*, 2002, **2**, 1117.

88. H. Takikawa, M. Ikeda, K. Hirahara, Y. Hibi, Y. Tao, P.A. Ruiz, T. Sakakibara, S. Itoh and S. Iijima, *Phys. B: Condensed Matter*, 2002, **323**, 277.

89. T. Guo, P. Nikolaev, A. Thess, D. T. Colbert and R. E. Smalley, *Chem. Phys. Lett.*, 1995, **243**, 49.

90. M. Yudasaka, R. Yamada, N. Sensui, T. Wilkins, T. Ichihashi and S. Iijima, *J. Phys. Chem. B*, 1999, **103**, 6224.

91. P. C. Eklund, B. K. Pradhan, U. J. Kim, Q. Xiong, J. E. Fischer, A. D. Friedman, B. C. Holloway, K. Jordan and M. W. Smith, *Nano Lett.*, 2002, **2**, 561.

92. (a) W. K. Maser, E. Munoz, A. M. Benito, M. T. Martinez, G. F. de la Fuente, Y. Maniette, E. Anglaret and J. L. Sauvajol, *Chem. Phys. Lett.*, 1998, **292**, 587; (b) A. P. Bolshakov, S. A. Uglov, A. V. Saveliev, V. I. Konov, A. A. Gorbunov, W. Pompe and A. Graff, *Diamond Relat. Mater.*, 2002, **11**, 927.

93. C. D. Scott, S. Arepalli, P. Nikolaev and R. E. Smalley, *Appl. Phys. A: Mater. Sci. Process.*, 2001, **72**, 573.

94. A. Thess, R. Lee, P. Nikolaev, H. Dai, P. Petit, J. Robert, C. Xu, Y. H. Lee, S. G. Kim, A. G. Rinzler, D. T. Colbert, G. E. Scuseria, D. Tomanek, J. E. Fischer and R. E. Smalley, *Science,* 1996, **273**, 483.

95. S. Arepalli, *J. Nanosci. Nanotechnol.*, 2004, **4**, 318.

96. (a) B. C. Satishkumar, A. Govindaraj, R. Sen and C. N. R. Rao, *Chem. Phys. Lett.*, 1998, **293**, 47; (b) S. Seraphin and D. Zhou, *Appl. Phys. Lett.*, 1994, **64**, 2087.

97. H. M. Cheng, F. Li, G. Su, H. Y. Pan, L. L. He, X. Sun and M. S. Dresselhaus, *Appl. Phys. Lett.*, 1998, **72**, 3282.

98. J. F. Colomer, C. Stefan, S. Lefrant, G. van Tendeloo, I. Willems, Z. Konya, A. Fonseca. Ch. Laurent and J. B. Nagy, *Chem. Phys. Lett.*, 2000, **317**, 83.

99. E. Flahaut, A. Govindaraj, A. Peigney, Ch. Laurent, A. Rousset and C. N. R. Rao, *Chem. Phys. Lett.*, 1999, **300**, 236.

100. H. D. Sun, Z. K. Tang, J. Chen and G. Li, *Appl. Phys. A.*, 1999, **69**, 381.

101. M. Endo, T. Hayashi, Y.A. Kim, H. Muramatu, M. Ezaka, P.C.P. Watts, K. Nishimura and T. Tsukada, *J. Nanosci. Nanotechnol.*, 2004, **4**, 132.

102. K. Hata, D. N. Futaba, K. Mizuno, T. Namai, M. Yumura and S. Iijima, *Science*, 2004, **306**, 1362.

103. Y. M. Li, D. Mann, M. Rolandi, W. Kim, A. Ural, S. Hung, A. Javey, J. Cao, D. W. Wang, E. Yenilmez, Q. Wang, J. E. Gibbons, Y. Nishi and H. J. Dai, *Nano Lett.*, 2004, **4**, 317.

104. (a) Y. Murakami, Y. Miyauchi, S. Chiashi and S. Maruyama, *Chem. Phys. Lett.*, 2003, **374**, 53; (b) S. Maruyama, S. Chiashi and Y. Miyauch, *Thermal Eng. Joint Conference* 6, 2003.

105. M. Su, B. Zheng and J. Liu, *Chem. Phys. Lett.*, 2000, **322**, 321.

106. B. Zheng, Y. Li and J. Liu, *Appl. Phys. A: Mater. Sci. & Process.*, 2002, **74**, 345.

107. R. Alexandrescu, A. Crunteanu, R. E. Morjan, I. Morjan, F. Rohmund, L. K. L. Falk, G. Ledoux and F. Huisken, *Infrared Phy. Technol.*, 2003, **44**, 43.

108. D. E. Resasco, W. E. Alvarez, F. Pompeo, L. Balzano, J. E. Herrera, B. Kitiyanan and A. Borgna, *J. Nanoparticle Res.*, 2002, **4**, 131; www.ou.edu/engineering/nanotube/comocat.html.

109. P. Nikolaev, M. Bronikowski, R. K. Bradley, F. Rohmund, D. T. Colbert, K. A. Smith and R. E. Smalley, *Chem. Phys. Lett.*, 1999, **313**, 91.

110. B. Kitiyanan, W. E. Alvarez, J. H. Harwell and D. E. Resasco, *Chem. Phys. Lett.,* 2000, **317**, 497.

111. R. E. Smalley and B. I. Yakobson, *Solid State Commun.*, 1998, **107**, 597.

112. M. J. Bronikowski, P. A. Willis, D. T. Colbert, K. A. Smith and R. E. Smalley, *J. Vac. Sci. Technol. A*, 2001, **19**, 1800.

113. P. Nikolaev, *J. Nanosci. Nanotechnol.*, 2004, **4**, 307.

114. R. L. V. Wal, L. J. Hall and G. M. Berger, *J. Phys. Chem. B*, 2002, **106**, 13122.

115. R. L. V. Wal, G. M. Berger and L. J. Hall, *J. Phys. Chem. B*, 2002, **106**, 3564.

116. D. Laplaze, P. Bernier, W. F. Maser, G. Flamant, T. Guillard and A. Loiseau, *Carbon*, 1998, **36**, 685.

117. S.-H. Jeong, J.-H. Ko, J.-B. Park and W. Park, *J. Am. Chem. Soc.*, 2004, **126**, 15982–15983.

118. (a) B. W. Smith, M. Monthioux and D. E. Luzzi, *Nature*, 1998, **396**, 323; (b) B. Burteaux, A. Claye, B. W. Smith, M. Monthioux, D. E. Luzzi and J. E. Fishcher, *Chem. Phys. Lett.*, 1999, **310**, 21–24.

119. (a) S. Bandow, K. Hirahara, T. Hiraoka, G. Chen, P. C. Eklund and S. Iijima, *MRS Bull.* 2004, April, 260 and the references therein; (b) K. Suenaga, T. Okazaki, C.-R. Wang, S. Bandow, H. Shinohara and S. Iijima, *Phys. Rev. Lett.*, 2003, **90**, 055506–1.

120. (a) S. Okada, S. Saito and A. Oshiyama, *Phys. Rev. Lett.*, 2001, **86**, 3835; (b) S. Berber, Y.-K. Kwon and D. Tomanek, *Phys.Rev. Lett.*, 2002, **88**, 185502.

121. J. L. Hutchison, N. A. Kiselev, E. P. Krinichnaya, A. V. Krestinin, R. O. Loutfy, A. P. Morawsky, V. E. Muradyan, E. D. Obraztsova, J. Sloan, S. V. Terekhov and D. N. Zakharov, *Carbon*, 2001, **39**, 761.

122. T. Sugai, H. Yoshida, T. Shimada, T. Okazaki and H. Shinohara, *Nano Lett.*, 2003, **3**, 769.

123. (a) S. C. Lyu, B.C. Liu, S. H. Lee, C. Y. Park, H. K. Kang, C.-W. Yang and C. J. Lee, *J. Phys. Chem. B*, 2004, **108**, 2192; (b) C.-J. Lee. (S. Korea), *PCT Int. Appl.*, 2004, 19 pp. (Patent No: WO 2004083113); (c) S. H. Lee, S. I. Jung, C. B. Kong, T. J. Lee, S. K. Choi, M. H. Park, C. W. Yang, C. Y. Park and C. J. Lee, *International Conference on the Science and Application of Nanotubes* San Luis Potosi, S. L. P. (Mexico), 2004, 115; (d) E. Flahaut, R. Bacsa, A. Peigney and Ch. Laurent, *Chem. Commun.*, 2003, **12**, 1442.

124. R. T. K. Baker and P. S. Harris, in *Chemistry and Physics of Carbon*, ed. P. L. Walker and P. A. Thrower, Marcel Dekker, New York, 1978, Vol. 14.

125. A. Oberlin, M. Endo and T. Koyama, *J. Cryst. Growth*, 1976, **32**, 335.

126. M. Endo and H. W. Kroto, *J. Phys. Chem.*, 1992, **96**, 6941.

127. R. E. Smalley, *Mat. Sci. Eng. B.*, 1993, **19**, 1.

128. T. W. Ebbesen, J. Tabuchi and K. Tanigaki, *Chem. Phys. Lett.*, 1992, **191**, 336.

129. Y. Saito, T. Yoshikawa, M. Inagaki, M. Tomita and T. Hayashi, *Chem. Phys. Lett.*, 1994, **204**, 277.

130. D. H. Robertson, D. W. Brenner and C. T. White, *J. Phys. Chem.*, 1992, **96**, 6133.

131. (a) S. Amelinckx, X. B. Zhang, D. Bernaerts. X. F. Zhang, V. Ivanov and J. B. Negy, *Science*, 1994, **265**, 635; (b) S. Amelinckx, D. Bernaerts. X. B. Zhang, G. Van Tendeloo and J. Van Landuyt, *Science*, 1995, **267**, 1334.

132. A. Maiti, C. T. Brabec and J. Bernholc, *Phys. Rev. B*, 1997, **55**, R6097.

133. A. Moisala, A. G. Nasibulin and E. I. Kaeuppinen, *J. Phys.: Condens. Matter.*, 2003, **15**, S3011.

134. K. Tohji, T. Goto, H. Takahashi, Y. Shinoda, N. Shimizu, B. Jeyadevan, I. Matsuoka, Y. Saito, A. Kasuya, T. Ohsuna, K. Hiraga and Y. Nishina, *Nature*, 1996, **383**, 679.

135. K. Tohji, T. Goto, H. Takahashi, Y. Shinoda, N. Shimizu, B. Jeyadevan, I. Matsuoka, Y. Saito, A. Kasuya, S. Ito and Y. Nishina, *J. Phys. Chem. B*, 1997, **101**, 1974.

136. S. Bandow, A. M. Rao, K. A. Williams, A. Thess, R. E. Smalley and P. C. Eklund, *Phys. Chem. B*, 1997, **101**, 8839.

137. M. T. Martinez, M. A. Callejas, A. M. Benito, W. K. Maser, M. Cochet, J. M. Andres, J. Schreiber, O. Chauvet and J. L. G. Fierro, *Chem. Commun.*, 2002, 1000.

138. A. G. Rinzler, J. Liu, P. Nikolaev, C. B. Huffman, M. Rodrígues, P. J. Boul, A. H. Lu, D. Heymann, D. T. Colbert, R. S. Lee, J. E. Fischer, A. M. Rao, P. C. Eklund and R. E. Smalley, *Appl. Phys. A*, 1998, **67**, 29.

139. A. C. Dillon, T. Gennett, K. M. Jones, J. L. Alleman, P. A. Parilla and M. J. Heben, *Adv. Mater.*, 1999, **11**, 1354.

140. I. W. Chiang, B. E. Brinson, R. E. Smalley, J. L. Margrave and R. H. Hauge, *J. Phys. Chem. B*, 2001, **105**, 1157.

141. I. W. Chiang, B. E. Brinson, A. Y. Huang, P. A. Willis, M. J. Bronikowski, J. L. Margrave, R. E. Smalley and R. H. Hauge, *J. Phys. Chem. B*, 2001, **105**, 8297.

142. R. Sen, S. M. Rickard, M. E. Itkis and R. C. Haddon, *Chem. Mater.*, 2003, **15**, 4273.

143. S. R. C. Vivekchand, A. Govindaraj, Md. Motin Seikh and C. N. R. Rao, *J. Phys. Chem. B*, 2004, **108**, 6935.

144. J. Liu, A. G. Rinzler, H. Dai, J. H. Hafner, R. K. Bradley, P. J. Boul, A. Lu, T. Iverson, K. Shelimov, C. B. Huffman, F. Rodriguez-Macias, Y. S. Shon, T. R. Lee, D. T. Colbert and R. E. Smalley, *Science*, 1998, **280**, 1253.

145. (a) G. S. Duesberg, J. Muster, V. Krstic, M. Burghard and S. Roth, *Appl. Phys. A*, 1998, **67**, 117; (b) G. S. Duesberg, W. Blau, H. J. Byrne, J. Muster, M. Burghard and S. Roth, *Synth. Metals*, 1999, **103**, 2484.

146. S. K. Doorn, I. R. F. Fields, H. Hu, M. Hamon, R.C. Haddon, J. P. Selegue and V. Majidi, *J. Am. Chem. Soc.*, 2002,**124**, 3169.

147. D. Chattopadhyay, S. Lastella, S. Kim and F. Papadimitrakopoulos, *J. Am. Chem. Soc.*, 2002,**124**, 728.

148. B. Chen and J. P. Selegue, *Anal. Chem.*, 2002, **74**, 4774.

149. E. Farkas, M. Elizabeth Anderson, Z. Chen and A. G. Rinzler, *Chem. Phys. Lett.*, 2002, **363**, 111.

150. D. Chattopadhyay, I. Galeska and F. Papadimitrakopoulos, *J. Am. Chem. Soc.*, 2003, **125**, 3370.

151. Ge. G. Samsonidze, S. G. Chou, A. P. Santos, V. W. Brar, G. Dresselhaus, M. S. Dresselhaus, A. Selbst, A. K. Swan, M. S. Ünlü, B. B. Goldberg, D. Chattopadhyay, S. N. Kim and F. Papadimitrakopoulos, *Appl. Phys. Lett.*, 2004, **85**, 1006.

152. Z. Chen, X. Du, M.-H. Du, C.D. Rancken, H.-P. Cheng and A.G. Rinzler, *Nano Lett.*, 2003, **3**, 1245.

153. M. J. O'Connell, S. M. Bachilo, C. B. Huffman, V. C. Moore, M. S. Strano, E. H. Haroz, K. L. Rialon, P. J. Boul, W. H. Noon, C. Kittrell, J. Ma, R. H. Hauge, B. R. Weisman and R. E. Smalley, *Science*, 2002, **297**, 593.

154. R. Krupke, F. Hennrich, H. v. Löhneysen and M. M. Kappes, *Science*, 2003, **301**, 344.

155. M. S. Strano, C. A. Dyke, M. L. Usrey, P. W. Barone, M. J. Allen, H. Shan, C. Kittrell, R. H. Hauge, J. M. Tour and R. E. Smalley, *Science*, 2003, **301**, 1519.

156. M. Zheng, A. Jagota, E. D. Semke, B. A. Diner, R. S. McLean, S. R. Lustig, R. E. Richardson and N. G. Tassi, *Nat. Mater.*, 2003, **2**, 338.

157. M. Zheng, A. Jagota, M. S. Strano, A. P. Santos, P. Barone, S. G. Chou, B. A. Diner, M. S. Dresselhaus, R. S. Mclean, G. B. Onoa, G. G. Samsonidze, E. D. Semke, M. Usrey and D. J. Walls, *Science*, 2003, **302**, 1545.

158. (a) R. C. Haddon, J. Sippel, A. G. Rinzler and F. Papadimitrakopoulos *MRS Bull./*2004 April, 252; (b) M. E. Itkis, D. E. Perea, R. Jung, S. Niyogi and R. C. Haddon, *J. Am. Chem. Soc.*, 2005, **127**, 3439.

159. S. Iijima, *MRS Bull.*, 1994, **19**, 43.

160. J. Tersoff and R. S. Ruoff, *Phys. Rev. Lett.*, 1994, **73**, 676.

161. D. Srivastava, D. W. Brenner, J. D. Schall, K. D. Ausman, M. Yu and R. S. Ruoff, *J. Phys. Chem. B*, 1999, **103**, 4330.

162. R. Martel, H. R. Shea and Ph. Avouris, *J. Phys. Chem. B*, 1999, **103**, 7551.

163. M. Ahlskog, E. Seynaeve, R. J. M. Vullers, C. Haesendonck, A. Fonseca, K. Hernadi and J. B. Nagy, *Chem. Phys. Lett.*, 1999, **300**, 202;

164. S. Huang, L. Dai and A. W. H. Mau, *J. Mater. Chem.*, 1999, **9**, 1221.

165. J. Sloan, R. E. Dunin-Borkowski, J. L. Hutchison, K. S. Coleman, V. C. Williams, J. B. Claridge, A. P. E. York, C. Xu, S. R. Baley, G. Brown, S. Friedrichs and M. L. H. Green, *Chem. Phys. Lett.*, 2000, **316**, 191.

166. R. Seshadri, A. Govindaraj, H. N. Aiyer, R. Sen, G. N. Subbanna, A. R. Raju and C. N. R. Rao, *Curr. Sci. (India)*, 1994, **66**, 839.

167. Y. Murakami, T. Shibata, K. Okuyama, T. Arai, H. Suematsu and Y. Yoshida, *J. Phys. Chem. Solids*, 1993, **54**, 1861.

168. W. Ruland, in *Chemistry and Physics of Carbon*, ed. P. L. Walker, Marcel Dekker, New York, 1968, Vol. 4.

169. 143. B. E. Warren, *Phys. Rev.*, 1941, **59**, 693.

170. T. W. Ebbesen, H. Hiura, J. Fujita, Y. Ochiai, S. Matsui and K.Tanigaki, *Chem. Phys. Lett.*, 1993, **209**, 83.

171. Y. Ando, *Jpn. J. Appl. Phys. Lett.*, 1993, **32**, L1342.

172. X. K. Wang, X. M. Lin, V. P. Dravid, J. B. Ketterson and R. P. H. Chang, *Appl. Phys. Lett.*, 1993, **62**, 1881.

173. Z. Zhang and C. M. Lieber, *Appl. Phys. Lett.*, 1993, **62**, 2792.

174. C. H. Olk and J. P. Haremans, *J. Mater. Res.*, 1994, **9**, 259.

175. R. Seshadri, H. N. Aiyer, A. Govindaraj and C. N. R. Rao, *Solid State Commun.*, 1994, **91**, 195.

176. D. L. Carroll, Ph. Redlich, P. M. Ajayan, J. C. Charlier, X. Blase, A. De-Vita and R. Car, *Phys. Rev. Lett.*, 1997, **78**, 2811.

177. L. C. Venema, V. Meunier, Ph. Lambin and C. Dekker, *Phys. Rev. B*, 2000, **61**, 2991.

178. M. Ashino, A. Schwarz, T. Behnke and R. Wiesendanger, *Phys. Rev. Lett.*, 2004, **93**, 136101.

179. J. M. Zuo, I. Vartanyants, M. Gao, R. Zhang and L. A. Nagahara, *Science*, 2003, **300**, 1419.

180. M. Zamkov, N.Woody, S. Bing, H. S. Chakraborty, Z. Chang, U. Thumm and P. Richard, *Phys. Rev. Lett.*, 2004, **93**, 156803–1.

181. R. A. Jishi, L. Venkataraman, M. S. Dresselhaus and G. Dresselhaus, *Chem. Phys. Lett.*, 1993, **209**, 77.

182. H. Hiura, T. W. Ebbesen, K. Tanigaki and H. Takahashi, *Chem. Phys. Lett.*, 1993, **202**, 509.

183. J. M. Holden, P. Zhou, X. X. Bi, P. C. Eklund, S. Bandow, R. A. Jishi, K. Das Chowdhury, G. Dresselhaus and M. S. Dresselhaus, *Chem. Phys. Lett.*, 1994, **220**, 186.

184. A. M. Rao, E. Richter, S. Bandow, P. C. Eklund, K. A. Williams, S. Fang, K. R. Subbaswamy, M. Menon, A. Thess, R. E. Smalley, G. Dresselhaus and M. S. Dresselhaus, *Science*, 1997, **275**, 187.

185. A. Kasuya, Y. Sasaki, Y. Saito, K. Tohji and Y. Nishina, *Phys. Rev. Lett.*, 1997, **78**, 4434.

186. A. M. Rao, A. Jorio, M. A. Pimenta, M. S. S. Dantas, R. Saito, G. Dresselhaus and M. S. Dresselhaus, *Phys. Rev. Lett.*, 2000, **84**, 1820.

187. A. Jorio, R. Saito, J. H. Hafner, C. M. Lieber, M. Hunter, T. McClure, G. Dresselhaus and M. S. Dresselhaus, *Phys. Rev. Lett.*, 2001, **86**, 1118.

188. M. S. Dresselhaus, G. Dresselhaus, A. Jorio, A. G. S. Filho and R. Saito, *Carbon*, 2002, **40**, 2043.

189. C. Fantini, A. Jorio, M. Souza, M. S. Strano, M. S. Dresselhaus and M. A. Pimenta, *Phys. Rev. Lett.*, 2004, **93**, 147406–1.

190. S. Bandow, G. Chen, G. U. Sumanasekera, R. Gupta, M. Yudasaka, S. Iijima and P. C. Eklund, *Phys. Rev. B*, 2002, **66**, 075416.

191. S. Ghosh, A. K. Sood and C. N. R. Rao, *J. Appl. Phys.*, 2002, **92**, 1165.

192. S. M. Bachilo, M. S. Strano, C. Kittrell, R. H. Hauge, R. E. Smalley and R. B. Weisman, *Science*, 2002, **298**, 2361.

193. A. Hartschuh, H. N. Pedrosa, L. Novotny and T. D. Krauss, *Science*, 2003, **301**, 1354.

194. J. Lefebvre, J. M. Fraser, P. Finnie and Y. Homma, *Phys. Rev. B*, 2004, **69**, 75403.

195. S. Zaric, G. N. Ostojic, J. Kono, J. Shaver, V. C. Moore, M. S. Strano, R. H. Hauge, R. E. Smalley and X. Wei, *Science*, 2004, **304**, 1129.

196. P. V. Teredesai, A. K. Sood, D. V. S. Muthu, R. Sen, A. Govindaraj and C. N. R. Rao, *Chem. Phys. Lett.*, 2000, **319**, 296.

197. A. K. Sood, P. V. Teredesai, D. V. S. Muthu, R. Sen, A. Govindaraj and C. N. R. Rao, *Phys. Stat. Sol. (b)*, 1999, **215**, 393.

198. S. Bandow, S. Asaka, Y. Saito, A. M. Rao, L. Grigorian. E. Richter and P. C. Eklund, *Phys. Rev. Lett.*, 1998, **80**, 3779.

199. U. D. Venkateswaran, A. M. Rao, E. Richter, M. Menon, A. Rinzler, R. E. Smalley and P. C. Eklund, *Phys. Rev. B*, 1999, **59**, 10928.

200. J. R. Wood, M. D. Frogley, E. R. Meurs, A. D. Prins, T. Peijs, D. J. Dunstan and H. D. Wagner, *J. Phys. Chem. B*, 1999, **103**, 10388.

201. H. D. Sun, Z. K. Tang, J. Chen and G. Li, *Solid State Commun.*, 1999, **109**, 365.

202. S. A. Chesnokov, V. A. Nalimova, A. G. Rinzler, R. E. Smalley and J. E. Fischer, *Phys. Rev. Lett.*, 1999, **82**, 343.

203. M. J. Peters, L. E. McNeil, J. P. Lu and D. Kahn, *Phys. Rev. B*, 2000, **61**, 5939.

204. M. Hanfland, H. Beister and K. Syassen, *Phys. Rev. B*, 1989, **39**, 12598.

205. D. W. Snoke, Y. P. Raptis and K. Syassen, *Phys. Rev. B*, 1992, **45**, 14419.

206. S. M. Sharma, S. Karmakar, S. K. Sikka, P. V. Teredesai, A. K. Sood, A. Govindaraj and C. N. R. Rao, *Phys. Rev. B*, 2001, **63**, 205417.

207. S. Iijima, C. Brabec, A. Maiti and J. Bernholc, *J. Chem. Phys.*, 1996, **104**, 2089.

208. J. W. Mintmire, B. I. Dunlap and C. T. White, *Phys. Rev. Lett.*, 1992, **68**, 631.

209. N. Hamada, S. Sawada and A. Yoshiyama, *Phys. Rev. Lett.*,1992, **68**, 1579.

210. R. Saito, M. Fujita, G. Dresselhaus and M. S. Dresselhaus, *Appl. Phys. Lett.*, 1992, **60**, 2204.

211. M. A. Hamon, M. E. Itkis, S. Niyogi, T. Alvaraez, C. Kuper, M. Menon and R. C. Haddon, *J. Am. Chem. Soc.*, 2001, **123**, 11292.

212. M. E. Itkis, S. Niyogi, M. Meng, M. Hamon, H. Hu and R.C. Haddon, *Nano Lett.*, 2002, **2**, 155.

213. E. Joselevich, *ChemPhysChem*, 2004, **5**, 619.

214. S. Aisebit, A. Karl, W. Eberhardt, J. E. Fischer, C. Sathe, A. Agui and J. Nordgren, *Appl. Phys. A*, 1998, **67**, 89.

215. Z. P. Tang, A. Kleinhammes, H. Shimoda, L. Fleming, K. Y. Bennoune, S. Sinha, C. Bower, O. Zhou and Y. Wu, *Science*, 2000, **288**, 492.

216. T. Hertel and G. Moos, *Chem. Phys. Lett.*, 2000, **320**, 359.

217. H. Ago, T. Kugler, F. Cacialli, W. K. Salaneck, M. S. P. Shaffer, A. H. Windle and R. H. Friend, *J. Phys. Chem. B*, 1999, **103**, 8116.

218. Y. Miyamoto, A. Rubio, M. L. Cohen and S. G. Louie, *Phys. Rev. B*, 1994, **50**, 4976.

219. B. C. Satishkumar, A. Govindaraj, K. R. Harikumar, J. P. Zhang, A. K. Cheetham and C. N. R. Rao, *Chem. Phys. Lett.*, 1999, **300**, 473.

220. O. Stephan, P. M. Ajayan, C. Colliex, Ph. Redlich, J. M. Lambert, P. Bernier and P. Lefin, *Science*, 1994, **266**, 1683.

221. Y. Zhang, H. Gu, K. Suenaga and S. Iijima, *Chem. Phys. Lett.*, 1997, **279**, 264.

222. M. Terrones, A. M. Benito, C. Mantega-Diego, W. K. Hsu, O. I. Osman, J. P. Hare, D. G. Reid, H. Terrones, A. K. Cheetham, K. Prassides, H. W. Kroto and D. R. M. Walton, *Chem. Phys. Lett.*, 1996, **257**, 576.

223. (a) R. Sen, B. C. Satishkumar, A. Govindaraj, K. R. Harikumar, G. Raina, J. P. Zhang, A. K. Cheetham and C. N. R. Rao, *Chem. Phys. Lett.*, 1998, **287**, 671; (b) R. Sen, B. C. Satishkumar, A. Govindaraj, K. R. Harikumar, M. K. Renganathan and C. N. R. Rao, *J. Mater. Chem.*, 1997, **7**, 2335.

224. C. Y. Zhi, X. D. Bai and E. G. Wang, *J. Nanosci. Nanotechnol.*, 2004, **4**, 35.

225. (a) M. Nath, B. C. Satishkumar, A. Govindaraj, C. P. Vinod and C. N. R. Rao, *Chem. Phys. Lett.*, 2000, **322**, 333; (b) S. L. Sung, S. H. Tsai, C. H. Tseng, F. K. Chiang, X. W. Liu and H. C. Shih, *Appl. Phys. Lett.*, 1999, **74**, 197.

226. (a) K. Suenaga, M. Yusadaka, C. Colliex and S. Iijima, *Chem. Phys. Lett.*, 2000, **316**, 365; (b) K. Suenaga, M. P. Johansson, N. Hellgren, E. Broitman, L. R. Wallenberg, C. Colliex, J.-E. Sundgren and L. Hultman, *Chem. Phys. Lett.*, 1999, **300**, 695.

227. D. Goldberg, Y. Bando, W. Han, K. Kurashima and T. Sato, *Chem. Phys. Lett.*, 1999, **308**, 337.

228. D. L. Carroll, Ph. Redlich, X. Blase, J.-C. Charlier, S. Curran, P. M. Ajayan, S. Roth and M. Rühle, *Phys. Rev. Lett.*, 1998, **81**, 2332.

229. K. Efsarjani, A. A. Farajin, Y. Hashi and Y. Kawazoe, *Appl. Phys. Lett.*, 1999, **74**, 79.

230. R. D. Antonov and A. T. Johnson, *Phys. Rev. Lett.*, 1999, **83**, 3274.

231. M. Terrones, A. Jorio, M. Endo, A. M. Rao, Y. A. Kim, T. Hayashi, H. Terrones, J.-C. Charlier, G. Dresselhaus and M. S. Dresselhaus, *Mater. Today*, 2004, 30.

232. O. Zhou, R. M. Fleming, D. W. Murphy, C. H. Chen, R. C. Haddon and A. P. Ramirez, *Science*,1994, **263**, 1744.

233. (a) V. Z. Mordkovitch, M. Baxendale, S. Yoshimura and R. P. H. Chang, *Carbon*, 1996, **34**, 1301; (b) V. Z. Mordkovitch, M. Baxendale, R. P. Chang and H. Yoshimura *Synth. Metals*, 1997, **86**, 2049.

234. P. M. Ajayan, *Condensed Matter News*, 1995, **4**, 9.

235. V. Z. Mordkovitch, *Mol. Cryst. Liq. Cryst.*, 2000, **340**, 775.

236. K. Metenier, L. Duclaux, H. Gaucher, J.P. Salvetat, P. Lauginie and S. Bonnamy, in *AIP Conference Proceedings*, American Instisute of Physics, 544, Woodbury, New York, 1998, pp. 51–54; L. Duclaux, K. Metenier, J. P. Salvetat, P. Lauginie, S. Bonnamy and F. Béguin, in: *AIP Conference Proceedings*, American Institute of Physics, 544, Woodbury, New York, 2000, pp. 408.

237. (a) E. Jouguelet, C. Mathis and P. Petit, *Chem. Phys. Lett.,* 2000, **318**, 561; (b) P. Petit, C. Mathis, C. Journet and P. Bernier, *Chem. Phys. Lett.*, 1999, **305**, 370.

238. A. Claye and J. E. Fischer *Mol. Cryst. Liq. Cryst.*, 2000, **340**, 743.

239. B. Gao, C. Bower, J.D. Lorentzen, L. Fleming, A. Kleinhammes and X. P. Tang, *Chem. Phys. Lett.*, 2000, **327**, 69.

240. R. M. Lago, S. C. Tsang, K. L. Lu, Y. K. Chen and M. L. H. Green, *J. Chem. Soc., Chem. Commun.*, 1995, 1355.

241. K. C. Hwang, *J. Chem. Soc., Chem. Commun.* 1995, 173.

242. H. Hiura, T. W. Ebbesen and K. Tanigaki, *Adv. Mater.* 1995, **7**, 275.

243. B. C. Satishkumar, A. Govindaraj, G. N. Subbanna, J. Mofokeng and C. N. R. Rao, *J. Phys. B, Atm. Mol. Opt. Phys.*, 1996, **29**, 4925.

244. C. Bower, S. Suzuki, K. Tanigaki and O. Zhou, *Appl. Phys. A*, 1998, **67**, 47.

245. A. Govindaraj, B. C. Satishkumar, M. Nath and C. N. R. Rao, *Chem. Mater.*, 2000, **12**, 202.

246. W. K. Hsu, W. Z. Li, Y. Q. Zhu, N. Grobert, M. Terrones, H. Terrones, N. Yao, J. P. Zhang, S. Firth, R. J. H. Clark, A. K. Cheetham, J. P. Hare, H. W. Kroto and D. R. M. Walton, *Chem. Phys. Lett.*, 2000, **317**, 77.
247. C. N. R. Rao, A. Govindaraj and B. C. Satishkumar, *Chem. Commun.*, 1996, 1525.
248. J. Sloan, J. Hammer, M. Zwiefka-Sibley and M. L. H. Green, *Chem. Commun.*, 1998, 347.
249. B. C. Satishkumar, E. M. Vogl, A. Govindaraj and C. N. R. Rao, *J. Phys. D. Appl. Phys.* 1996, **29**, 3173.
250. C. Bower, A. Kleinhammes, Y. Wu and O. Zhou, *Chem. Phys. Lett.*, 1998, **288**, 481.
251. G. U. Sumanasekera, J. L. Allen, S. L. Fang, A. L. Loper, A. M. Rao and P. C. Eklund, *J. Phys. Chem. B*, 1999, **103**, 4292.
252. S. Kazaoui, N. Minami, R. Jacquemin, H. Kataura and Y. Achiba *Phys. Rev. B*, 1999, **60**, 13339.
253. X. Fan, E. C. Dickey, P. C. Eklund, K. A. Williams, L. Grigorian, R. Buczko, S. T. Pantelides and S. J. Pennycook, *Phys. Rev. Lett.*, 2000, **84**, 4621.
254. J. Sloan, D. E. Luzzi, A. I. Kirkland, J. L. Hutchison and M. L. H. Green, *MRS Bull.*, 2004 (April), 267 and the references cited therein.
255. S. Niyogi, M. A. Hamon, H. Hu, B. Zhao, P. Bhowmik, R. Sen, M. E. Itkis and R. C. Haddon, *Acc. Chem. Res.,* 2002, **35**, 1105.
256. Y. Chen, R. C. Haddon, S. Fang, A. M. Rao, P. C. Eklund, W. H. Lee, E. C. Dickey, E. A. Grulke, J. C. Pendergrass, A. Chavan, B. E. Haley and R. E. Smalley, *J. Mater. Res.*, 1998, **13**, 2423.
257. E. Dujardin, T. W. Ebbesen, A. Krishnan and M. M. J. Treacy, *Adv. Mater.*, 1998, **10**, 1472.
258. H. Hu, P. Bhowmik, B. Zhao, M. A. Hamon, M. E. Itkis and R. C. Haddon, *Chem. Phys. Lett.*, 2001, **345**, 25.
259. Z. Yu and L. E. Brus, *J. Phys. Chem. A*, 2000, **104**, 10995; C. Bower, A. Kleinhammes, Y. Wu and O. Zhou, *Chem. Phys. Lett.*, 1998, **288**, 481.
260. S. Bandow, S. Asaka, X. Zhao and Y. Ando, *Appl. Phys. A*, 1998, **67**, 23.
261. E. Dujardin, T. W. Ebbesen, A. Krishnan and M. M. J. Treacy, *Adv. Mater.*, 1998, **10**, 611.
262. J. Chen, M. A. Hamon, H. Hu, Y. Chen, A. M. Rao, P. C. Eklund and R. C. Heddon, *Science*, 1998, **282**, 95.
263. (a) E. T. Mickelson, I. W. Chiang, J. L. Zimmerman, P. J. Boul, J. Lozano, J. Liu, R. E. Smalley, R. H. Hauge and J. L. Margrave, *J. Phys. Chem. B*, 1999, **103**, 4318; (b) H.F. Bettinger, *ChemPhysChem*, 2003, **4**, 1283.
264. K. F. Kelly, I. W. Chiang, E. T. Mickelson, R. H. Hauge, J. L. Margrave, X. Wand, G. E. Scuseria, C. Radloff and N. J. Halas, *Chem. Phys. Lett.*, 1999, **313**, 445.
265. P. J. Boul, J. Liu, E. T. Mickelson, C. B. Huffman, L. M. Ericson, I. W. Chiang, K. A. Smith, D. T. Colbert, R. H. Hauge, J. L. Margrave and R. E. Smalley, *Chem. Phys. Lett.*, 1999, **310**, 367.
266. M. A. Hamon, J. Chen, H. Hu, Y. Chen, A. M. Rao, P. C. Eklund and R. C. Haddon, *Adv. Mater.*, 1999, **11**, 834.

267. J. Chen, A. M. Rao, S. Lyuksyutov, M. E. Itkis, M. A. Hamon, H. Hu, R. W. Cohn, P. W. Eklund, D. T. Colbert, R. E. Smalley and R. C. Haddon, *J. Phys. Chem. B*, 2001, **105**, 2525.

268. J. E. Riggs, Z. Guo, D. L. Carroll and Y.-P. Sun, *J. Am. Chem. Soc.*, 2000, **122**, 5879.

269. Y. P. Sun, W. Huang, Y. Lin, Y. Kefu, A. Kitaygorodskiy, L. A. Riddle, Y. Yu and D. L. Caroll, *Chem. Mater.*, 2001, **13**, 2864.

270. S. Banerjee and S. S. Wong, *Nano Lett.*, 2002,**2**, 195.

271. W. Huang, S. Taylor, K. Fu, Y. Lin, D. Zhang, T. W. Hanks, A. M. Rao and Y.-P. Sun, *Nano Lett.*, 2002,**2**, 311.

272. M. Sano, A. Kamino, J. Okamura and S. Shinkai, *Langmuir*, 2001, **17**, 5125.

273. F. Pompeo and D. E. Resasco, *Nano Lett.*, 2002, **2**, 369.

274. W. Huang, Y. Lin, S. Taylor, J. Gaillard, A. M. Rao and Y.-P. Sun, *Nano Lett.*, 2002, **2**, 231.

275. E. V. Basiuk, V. A. Basiuk, J.-G. Banuelos, J.-M. Saniger-Blesa, V. A. Pokrovskiy, T. Y. Gromovoy, A. V. Mischanchuk and B. G. Mischanchuk, *J. Phys. Chem. B*, 2002, **106**, 1588.

276. J. L. Bahr, J. Yang, D. V. Kosynkin, M. J. Bronikowski, R. E. Smalley and J. M. Tour, *J. Am. Chem. Soc.*, 2001, **123**, 6536.

277. J. L. Bahr and J. L. Tour, *Chem. Mater.*, 2001, **13**, 3823.

278. M. Holzinger, O. Vostrowsky, A. Hirsch, F. Hennrich, M. Kappes, R. Weiss and F. Jellen, *Angew. Chem., Int. Ed.*, 2001, **40**, 4002; V. Georgakilas, K. Kordatos, M. Prato, D. M. Guldi, M. Holzinger and A. Hirsch, *J. Am. Chem. Soc.*, 2002, **124**, 760.

279. J. L. Bahr, E. T. Mickelson, M. J. Bronikowski, R. E. Smalley and J. M. Tour, *Chem. Commun.*, 2001, 193.

280. Z. Liu, Z. Shen, T. Zhu, S. Hou, L. Ying, Z. Shi and Z. Gu, *Langmuir*, 2000, **16**, 3569.

281. H. Shimoda, S. J. Oh, H. Z. Geng, R. J. Walker, X. B. Zhang, L. E. McNeil and O. Zhou, *Adv. Mater.*, 2002, **14**, 899.

282. S. S. Wong, A. T. Wooley, E. Joselevich and C. M. Lieber, *Chem. Phys. Lett.*, 1999, **306**, 219.

283. J. J. Davis, K. S. Coleman, B. R. Azamian, C. B. Bagshaw and M. L. H. Green, *Chem. Eur. J.*, 2003, **9**, 3732.

284. C. Richard, F. Balavoine, P. Schultz, T. W. Ebbesen and C. Mioskowski, *Science*, 2003, **300**, 775.

285. E. Katz and I. Willner, *ChemPhysChem*, 2004, **5**, 1084.

286. W. Song, I. A. Kinloch and A. H. Windle, *Science*, 2003, **302**, 1363.

287. T. W. Odom, J. L. Huang, P. Kim and C. M. Lieber, *J. Phys. Chem. B.*, 2000, **104**, 2794.

288. M. Ouyang, J.-L. Huang and C. M. Lieber, *Acc. Chem. Res.*, 2002, **35**, 1018.

289. P. L. McEuen and J.-Y. Park, *MRS Bull.*, 2004, 72.

290. H. Kataura, Y. Kumazawa, Y. Maniwa, I. Umezu, S. Suzuki,Y. Ohtsuka and Y. Achiba, *Synth. Metals*, 1999, **103**, 2555.

291. M. Bockrath, D. V. Cobden, P. L. McEuen, N. G. Chopra, A. Zettl, A. Thess and R. E. Smalley, *Science*,1997, **275**, 1922.

292. S. J. Tans, M. H. Devoret, H. Dai, A. Thess, R. E. Smalley, L. J. Greeligs and C. Dekker, *Nature*, 1997, **386**, 474.

293. L. C. Venema, J. W. G. Wildoer, J. W. Janssen, S. J. Tans, H. L. J. T. Tuinstra, L. P. Kouwenhoven and C. Dekker, *Nature*, 1999, **283**, 52.

294. L. Chico, M. P. Lopez Sancho and M. C. Munoz, *Phys. Rev. Lett.*, 1998, **81**, 1278.

295. C. L. Kane, E. J. Mele, R. S. Lee, J. E. Fischer, P. Petit, H. Dai, A. Thess, R. E. Smalley, A. R. M. Verschuren, S. J. Tans and C. Dekker, *Europhys. Lett.*, 1998, **41**, 683.

296. J. Tersoff, *Appl. Phys. Lett.*, 1999, **74**, 2122.

297. A. D. Bozko, D. E. Slovsky, V. A. Nalimova, A. G. Rinzler, R. E. Smalley and J. E. Fischer, *Appl. Phys. A*, 1998, **67**, 75.

298. J. Kong, C. Zhou, A. Morpurgo, H. T. Soh, C. F. Quate, C. Marcus and H. Dai, *Appl. Phys. A.*, 1999, **69**, 305.

299. T. W. Tombler, C. Zhou, J. Kong and H. Dai, *Appl. Phys. Lett.*, 2000, **76**, 2412.

300. Ph. Avouris, T. Martel, T. Schmidt, H. R. Shea and R. E. Walkup, *Appl. Surf. Sci.*, 1999, **141**, 201.

301. S. J. Tans, A. R. M. Verschueren and C. Dekker, *Nature*, 1998, **393**, 49.

302. R. Martel, T. Schmidt, H. R. Shea, T. Hertel and Ph. Avouris, *Appl. Phys. Lett.*, 1998, **73**, 2447.

303. A. Javey, J. Guo, M. Paulsson, Q. Wang, D. Mann, M. Lundstrom and H. Dai, *Phys. Rev. Lett.*, 2004, **92**, 106804–1.

304. U. C. Coskun, T.-C. Wei, S. Vishveshwara, P. M. Goldbart and A. Bezryadin, *Science*, 2004, **304**, 1132.

305. W. Liang, M. Bockrath, D. Bozovic, J. H. Hafner, M. Tinkham and H. Park, *Nature*, 2001, **411**, 665.

306. Z. Yao, C. L. Kane and C. Dekker, *Phys. Rev. Lett.*, 2000, **84**, 2941.

307. Z. Yao, C. Dekker and Ph. Avouris, *Top. Appl. Phys.*, 2001, **80**, 147.

308. J. Hone, I. Ellwood, M. Muno, A. Mizel, M. L. Cohen, A. Zettl, A. G. Rinzler and R. E. Smalley, *Phys. Rev. Lett.*, 1998, **80**, 1042.

309. L. Grigorian, G. U. Sumanasekera, A. L. Loper, S. L. Fang, J. L. Allen and P. C. Eklund, *Phys. Rev. B*, 1999, **60**, R11309.

310. G. T. Kim, E. S. Choi, D. C. Kim, D. S. Suh, Y. W. Park, K. Liu, G. Duesberg and S. Roth, *Phys. Rev. B*, 1998, **58**, 16064.

311. A. A. Farajian, K. Esfarjani and Y. Kawazoe, *Phys. Rev. Lett.*, 1999, **82**, 5084.

312. A. Bachtold, C. Strunk, J. P. Salvetat, J. M. Bonard, L. Forro, T. Nussbaumer and C. Schonenberger, *Nature*, 1999, **397**, 673.

313. M. Bockrath, J. Hone, A. Zettl, P. L. McEuen, A. G. Rinzler and R. E. Smalley, *Phys. Rev. B*, 2000, **61**, R10606.

314. R. Gaal, L. Thien-Nga, R. Basca, J. P. Salvetat, L. Forro, B. Ruzicka and L. Degiorgi, *Phys. Rev. B*, 2000, **61**, R2468.

315. R. S. Lee, H. J. Kim, J. E. Fischer, J. Lafebvre, M. Radosavljevic, J. Hone and A. T. Johnson, *Phys. Rev. B*, 2000, **61**, 4526.

316. S. Frank, P. Poncharal, Z. L. Wang and W. A. de Heer, *Science*, 1998, **280**, 1744.

317. P. Kim, L. Shi, A. Majumdar and P. L. McEuen, *Phys. Rev. Lett.*, 2001, **87**, 215502–1.

318. M. Kociak, A. Yu. Kasumov, S. Guéron, B. Reulet, I. I. Khodos, Yu. B. Gorbatov, V. T. Volkov, L. Vaccarini and H. Bouchiat, *Phys. Rev. Lett.*, 2001, **86**, 2416.

319. Z. K. Tang, L. Zhang, N. Wang, X. X. Zhang, G. H. Wen, G. D. Li, J. N. Wang, C. T. Chan and P. Sheng, *Science*, 2001, **292**, 2462.

320. M.-F. Yu, B. S. Files, S. Arepalli and R. S. Ruoff, *Phys. Rev. Lett.*, 2000, **84**, 5552.

321. M. R. Falvo, G. J. Clary, R. M. Taylor, V. Chi, F. P. Brooks, S. Washburn and R. Superfine *Nature*, 1997, **389**, 582.

322. E. T. Thostenson, Z. F. Ren and T. W. Chou *Compos. Sci. Technol.*, 2001, **61**, 1899; K. T. Lau and D. Hui, *Compos. Part B, Eng.*, 2002, **33**, 263.

323. M. S. P. Shaffer, X. Fan and A. H. Windle, *Carbon*, 1998,**36**, 1603.

324. M. S. P. Shaffer and A. H. Windle: *Adv. Mater.*, 1999, **11**, 937.

325. D. E. Hill, Y. Lin, A. M. Rao, L. F. Allard and Y.-P. Sun, *Macromolecules*, 2002, **35**, 9466.

326. D. Qian, E. C. Dickey, R. Andrews and T. Rantell *Appl. Phys. Lett.*, 2000, **76**, 2868.

327. R. Andrews, D. Jacques, D. Qian and T. Rantell, *Acc. Chem. Res.*, 2002, **35**, 1008.

328. R. Haggenmueller, H. H. Gommans, A. G. Rinzler, J. E. Fischer and K. I. Winey, *Chem. Phys. Lett.*, 2000, **330**, 219.

329. P. Potschke, T. D. Fornes and D. R. Paul, *Polymer*, 2002, **43**, 3247.

330. M. Sennett, E. Welsh, J. B. Wright, W. Z. Li, J. G. Wen and Z. F. Ren *Appl. Phys. A*, 2003, **76**, 111.

331. (a) P. M. Ajayan, O. Stephan, C. Colliex and D. Trauth, *Science*, 1994, **265**, 1212; (b) J. Sandler, M. S. P. Shaffer, T. Prasse, W. Bauhofer, K. Schulte and H. Windle, *Polymer*, 1999, **40**, 5967; (c) A. Allaoui, S. Bai, H. M. Cheng and J. B. Bai, *Compos. Sci. Technol.*, 2002, **62**, 1993; (d) X. Xu, M. M. Thwe, C. Shearwood and K. Liao, *Appl. Phys. Lett.*, 2002, **81**, 2833; (e) D. Puglia, L. Valentini and J. M. Kenny, *J. Appl. Polym. Sci.*, 2003, **88**, 452.

332. M. S. P. Shaffer and K. Koziol, *Chem. Commun.*, 2002, 2074.

333. C. Park, Z. Ounaies, K. A. Watson, R. E. Crooks, J. Smith, S. E. Lowther, J. W. Connell, E. J. Siochi, J. S. Harrison and T. L. St Clair, *Chem. Phys. Lett.*, 2002, **364**, 303.

334. K. Jurewicz, S. Delpeux, V. Bertagna, F. Béguin and E. Frackowiak, *Chem. Phys. Lett.*, 2001, **347**, 36; M. Hughes, G. Z. Chen, M. S. P. Shaffer, D. J. Fray and A. H. Windle, *Chem. Mater.*, 2002, **14**, 1610.

335. A. A. Mamedov, N. A. Kotov, M. Prato, D. M. Guldi, J. P. Wicksted and A. Hirsch, *Nat. Mater.*, 2002, **1**, 190.

336. J. H. Rouse and P. T. Lillehei, *Nano Lett.*, 2003, **3**, 59.

337. P. C. P. Watts and W. K. Hsu, *Nanotechnology*, 2003, **14**, L7.

338. F. H. Gojny, J. Nastalczyk, Z. Roslaniec and K. Schulte, *Chem. Phys. Lett.*, 2003, **370**, 820.

339. K. Liao and S. Li, *Appl. Phys. Lett.*, 2001, **79**, 4225.
340. A. Star, J. F. Stoddart, D. Steuerman, M. Diehl, A. Boukai, E. W. Wong, X. Yang, S. W. Chung, H. Choi and J. R. Heath, *Angew. Chem.-Int. Ed.*, 2001, **40**, 1721; A. Star, Y. Liu, K. Grant, L. Ridvan, J. F. Stoddart, D. W. Steuerman, M. R. Diehl, A. Boukai and J. R. Heath, *Macromolecules*, 2003, **36**, 553.
341. M. J. Biercuk, M. C. Llaguno, M. Radosavljevic, J. K. Hyun, A. T. Johnson and J. E. Fischer, *Appl. Phys. Lett.*, 2002, **80**, 2767.
342. T. Fukushima, A. Kosaka, Y. Ishimura, T. Yamamoto, T. Takigawa, N. Ishii and T. Aida, *Science*, 2003, **300**, 2072.
343. Y-L. Li, I. A. Kinloch and A. H. Windle, *Science*, 2004, **204**, 276.
344. M. Zhang, K. R. Atkinson and R. H. Baughman, *Science*, 2004, **306**,1358.
345. M. Cochet, W. K. Maser, A. M. Benito, M. A. Callejas, M. T. Martýnez, J.-M. Benoit, J. Schreiber and O. Chauvet, *Chem. Commun.*, 2001, 1450.
346. S. R. C. Vivekchand, L. Sudheendra, A. Govindaraj and C. N. R. Rao, *J. Nanosci. Nanotechnol.*, 2002, **2**, 631.
347. V. Jamieson, *New Sci.*, 2003 15 Mar, 30.
348. S. A. Curran, P. M. Ajayan, W. J. Blau, D. L. Carroll, J. N. Coleman, A. B. Dalton, A. P. Davey, A. Drury, B. McCarthy, S. Maier and A. A. Strevens, *Adv. Mater.*, 1998, **10**, 1091.
349. E. Kymakis and G. A. J. Amaratunga, *Appl. Phys. Lett.*, 2002, **80**, 112.
350. Y.-C. Chen, N. R. Raravikar, L. S. Schadler, P. M. Ajayan, Y.-P. Zhao, T.-M. Lu, G.-C. Wang and X.-C. Zhang, *Appl. Phys. Lett.*, 2002, **81**, 975.
351. S. A. O'Flaherty, R. Murphy, S. V. Hold, M. Cadek, J. N. Coleman and W. J. Blau, *J. Phys. Chem. B*, 2003, **107**, 958.
352. (a) H. Han, S. Vijayalakshmi, A. Lan, Z. Iqbal, H. Grebel, E. Lalanne and A. M. Johnson, *Appl. Phys. Lett.*, 2003, **82**, 1458; (b) W. A. Curtin and B. W. Sheldon, *Mater. Today,* 2004, **7**, 44.
353. W. A. de Heer, *MRS Bulle.*, 2004, 281.
354. J. Robertson, *Mater. Today* 2004 (October), 46.
355. (a) L. Chico, V. H. Crespi, L. X. Benedict, S. G. Louie and M. L. Cohen, *Phys. Rev. Lett.*, 1996, **76**, 971; (b) L. Kouwenhoven, *Science*, 1997, **275**, 1896; (c) P. L. McEuen, *Nature*, 1998, **393**, 15.
356. Y. Zhang, T. Ichihashi, E. Landree, F. Nihey and S. Iijima, *Science*, 1999, **285**, 1719.
357. Z. Yao, H. W. Ch. Postma, L. Balents and C. Dekker, *Nature*, 1999, **402**, 273.
358. M. Menon and D. Srivastava, *Phys. Rev. Lett.*, 1997, **79**, 4453.
359. M. Menon and D. Srivastava, *J. Mater. Res.*,1998, **13**, 2357.
360. M. S. Fuhrer, J. Nygard, L. Shih, M. Forero, Y. G. Yoon, M. S. C. Mazzoni, H. J. Choi, J. Ihm, S. G. Louie, A. Zettl and P. L. McEuen, *Science*, 2000, **288**, 494.
361. J. Li, C. Papadopoulos and J. Xu, *Nature*, 1999, **402**, 254.
362. B. C. Satishkumar, P. J. Thomas, A. Govindaraj and C. N. R. Rao, *Appl. Phys. Lett.*, 2000, **77**, 2530; F. L. Deepak, A. Govindaraj and C. N. R. Rao, *Chem. Phys. Lett.*, 2001, **345**, 5.

363. M. Menon, A. N. Andriotis, D. Srivastava, I. Ponomarevax and L. A. Chernozatonskiik, *Phys. Rev. Lett.*, 2003, **91**, 145501–1.

364. J. Hu, M. Ouyang, P. Yang and C. M. Lieber, *Nature*, 1999, **399**, 48.

365. H. Nishijima, S. Akita and Y. Nakayama, *Jpn. J. Appl. Phys.*, 1999, **38**, 7247.

366. J. Lafebvre, J. F. Lynch, M. Llaguno, M. Radosavljevic and A. T. Johnson, *Appl. Phys. Lett.*, 1999, **75**, 3014.

367. E. B. Cooper, S. R. Manalis, H. Fang, H. Dai, K. Matsumoto, S. C. Minne, T. Hunt and C. F. Quate, *Appl. Phys. Lett.*, 1999, **75**, 3566.

368. K. Tsukagoshi, B. W. Alphenaar and H. Ago, *Nature*, 1999, **401**, 572–4.

369. N. Mason, M. J. Biercuk and C. M. Marcus, *Science*, 2004, **303**, 655.

370. I. Dierking, G. Scalia, P. Morales and D. LeClere, *Adv. Mater.*, 2004, **16**, 865.

371. T. Rueckes, K. Kim, E. Joseluich, G. Y. Tsang, C. L. Cheung and C. M. Leiber, *Science*, 2000, **289**, 94.

372. A. Bachtold, P. Hadley, T. Nakanishi and C. Dekker, *Science*, 2001, **294**, 1317.

373. P. G. Collins, M. S. Arnold and Ph. Avouris, *Science*, 2001, **292**, 706.

374. Y. Yaish, J.-Y. Park, S. Rosenblatt, V. Sazonova, M. Brink and P. L. McEuen, *Phys. Rev. Lett.*, 2004, **92**, 046401.

375. S. J.Wind, J. Appenzeller and Ph. Avouris, *Phys. Rev. Lett.*, 2003, **91**, 058301–1.

376. J. Appenzeller, J. Knoch, M. Radosavljević and Ph. Avouris, *Phys. Rev. Lett.*, 2004, **92**, 226802–1.

377. K. Keren, R. S. Berman, E. Buchstab, U. Sivan and E. Braun, *Science*, 2003, **302**, 1380.

378. A. Star, J.-C. P. Gabriel, K. Bradley and G. Gruner, *Nano Lett.*, 2003, **3**, 459.

379. K. Bradley, M. Briman, A. Star and G. Gruner, *Nano Lett.*, 2004, **4**, 253.

380. N. de Jonge and J.-M. Bonard, *Phil. Trans. R. Soc. Lond. A*, 2004, **362**, 2239.

381. W. I. Milne, K. B. K. Teo, G. A. J. Amaratunga, P. Legagneux, L. Gangloff, J.-P. Schnell, V. Semet, V. Thien Binh and O. Groening, *J. Mater. Chem.*, 2004, **14**, 933.

382. W. A. de Heer, A. Châtelain and D. Ugarte, *Science*, 1995, **270**, 1179.

383. A. G. Rinzler, J. H. Hafner, P. Nikolaev, L. Lou, S. G. Kim, D. Tomanek, P. Nordlander, D. T. Colbert and R. E. Smalley, *Science*, 1995, **269**, 1550.

384. N. S. Lee, D. S. Chung, I. T. Han, J. H. Kang, Y. S. Choi, H. Y. Kim, S. H. Park, Y. W. Jin, W. K. Yi, M. J. Yun, J. E. Jung, C. J. Lee, J. H, You, S. H. Jo, C. G. Lee and J. M. Kim, *Diamond Relat. Mater.*, 2001, **10**, 265.

385. Y. Saito and S. Uemura, *Carbon*, 2000, **38**, 169.

386. R. Rosen, W. W. Simendinger, C. Debbault, H. Shimoda, L. Fleming, B. Stoner and O. Zhou, *Appl. Phys. Lett.*, 2000, **76**, 1668.

387. H. Sugie, M. Tanemura, V. Filip, K. Iwata, K. Takahashi and F. Okuyama, *Appl. Phys. Lett.*, 2001, **78**, 2578.

388. J.-M. Bonard, T. Stöckli, F. Maier, W. A. de Heer, A. Châtelain, J.-P. Salvetat and L. Forró, *Phys. Rev. Lett.*, 1998, **81**, 1441.

389. J.-L. Kwo, C. C. Tsou, M. Yokoyama, I. N. Lin, C. C. Lee, W. C. Wang and F. Y. Chuang, *J. Vac. Sci. Technol. B*, 2001, **19**, 23.

390. W. Zhu, C. Bower, O. Zhou, G. Kochanski and S. Jin, *Appl. Phys. Lett.*, 1999, **75**, 873.

391. Y. Saito, K. Hamaguchi, S. Uemura, K. Uchida, Y. Tasaka, F. Ikazaki, M. Yumura, A. Kasuya and Y. Nishina, *Appl. Phys. A.*, 1998, **67**, 95.

392. (a) W. B. Choi, D. S. Chung, J. H. Kang, H. Y. Kim, Y. W. Jin, I. T. Han, Y. H. Lee, J. E. Jung, N. S. Lee, G. S. Park and J. M. Kim, *Appl. Phys. Lett.*, 1999, **75**, 3129; (b) C. J. Lee, J. Park, S. Y. Kang and J. H. Lee, *Chem. Phys. Lett.*, 2000, **326**, 175.

393. D. Lovall, M. Buss, E. Graugnard, R. P. Andres and R. Reifenberger, *Phys. Rev. B*, 2000, **61**, 5683.

394. O. Groning, O. M. Kuttel, Ch. Emmenegger, P. Groning and L. Schlapbach, *J. Vac. Sci. Technol. B*, 2000, **18**, 665.

395. R. B. Sharma, V. N. Tondare, D. S. Joag, A. Govindaraj and C. N. R. Rao, *Chem. Phys. Lett.*, 2001, **344**, 283.

396. J.-C. Charlier, M. Terrones, M. Baxendale, V. Meunier, T. Zacharia, N. L. Rupesinghe, W. K. Hsu, N. Grobert, H. Terrones and G. A. J. Amaratunga, *Nano Lett.*, 2002, **2**, 1191.

397. D. Golberg, P. S. Dorozhkin, Y. Bando, Z.-C. Dong, C. C. Tang, Y. Uemura, N. Grobert, M. Reyes-Reyes, H. Terrones and M. Terrones, *Appl. Phys. A, Mater. Sci. Process.*, 2003, **76**, 499.

398. K. H. An, W. S. Kim, Y. S. Park, J.-M. Moon, D. J. Bae, S. C. Lim, Y. S. Lee and Y. H. Lee, *Adv. Funct. Mater.*, 2001, **11**, 387.

399. C. Niu, E. K. Sickel, R. Hoch, D. Moy and H. Tennent, *Appl. Phys. Lett.*, 1997, **70**, 1480.

400. C. Niu, J. Kupperschmidt and R. Hock, in *Proceedings of the 39th Power Sources Conference* Maple Hill, NJ, USA, 2000, pp 314–317.

401. R. H. Baughman, C. Cui, A. A. Zakidov, Z. Iqbal, J. N. Berisci, G. M. Spinks, G. G. Wallace, A. Mazzoldi, D. De Rossi, A. G. Rinzler, O. Jaschinski, S. Roth and M. Kertesz, *Science,* 1999, **284**, 1340.

402. A. M. Fennimore, T. D. Yuzvinsky, W.-Q. Han, M. S. Fuhrer, J. Cumings and A. Zettl, *Nature*, 2003, **424**, 408.

403. P. G. Collins, K. Bradley, M. Ishigami and A. Zettl, *Science*, 2000, **287**, 1801.

404. J. Kong, N. R. Franklin, C. Zhou, M. G. Chapline, S. Peng, K. Cho and H. Dai, *Science*, 2000, **287**, 622.

405. S. Ghosh, A. K. Sood and N. Kumar, *Science*, 2003, **299**, 1042.

406. S. S. Wong, E. Joselevich, A. T. Woolley, L. C. Chin and C. M. Lieber, *Nature*, 1998, **394**, 52.

407. F. Villalpando-Páez, A. H. Romero, E. Muñoz-Sandoval, L. M. Martínez, H. Terrones and M. Terrones, *Chem. Phys. Lett.*, 2004, **386**, 137.

408. S. Peng and K. Cho, *Nano Lett.*, 2003, **3**, 513.

409. B. R. Azamian, J. J. Davis, K. S. Coleman, C. Bagshaw and M. L. H. Green, *J. Am. Chem. Soc.*, 2002, **124**, 12664; J. J. Davis, K. S. Coleman, B. R. Azamian, C. B. Bagshaw and M. L. H. Green, *Chem. Eur. J.*, 2003, **9**, 3732.

410. J. Wang, G. Liu and M. R. Jan, *J. Am. Chem. Soc.*, 2004, **126**, 3010.

411. J. Cao, Q. Wang and H. Dai, *Phys. Rev. Lett.*, 2004, **90**, 157601–1.

412. J. H. Hafner, C. L. Cheung and C. M. Leiber, *Nature*, 1999, **398**, 761.

413. H. Dai, J. H. Hafner, A. G. Rinzler, D. T. Colbert and R. E. Smalley, *Nature*, 1996, **384**, 147.

414. P. Kim and C. L. Lieber, *Science*, 1999, **286**, 2148.

415. M. Endo, Y. A. Kim, T. Hayashi, K. Nishimura, T. Matusita, K. Miyashita and M. S. Dresselhaus, *Carbon*, 2001, **39**, 1287.

416. I. Mukhopadhyay, N. Hoshino, S. Kawasaki, F. Okino, W. K. Hsu and H. Touhara, *J. Electrochem. Soc.*, 2002, **149**, A39.

417. (a) B. Gao, A. Kleinhammes, X. P. Tang, C. Bower, L. Fleming, Y. Wu and O. Zhou, *Chem. Phys. Lett.*, 1999, **307**, 153; (b) F. Salver-Disma, C. Lenain, B. Beaudoin, L. Aymard and J.-M. Tarascon, *Solid State Ionics*, 1997, **98**, 145; (c) D. Y. Zhong, G. Y. Zhang, S. Liu, and E. G. Wang, Q. Wang, H. Li and X. J. Huang, *Appl. Phys. Lett.*, 2001, **79**, 3500.

418. M. Eswaramoorthy, R. Sen and C. N. R. Rao, *Chem. Phys. Lett.*,1999, **304**, 207.

419. (a) G. Stan, J. H. Hartman, V. H. Crespi, S. M. Gatica and M. W. Cole, *Phys. Rev. B*, 2000, **61**, 7288; (b) W. Teizer, R. B. Hallock, E. Dujardin and T. W. Ebbesen, *Phys. Rev. Lett.*, 1999, **82**, 5305.

420. A. Kuznetsova, D. B. Mawhinney, V. Naumenko, J. T. Yates Jr, J. Liu and R. E. Smalley, *Chem. Phys. Lett.*, 2000, **321**, 292.

421. M. S. Dresselhaus, K. A. Williams and P. C. Eklund, *MRS Bull.*, 2000, **24**, 45.

422. G. G. Tibbetts, G. P. Meisner and C. H. Olk, *Carbon*, 2001, **39**, 2291.

423. M. Hirscher, M. Becher, M. Haluska, A. Quintel, V. Skakalova, Y.-M. Choi, U. Dettlaff-Weglikowska, S. Roth, I. Stepanek, P. Bernier, A. Leonhardt and J. Fink, *J. Alloys Comp.*, 2002, **330–332**, 654.

424. C. Zandonella, *Nature*, 2001, **410**, 734.

425. Y. Ye, C. C. Ahn, C. Witham, B. Fultz, J. Liu, A. G. Rinzler, D. Colbert, K. A. Smith and R. E. Smalley, *Appl. Phys. Lett.*, 1999, **74**, 2307.

426. A. C. Dillon, K. M. Jones, T. A. Bekkedahl, C. H. Kiang, D. S. Bethune and M. J. Heben, *Nature*, 1997, **386**, 377.

427. C. Liu, Y. Y.Fan, M. Liu, H. T. Cong, H. M. Cheng and M. S. Dresselhaus, *Science*, 1999, **286**, 1127.

428. G. Gundiah, A. Govindaraj, N. Rajalakshmi, K. S. Dhathathreyan and C. N. R. Rao, *J. Mater. Chem.*, 2003, **13**, 209.

429. S. R. Mishra, H. S. Rawat, S. C. Mehendale, K. C. Rustagi, A. K. Sood, R. Bandyopadhyay, A. Govindaraj and C. N. R. Rao, *Chem. Phys. Lett.*, 2000, **317**, 510.

430. K. Yoshino, H. Kajii and H. Araki, *Fuller. Sci. Technol.*, 1999, **7**, 695.

431. J. R. Wood and H. D. Wagner, *Appl. Phys. Lett.*, 2000, **76**, 2883.

432. A. K. Sood and S. Ghosh, *Phys. Rev. Lett.*, 2004, **93**, 086601.

433. (a) P. Karl and D. Tomanek, *Phys. Rev. Lett.*, 1999, **82**, 5373; (b) B. C. Regan, S. Aloni, R. O. Ritchle, U. Dahmen and A. Zettl, *Nature*, 2004, **428**, 924.

CHAPTER 2

Inorganic Nanotubes

1 Introduction

The discovery of carbon nanotubes by Iijima,[1] following the discovery of fullerenes,[2] prompted the investigation of other layered materials that may form similar tubular structures. Many inorganic compounds possess structures comparable to that of graphite, typical examples being metal dichalcogenides (sulphides and selenides), halides, oxides and hydroxides. Metal dichalcogenides, such as MoS_2, WS_2, $MoSe_2$, NbS_2 and HfS_2 contain a metal layer sandwiched between two chalcogen layers, with the metal in trigonal pyramidal or octahedral coordination. The MX_2 (M = metal, X = S or Se) layers are stacked along the c-direction in the ABAB fashion, making them analogous to the single graphene sheets in the graphite structure (Figure 2.1). The layers show the presence of dangling bonds due to the absence of an X or M atom at the edges when viewed parallel to the c-axis. While the metal atom in the MX_2 layer is six-fold coordinated to the chalcogen atoms in the bulk, it is four-fold coordinated at the edge of the MS_2 nanocluster. The chalcogen atom in the rim is bonded only two-fold to the metal atoms instead of being three-fold coordinated as in the bulk (Figure 2.1). Such unsaturated bonds at the edges of the layers occur in graphite as well. With increasingly smaller molecular sheets the relative number of rim atoms with unsaturated bonds increases, compared to the fully bonded bulk atoms. The dichalcogenide layers, therefore, become unstable towards bending and have a high propensity to roll into curved structures, forming hollow fullerene-like structures.

Folding in the layered transition metal chalcogenides was identified by Chianelli *et al.*[3] and Sanders[4] even before the discovery of the carbon fullerenes and nanotubes. Chianelli *et al.* reported rag-like and tubular structures of MoS_2 and studied their usefulness in catalysis. The folded sheets appear as crystalline needles in low-magnification TEM images, and were described as layers that fold onto themselves. These structures indeed represent those of nanotubes or, more probably, those of nanoscrolls. Tenne and co-workers[5–8] first recognized that the nanosheets of Mo and W dichalcogenides are unstable against folding and closure and that they can form fullerene-like nanoparticles as well as nanotubes (Figure 2.2). These new nanoparticles were designated inorganic fullerenes (IF). The dichalcogenide structures contain concentrically nested fullerene cylinders,

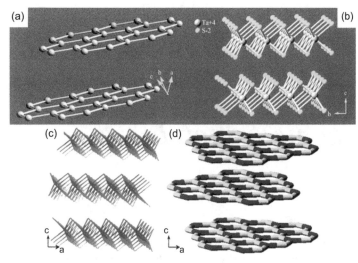

Figure 2.1 *Comparison of the structures of* (a) *graphite and inorganic layered compounds such as* (b) *NbS$_2$/TaS$_2$;* (c) *MoS$_2$;* (d) *BN. In the layered dichalcogenides, the metal is in trigonal prismatic (TaS$_2$) or octahedral coordination (MoS$_2$). Note that the rim atoms have lower coordination to the neighbouring atoms than with bulk atoms*
(Reproduced from ref. 11)

with a less regular structure than in the carbon nanotubes. Nearly defect-free MX$_2$ nanotubes are rigid as a consequence of their structure and do not permit plastic deformation. The folding of a MS$_2$ layer in the process of forming a nanotube is shown schematically in Figure 2.3. There has been some speculation on the cause of folding and curvature in layered metal chalcogenides. Stoichiometric chains and layers such as those in TiS$_2$ possess an inherent ability to bend and fold, as observed in intercalation reactions. The existence of alternate coordination and stoichiometry in the layered chalcogenides may also cause folding. A change in the

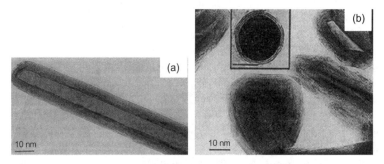

Figure 2.2 *TEM images of* (a) *a multi-walled nanotube of WS$_2$ and* (b) *hollow particles (inorganic fullerenes, IF) of WS$_2$*
(Reproduced from refs. 5 and 6)

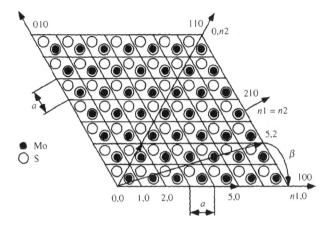

Figure 2.3 *Illustration of the bending of a MoS$_2$ layer; n1 and n2 are the lattice para-meters of the nanotube*
(Reproduced from ref. 9)

stoichiometry within the material would give rise to closed rings. There has been considerable progress recently in the synthesis and characterization of the nano-tubes of Mo and W dichalcogenides. Detailed reviews on the synthesis, structural characterization and applications of inorganic nanotubes have appeared.[9–12]

Transition metal chalcogenides possess a wide range of interesting physical properties. They are widely used in catalysis and as lubricants. They exhibit semiconducting and superconducting properties. With the synthesis and character-ization of fullerene-like structures and nanotubes of MoS$_2$ and WS$_2$, a wide field of research opened up, resulting in the successful synthesis and characterization of nanotubes of other metal chalcogenides and 2D layered compounds. Indeed, the dichalcogenides of many Group 4 and 5 metals have layered structures suitable for forming nanotubes. Curved structures are not limited to carbon and the dichalco-genides of Mo and W alone. A well-known example of a tube-like structure with diameters in the nanometer range is that of asbestos minerals (*kaolinite, chrysotile*) whose fibrous characteristics are determined by the tubular structure of the fused tetrahedral and octahedral layers.[13]

Nanotubes of oxides of transition metals as well as of other metals have been synthesized employing different methodologies.[12,14–20] Polycrystalline silica nano-tubes are formed during the synthesis of spherical silica particles by the hydrolysis of tetraethylorthosilicate in a mixture of water, ammonia, ethanol and D,L-tartaric acid.[14] Templated reactions have been employed, using carbon nanotubes, to obtain nanotube structures of metal oxides.[16–18] Oxides such as V$_2$O$_5$ have good catalytic activity in the bulk phase and the activity is retained in the nanotubular structure. V$_2$O$_5$ nanotubes have been prepared by chemical methods.[19] Boron nitride (BN) crystallizes in a graphite-like structure and can be simply viewed as replacing a C-C pair in the graphene sheet with the iso-electronic B-N pair. It can, therefore, be considered as an ideal precursor for the formation of BN nanotubes. Replacement of the C-C pairs, for example in C$_{60}$, partly or entirely by B-N pairs in the hexagonal network of graphite affords a wide array of two-dimensional phases that can form hollow cage structures and nanotubes.[21-24] BN nanotubes of varying wall thickness and morphology have been generated by employing several procedures.[25-28] Unsurprisingly, therefore, nanotube structures of other layered materials can also be prepared. For example, metal halides such as $NiCl_2$, oxides such as Tl_2O, nitrides such as GaN and chalcogenides such as GaSe crystallize in layered structures and nanotubes of many of these materials have been characterized. Nanotubes of elemental materials such as Ag and Ni and Te have also been prepared.[29,30] There is continuing interest in preparing exotic nanotubes and to study their properties.

2 Synthetic Methods

Carbon nanotubes have been prepared by arc evaporation and pyrolysis methods, laser ablation of graphite as well as electrochemical and templating techniques.[31] These methods fall under two categories. Methods such as arc evaporation and laser ablation employ processes, that are far from equilibrium. Chemical routes are generally closer to equilibrium conditions. Nanotubes of metal chalcogenides and boron nitride can also be prepared by employing techniques similar to those of carbon nanotubes, although there is an inherent difference in that the nanotubes of inorganic materials such as MoS_2 or BN would require reactions involving the component elements or compounds containing the elements. Decomposition of precursor compounds containing the elements is another route.

Nanotubes and fullerene-like nanoparticles of dichalcogenides such as MoS_2, $MoSe_2$ and WS_2 have been prepared by employing processes far from equilibrium such as arc discharge[32] and laser ablation.[33,34] By far, the most successful routes employ appropriate chemical reactions.[9,35] Thus, MoS_2 and WS_2 nanotubes are conveniently obtained from the stable oxides, MoO_3 and WO_3. The oxides are first heated at high temperatures in a reducing atmosphere and then reacted with H_2S. Reaction with H_2Se is used to obtain the selenides.[7,36] Recognizing that the trisulphides MoS_3 and WS_3 are likely to be intermediates in the formation of the disulfide nanotubes, the trisulphides have been directly decomposed to obtain the disulphide nanotubes.[37] Diselenide nanotubes have been obtained by the decomposition of metal triselenides.[38] The trisulphide route provides a general route for the synthesis of the nanotubes of many metal disulphides, such as NbS_2 and HfS_2.[39,40] For Mo and W dichalcogenides, it is possible to use the decomposition of precursor ammonium salts, $(NH_4)_2MX_4$ (X = S, Se; M = Mo, W), to prepare the nanotubes.[37] Other methods employed for the synthesis of dichalcogenide nanotubes include hydrothermal methods where an organic amine is taken as one of the components in the reaction mixture.

The hydrothermal route has been used to synthesize nanotubes and related structures of various other inorganic materials. Thus, nanotubes of several metal oxides (*e.g.* SiO_2, V_2O_5, ZnO) have been produced hydrothermally.[19,41,42] Nanotubes of oxides such as V_2O_5 are also prepared from an appropriate oxide precursor in the presence of an organic amine or a surfactant.[43] Surfactant-assisted synthesis of CdSe and CdS nanotubes has been reported. Here, the metal oxide reacts with the sulphidizing or selenidizing agent in the presence of a surfactant such as Triton X.[44]

Sol–gel chemistry is widely used in the synthesis of metal oxide nanotubes, a good example being that of silica and TiO_2.[14,45] Oxide gels in the presence of surfactants or suitable templates form nanotubes. For example, by coating carbon nanotubes (CNTs) with oxide gels and then burning off the carbon, one obtains nanotubes and nanowires of various metal oxides, including ZrO_2, SiO_2 and MoO_3.[17] Sol–gel synthesis of oxide nanotubes is also possible in the pores of alumina membranes. Notably, that MoS_2 nanotubes are also prepared by the decomposition of a precursor in the pores of an alumina membrane.[46] A layer-by-layer film-forming method for producing nanotubes in the pores of alumina membranes, involving the alternate immersion of the template with a diorganodiphosphonate and then into a solution of ZrO^{2+}, has been developed.[47]

High-temperature laser ablation of $NiCl_2$ in the presence of CCl_4 at 700–940 °C gives $NiCl_2$ IF structures and nanotubes.[48] To form germanium-filled silica nanotubes, Hu *et al.*[49] used a two-stage process by combining simultaneous thermal evaporation of SiO powder and laser ablation of the Ge target (KrF excimer laser).

Boron nitride nanotubes have been obtained by striking an electric arc between HfB_2 electrodes in a N_2 atmosphere.[50] BCN and BC nanotubes are obtained by arcing between B/C electrodes in an appropriate atmosphere. Considerable effort has gone into the synthesis of BN nanotubes, starting with different precursor molecules containing B and N. Decomposition of borazine in the presence of transition metal nanoparticles or of the 1:2 melamine-boric acid addition compound yields BN nanotubes.[25,26] Reaction of boric acid or B_2O_3 with N_2 or NH_3 at high temperature in the presence of carbon or catalytic metal particles has been employed to synthesize BN nanotubes.[28] Goldberger *et al.*[51] have synthesized single-crystal GaN nanotubes with inner diameters of 30–200 nm and wall thicknesses of 5–50 nm by employing an epitaxial casting-approach. They used hexagonal ZnO nanowires as templates for the epitaxial overgrowth of thin GaN layers in a chemical vapour deposition (CVD) system. In a typical experiment, they used trimethylgallium and ammonia as precursors with argon or nitrogen as carrier gas and maintained the deposition temperature at 600–700 °C. The ZnO nanowire templates were subsequently removed by thermal reduction and evaporation, resulting in ordered arrays of GaN nanotubes on the substrates. A carbon-free CVD process using B_2O_3–Ga_2O_3 mixtures and NH_3 has also been employed for BN nanotubes.[52a] A template-free synthesis of InP nanotubes starting with In_2O_3, In metal and red P has been reported.[52b]

The use of organic templates (organogels) to form morphologically interesting inorganic materials has been reviewed by Shinkai and co-workers.[53] Organogels form fibrous, tubular, ribbon-like, lamellar and hollow spherical morphologies. The organogels consist of an organic liquid and low concentrations (<0.5 wt%) of relatively low molecular weight gelator molecules.[54] These are usually prepared by heating a mixture of a gelator and a solvent until the solid dissolves; upon cooling, the solution (sol) thickens to form a gel. Gelation of compounds such as tetraethylorthosilicate (TEOS) along with a cholesterol-based gelator, under acidic pH conditions, followed by polycondensation exhibits a network of fibres with diameters in the 50–200 nm range.[53] Subsequent drying and calcination result in silica tubes without the presence of the original organic template. A hydrogel has also been employed to prepare various inorganic nanotubes.[55]

3 Specific Cases

Some of the important inorganic nanotubes synthesized and characterized in the last few years are:[35,11]

Metals:	Ni, Ag, Cu, Te, Co, Fe, Bi
Chalcogenides:	MoS_2, WS_2, $MoSe_2$, WSe_2, NbS_2, $NbSe_2$, HfS_2, ZrS_2, TiS_2, $TiSe_2$
Oxides:	TiO_2, $H_2Ti_3O_7$, ZrO_2, VO_x, SiO_2, IrO_2, ZnO, Ga_2O_3, $BaTiO_3$, $PbTiO_3$, $K_4Nb_6O_{17}$
Nitrides:	BN, AlN, GaN
Phosphides:	InP
Halides:	$NiCl_2$

Metal nanotubes and their arrays have been prepared using porous alumina membranes.[56] The design of a process for fabricating uniform nanotube arrays *via* a multi-step template replication and electrodeposition is shown schematically in Figure 2.4(a). The nanotubes so obtained are highly ordered and uniform in wall thickness and diameter along the entire nanotubes (Figure 2.4b and 2.4c). Electroless deposition has been carried out in the pores of alumina menbranes to obtain good yields of Ag and Cu nanotubes.[57,58] Thermal decomposition of $AgNO_3$ deposited in the pores of the membranes yields Ag nanotubes.[59] Bismuth nanotubes are obtained by the glycol reduction of Bi_2O_3.[60] Nanotubes of Pt, Pd and Ag have been prepared by employing lyotropic mixed surfactant liquid crystal templates,[61] while copper nanotube arrays have been obtained by electrodeless deposition using porous anodic alumina membrane templates.[62] Single-walled metal nanotubes are prepared by the galvanic replacement reaction between Ag nanowires and various metal ions.[63] By combining this with electroless plating of Ag, multiple-walled metal nanotubes could be obtained.

Other examples of elemental nanotubes are those of B, Se and Te. Boron SWNTs are generated by the reaction of BCl_3 with H_2 over a Mg-MCM-4 catalyst.[64] Under hydrothermal conditions, selenium nanotubes are obtained, along with nanowires, wherein Se powder and ammonia are reacted at 200 °C.[65] Pyrolysis of CdS in an inert gas atmosphere yields single-crystalline Cd nanotubes.[66]

Figure 2.5, shows TEM images of MoS_2 nanotubes prepared by the direct thermal decomposition of ammonium thiomolybdate in H_2.[37] The MoS_2 nanotubes consist of disulphide layers stacked along the *c*-direction. This implies that the S–S interaction between the MoS_2 slabs is weaker than the intralayer interactions. The S–S interlayer distances are, therefore, susceptible to distortions during the folding of the layers. This is exemplified by the slight expansion of the *c*-axis (2%) in the MoS_2 nanotubes.[9] HREM images of the disulfide nanotubes show stacking of

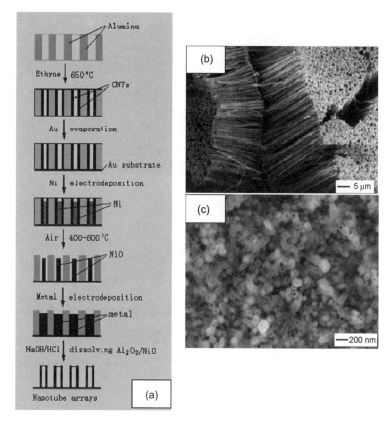

Figure 2.4 (a) *Schematic of the multi-step template replication process for preparing uniform metal nanotube arrays.* (b) *and* (c) *Metal nanotube arrays after dissolution of the template*
(Reproduced from ref. 56)

the (002) planes parallel to the tube axis; the distance between the layer fringes corresponds to the (002) spacing. A SEM image of nanotubes of HfS_2 obtained by the decomposition of HfS_3^{40} is shown in Figure 2.6(a). The nanotubes are quite long, some being over a micron. TEM images of these nanotubes are displayed in Figures 2.6(b), 2.6(c) and 2.6(d). Nanotubes of II–VI semiconductor compounds such as CdS and CdSe have been obtained by a soft chemical route involving surfactant-assisted synthesis.[44] Both the CdSe and CdS nanotubes (Figure 2.7) seem to be polycrystalline, formed by aggregates of nanoparticles.[44,67] Aligned CdS nanotubes have been prepared in the channels of porous alumina membranes by the pyrolysis of a dithiocarbamate precursor.[68] Nanotubes of Bi_2Se_3 are obtained by hydrothermal co-reduction, starting from $BiCl_3$, H_2SeO_3 and hydrazine.[69]

Exhaustive studies have been carried out on the synthesis of BN nanotubes and nanowires by various CVD techniques.[28] BN nanotubes with different structures are obtained on heating boric acid and iron particles in the presence of NH_3

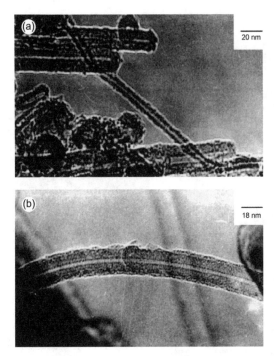

Figure 2.5   (a) *Low-resolution TEM images of MoS₂ nanotubes grown by the decomposition of ammonium thiomolybdate;* (b) *HREM image of the MoS₂ nanotube*
(Reproduced from ref. 37)

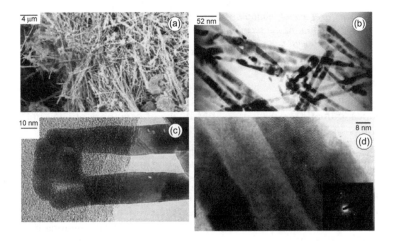

Figure 2.6   (a) *SEM image of the HfS₂ nanostructures;* (b) *and* (c) *low-resolution TEM images showing hollow nanotubes. The tube in* (c) *has a flat tip;* (d) *HREM image of the HfS₂ nanotubes, showing a layer separation of ~0.6 nm in the walls. Inset shows a typical ED pattern*
(Reproduced from ref. 40)

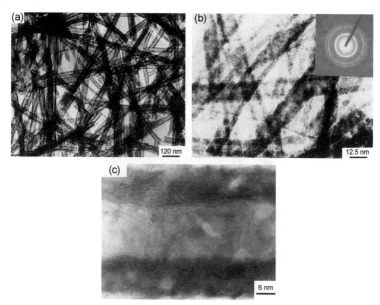

Figure 2.7 (a) *and* (b) *low-resolution TEM images of CdSe nanotubes. Inset shows a typical ED pattern;* (c) *HREM image of the CdSe nanotube, showing walls containing several nano-crystallites* (Reproduced from ref. 44)

(Figure 2.8). Aligned BN nanotubes are formed when aligned multi-walled carbon nanotubes are used as the templates.

Needle-shaped TiO_2 (anatase) nanotubes are readily precipitated from a gel containing a mixture of SiO_2 and TiO_2.[45a] The nanotubes formed by this method have a diameter of ~8 nm and lengths of up to 100 nm (Figure 2.9(a)). Titania nanotubes are also obtained by the direct deposition of titanium tetrafluoride (TiF_4) in the pores of an alumina membrane.[45b] Titania nanotubes prepared in the nanochannels of a membrane are shown in Figure 2.9(b) and 2.9(c). The tubes have straight channels, ~100–150 nm in diameter, and consist of small particles (10–20 nm). The pore diameter of the tubes was controllable over the range 50–150 nm by changing the deposition time. Much smaller TiO_2 nanotubes have been prepared by a simple procedure.[70] TiO_2 with anatase or rutile structure was treated with NaOH and subsequently with HCl. The resulting TiO_2 nanotubes are 50–200 nm long and about 10 nm in diameter. The HREM image of such TiO_2 nanotubes shows the presence of lattice fringes, indicating the crystalline structure of TiO_2 nanotubes. The structure of TiO_2 nanotubes has been greatly debated ever since Kasuga *et al.*[45a] reported the formation of such nanotubes by hydrothermal treatment of anatase TiO_2 powder in NaOH solution at mild temperatures. While some researchers propose that these nanotubes are nanoscrolls made of exfoliated layers of anatase TiO_2, others argue that they are made of the layered $H_2Ti_3O_7$ or its sodium salt. TiO_2 nanotube aggregates with different porosities are also reported

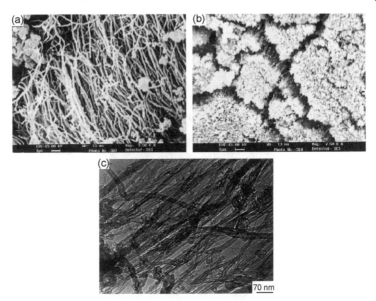

Figure 2.8 *SEM and TEM images of aligned BN nanotubes:* (a) *and* (b) *give side and top view SEM images, respectively;* (c) *TEM image of pure BN nanotube* (Reproduced from ref. 28)

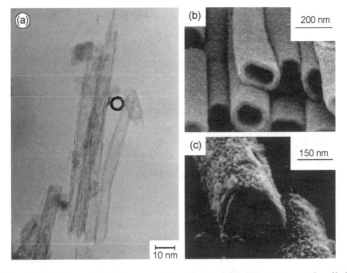

Figure 2.9 (a) *TEM image of TiO₂ nanotubes.* (b) *and* (c) *SEM images of well-developed TiO₂ nanotubes deposited from TiF₄ solution* (Part (a) reproduced from refs. 45a, 70; part (b) and (c) reproduced from ref. 45b)

from the hydrothermal reaction of TiO_2 in $NaOH$.[71] The temperature of the reaction and HCl washing after the preparation affect the structure and nature of the aggregates. Some studies revealed that this procedure leads to the exfoliation of anatase TiO_2 crystallites and the formation of nanoscrolls of the layered titanate, $Na_xH_{2-x}Ti_3O_7$.[72,73] Hydrothermal treatment of TiO_2 with $NaOH$ appears to yield layered $H_2Ti_3O_7 \cdot nH_2O$ first, which then transforms into B-TiO_2 nanotubes.[74]

Thermally stable metal silicate nanotubes with uniform pores have been obtained hydrothermally.[75] Well-aligned IrO_2 nanotubes can be grown on $LiTaO_3$ substrates by metal–organic vapour deposition.[76] Eu_2O_3 nanotube arrays have been generated by the sol–gel method.[77]

2D nanosheets have been transformed directly into nanotubes, in the case of oxides such as TiO_2 and MnO_2, by intercalation and deintercalation at room temperature.[78] A new nanostructure, well aligned nano-"box-beams" (nanotubes with square or rectangular cross-sections) of tin oxide, has been synthesized by combustion CVD of tin(II) 2-ethylhexanoate in absolute ethanol above 900 °C.[79] NbC nanotubes with a superconducting transition temperature of 11.2 K have been synthesized at 650 °C in an autoclave by reacting carbon nanotubes with $NbCl_5$ and Na.[80] The length of the NbC nanotubes ranges from hundreds of nanometers to several micrometres, with diameters of 20–60 nm.

Nesper and co-workers[19,81] have synthesized nanotubes of alkylammonium intercalated VO_x by hydrothermal means. The vanadium alkoxide precursor was hydrolysed in the presence of hexadecylamine and the hydrolysis product (lamellar structured composite of the surfactant and the vanadium oxide) yielded VO_x nanotubes along with the intercalated amine under hydrothermal conditions (Figures 2.10a and 2.10b). The interesting feature of this vanadium oxide nanotube is the presence of vanadium in the mixed valent state, thereby rendering it redox-active. The alkylamine intercalated in the intertubular space could be exchanged with other alkylamines of varying chain lengths as well as α and ϖ-diamines.[81] The distance between the layers in the VO_x nanotubes can be controlled by the length of the $-CH_2-$ chain in the amine template. Most VO_x nanotubes obtained by the hydrothermal method are open-ended. Very few closed tubes had flat or pointed conical tips. Cross-sectional TEM images of the nanotubular phases show that, instead of concentric cylinders (*i.e.*, layers that fold and close within themselves), the tubes are made up of single or double layer scrolls, providing a serpentine-like morphology.[81,82] The scrolls are seen as circles that do not close in the images (Figure 2.10c). Non-symmetric fringe patterns in the tube walls exemplify that most of the nanotubes are not rotationally symmetric and carry depressions and holes in the walls. Diamine-intercalated VO_x nanotubes are multilayer scrolls with narrow cores and thick walls, composed of packs of several vanadium oxide layers (Figure 2.10d). Notably, most of the low temperature (<300 °C) chemie douce (soft chemistry) processes, like intercalation–exfoliation, sol–gel processes, and template induced crystallization in the presence of a structure directing compounds, afford open-ended nanoscrolls, rather than the perfectly crystalline closed-ended nanotubes. Notwithstanding the lower energy of the fully crystalline nanotubes, the kinetic barrier is too high to afford their formation at typical chemie douce temperatures (around 200 °C). Consequently, the somewhat less stable, but kinetically more

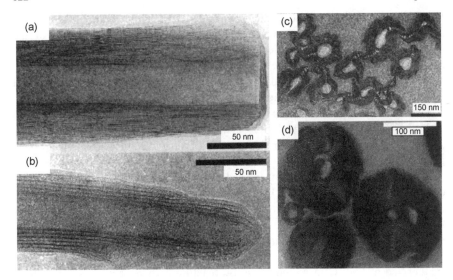

Figure 2.10   *TEM images of VO$_x$ nanotubes with intercalated amine having varying chain lengths; (a) C$_4$VO$_x$–NT; (b) C$_{16}$–VO$_x$–NT. (c) Cross-sectional TEM images of monoamine-intercalated VO$_x$ nanotubes, showing serpentine-like scrolls. (d) Cross-sectional TEM images of diamine-intercalated VO$_x$ nanotubes, showing a larger thickness of the tube walls and a smaller inner core (Reproduced from ref. 81)*

affordable nanoscroll structures are favoured with these low-temperature processes. Raman spectroscopy has proved useful in probing the structure of vanadate nanotubes.[83]

Single-crystalline Mg(OH)$_2$ nanotubes can be obtained from Mg$_{10}$(OH)$_{18}$Cl$_2$.5H$_2$O under solvothermal conditions.[84] Ni(OH)$_2$ nanotubes, which are potential positive-electrode materials in alkaline rechargeable batteries, have been prepared by depositing Ni ions and NH$_3$ in alumina membranes.[85] Nanotubes of certain complex inorganic materials have also been prepared. For example, a solution-phase method yields nanotubes of Ni(NH$_3$)$_6$Cl$_2$.[86] A 3.5 nm coordination nanotube has been synthesized by making use of flexible molecular tape containing Py-Py$'$-Py units and Pd ions.[87] K$_4$Nb$_6$O$_{17}$-type nanotubes have been prepared by the intercalation and exfoliation method.[88]

4 Properties

Physical properties of inorganic nanotubes constitute a fertile field of exploration. Studies carried out hitherto have been limited to a small group of inorganic nanotubes, where the synthesis is sufficiently advanced to provide adequate amounts of nanotubes – metal dichalcogenides and BN being important examples.

Mechanical Properties

The mechanical properties of inorganic nanotubes have been studied to some extent, using both experiment and theory. The Young's modulus of BN nanotubes,

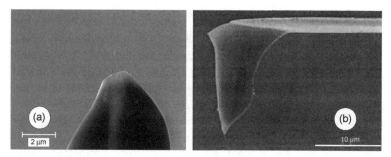

Figure 2.11 *AFM cantilever with WS$_2$ nanotube glued to its tip*
(Reproduced from ref. 91)

determined by measuring the amplitude of a vibrating nanotube in the TEM, is 1 GPa,[89] similar to that of carbon nanotubes. Theory predicts somewhat smaller Young's moduli for the BN nanotubes.[90] Using AFM tips of WS$_2$ nanotubes (Figure 2.11), the Young's modulus was determined by pushing the nanotube tip, attached to the cantilever, against the surface of a silicon substrate.[91] Figure 2.12 shows the force–distance trace of the nanotubes, where a kink associated with the buckling of the nanotubes is observed. The modulus determined thus was 170 MPa, which is close to the modulus of the bulk material calculated from first principles (150 MPa).

Tensile tests on individual multi-walled WS$_2$ nanotubes within a high-resolution SEM have provided the Young's modulus (*ca.* 140 GPa), strength (13 GPa) and the critical strain for failure (12–19%).[91] Theoretical studies show that the Young's modulus of zigzag nanotubes is ~10% higher than that of armchair nanotubes.

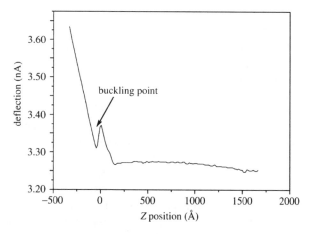

Figure 2.12 *Deflection of the silicon cantilever versus* its *Z position as recorded by an AFM*
(Reproduced from ref. 91)

Electronic, Magnetic, Optical and Related Properties

The electronic and optical properties of inorganic nanotubes have been studied by a few authors, especially through theoretical calculations.[92–95] For example, the electronic structures of BN, BC_2N and BC_3 nanotubes have been theoretically investigated,[21,96,97] along with BN fullerenes.[98] The calculations show that N-N and B-B nearest neighbours do not provide a stable nanotubular structure. The folding of the hexagonal BN network appears to be produced by disposing $(B-N)_2$ squares, full closure of the polyhedron being obtained by six such squares. If the six squares are symmetrically disposed, a stable structure of a truncated octahederon, like in $B_{12}N_{12}$, is obtained.

The electronic structure of metal-dichalcogenide nanotubes has been studied using tight-binding density functional theory (TB-DFT).[99–101] This work shows that nanotubes of materials such as MoS_2 retain the semiconducting behaviour. Nanotubes of materials with metallic character, such as NbS_2, or $NbSe_2$, with a high density of states near the Fermi level are likely to be metallic, irrespective of the diameter and chirality.[102] The elastic energy of bending of these nanotubes is appreciably higher than that of either carbon or BN nanotubes with the same diameter. This behaviour is not surprising, notwithstanding the bulky structure of MS_2 nanotubes, which imposes severe steric hindrance on the folding. In addition, the energy gap decreases with the shrinking diameter of the (semiconducting) nanotubes, an effect that is counterintuitive, in the light of quantum size effects in semiconductor nanoparticles. This behaviour has been confirmed through optical measurements.[103] The energy gap of individual MoS_2 nanotubes has been determined as a function of diameter using the *I–V* curves of a STM tip.[104] Arm-chair (n,n) MoS_2 nanotubes remain as indirect bandgap semiconductors while the zigzag $(n,0)$ nanotubes are direct bandgap semiconductors. The electronic structure of twisted MoS_2 nanotubes has been investigated.[105] The electronic structure of single-walled MoS_2 nanotubes and bundles of iodine intercalated nanotubes has been studied using STM.[106] Isolated nanotubes are found to be metallic, with a conductivity as high as graphite.

Bundles of iodine-doped single-walled MoS_2 nanotubes give stable field-emission currents.[107] Field emission from manganese oxide nanotubes synthesized by cyclic voltammetric electrodeposition method showed a turn-on field of 8.4 V μm^{-1} at a current density of 1 μA cm^{-2}.[108] These nanotubes may be useful for field-emission applications because they can be easily and economically synthesized at low temperatures. The electronic field emission of NbS_2 nanowires produced by direct heating of Nb and S powders in the presence of I_2 at 800 °C gave an emission current of 5 μA cm^{-2} at a threshold field of 5.5 V μm^{-1}.[109] The behaviour of these nanowires is comparable to that of carbon nanotubes.

The α-form of bismuth is a polytype with puckered layered structure, whereby each Bi atom is bonded to three neighbouring bismuth atoms, and it exhibits metallic behaviour. Using first principles theory, the structure and band structures of a single sheet of α-Bi as well as of armchair and zigzag nanotubes have been determined.[110] In the absence of layer-to-layer interaction, the single sheet becomes a direct bandgap semiconductor with a forbidden gap of 0.63 eV. Single-layer nanotubes are also semiconductors with a variable gap. This work indicates that new electronic and optical behaviour could be anticipated by controlling the number of layers of such nanostuctures. The crystalline and electronic structure of nanotubes of B_2O and BeB_2, both of which are isoelectronic with graphite,[111] show two of the configurations of B_2O to be metallic, while the most stable structure is a semiconductor with a bandgap of 2.8 eV. The (3,0) zigzag nanotube is a semiconductor with a direct gap of 1.63 eV, which is appreciably smaller than that of the bulk material. BeB_2 nanotubes are metallic.

The stability and electronic structure of AlB_2 nanotubes has been studied theoretically. These nanotubes tend to self-assemble into stable bundles having a metallic character.[112] Zigzag nanotubes are less stable than their armchair counterparts,[113] but both kinds are metallic. Ivanovskaya and co-workers[101a,114] studied the transformation of bulk MgB_2 into a sheet (2D), nanotubes (1D) and fullerene-like structures using tight-binding model calculations. In MgB_2 (space group *P6/mmm*) the boron atoms are arranged in a graphitic network with strong B–B sigma (σ) bonds. The Mg–B bonds are weaker by one order of magnitude. The Mg atoms are sandwiched between the pseudo-graphitic B layers, but with no Mg–Mg bonds. When going to the 2D layer, and more so when nanotubes (1D) and fullerene-like $Mg_{30}B_{60}$ (0D) structures are formed, the strength of the B–B is preserved, while the Mg–B is weakened and somewhat stronger Mg–Mg bonds are formed. In the nanotubes, a stable configuration is obtained by placing the Mg atoms inside, towards the hollow core, and the boron atoms on the outside perimeter. The same picture holds for the fullerene-like $Mg_{30}B_{60}$ structure. The Fermi energy (E_F) is in the vicinity of the B_{2p} states for the bulk as well as for the nanotubes. However, the density of states is higher for zigzag than for armchair nanotubes, which may indicate a higher T_c for the former. Theoretical investigations of TiB_2 nanotubes show that the (6,0) zigzag TiB_2 nanotube is a semiconductor with an energy gap of 1.32 eV; addition of a (12,0) B (outer) layer renders the nanotube metallic.[115]

Superconductivity has been found in carbon nanotubes. The critical temperature in ropes of single–CNTs (SW-CNTs) is 0.55 K.[116] In another study, a T_c of 15 K was reported for SW-CNTs 0.4 nm in diameter.[117] Measurements on $NbSe_2$ nanotubes and nanorods did not show any significant difference between the superconducting transition temperature compared to the bulk 2H platelets of the compound.[118]

Single-walled V_2O_5 nanotubes have been studied using tight-binding theory.[119] The strongest covalent bond is found between the V–O pairs, and weak bonding is found for the V–V pair. Furthermore, zigzag nanotubes were more stable than the armchair nanotubes. Both kinds of nanotubes were semiconductors. The bandgap was shown to vanish for the nanotubes with a small diameter due to the large strains accorded by the folding of such nanotubes.

Nanotubes and nanowires with both elemental (carbon or silicon) and multi-element compositions (such as compound semiconductors or oxides), and exhibiting electronic properties ranging from metallic to semiconducting, are being extensively investigated for use in device structures designed to control electron charge. However, another important degree of freedom – electron spin, the control of which underlies the operation of ''spintronic'' devices – has been much less explored. This is probably due to the relative paucity of nanometre-scale ferromagnetic building blocks (in which electron spins are naturally aligned) from which spin-polarized electrons can be injected. Elbaun *et al.*[120] describe VO$_x$ nanotubes, formed by controllable self-assembly, that are ferromagnetic at room temperature. Thus, the emerging picture is that of open-net layered-wall vanadium oxide nanotubes that can be spin-tuned by charge doping. In doping, the Fermi level is swept through the Mott gap, removing the frustration in the undoped system responsible for the spin gap, and promoting ferrimagnetism. Itinerant carriers under spin control are produced in the V(1) band, which interacts with the other more localized vanadium spins. Although manipulating and achieving doping selectivity of the tubes deserve further effort towards the practical implementation of this effect, their findings suggest a path to new spin-aligned nanoscale building blocks, in which the Fermi-level sweep can be accomplished by applied voltage. In other words the as-formed nanotubes are transformed from spin-frustrated semiconductors to ferromagnets by doping with either electrons or holes, potentially offering a route to spin control in nanotube-based heterostructures.

Optical-limiting behaviour of vanadium oxide nanotubes has been studied in the visible and IR spectral ranges using short pulses from a Nd:YAG laser.[121a,b] Vanadium oxide nanotube dispersions in water suspensions and embedded in solid poly(methyl methacrylate) films exhibit strong optical limiting at 532 nm, while the carbon nanotubes only exhibited a relatively small effect. No nonlinear behaviour was observed at 1064 nm. This suggests that a two-photon or excited state absorption mechanism is responsible for the observed nonlinearity. In addition, the effect of sheet distance on the optical properties of mixed-valent vanadium oxide nanotubes was studied.[121b] The sheet distance was adjusted with various amine templates, allowing the tube size to be altered. The sheet distance dependence of the optical gap is predicted to decrease from the bulk value with decreasing tube diameter. These modifications experimentally showed no systematic effect on the optical gap, but the 5 eV excitation showed a modest sheet distance dependence, red-shifting with increasing sheet distance.

Ivanovskaya and co-workers[122] have studied the electronic structure of single-wall TiO$_2$ nanotubes using tight-binding theory. They find the nanotubes to be stable semiconductors, with the bandgap decreasing with the decreasing diameter. Conversely, single-wall VO$_2$ nanotubes with rutile structure are metallic. Adachi and co-workers[123] have demonstrated that such nanotubular phases may play an important role in the charge-transfer process of dye-sensitized solar cells.

Tribological Properties

Tribological behaviour of fullerene-like WS$_2$ and MoS$_2$ nanoparticles and nanotubes has been studied.[32,124,125] Under severe loads, where fluids or greases are unable to effectively support the load and are squeezed away from the contact area, the elastic and quasi-spherical IF nanoparticles keep the asperities of the matting pairs apart and provide easy shear for the metal pairs. Furthermore, the "onion-like" structure of the nested particles leads to their gradual exfoliation. Thereby, mono- or bimolecular nanosheets of WS$_2$ are transferred onto the metal surface, further alleviating the friction and wear between the matting pairs. The presently available data indicate that these materials are likely to be used commercially. Impregnating the nanoparticles in various metal, ceramic and polymer coatings has been accomplished, bringing about substantial reduction in friction and wear of the tribological contacts. The impact resistance of WS_2 nanotubes against shock-waves has been tested.[126] They can withstand shock-waves as powerful as 21 GPa with minor damage to their framework, in contrast to carbon nanotubes, which transform under similar conditions into diamond under much weaker impacts (*ca.* 9 GPa). Potential applications for these findings are in the car and defense industries.

Chemical Modifications and Applications

Intercalation of fullerene-like MoS_2 and WS_2 nanoparticles and WS_2 nanotubes with alkali metal atoms by exposing the nanoparticles to the alkali-metal vapour has been accomplished recently.[127] Single-walled MoS_2 (3,3) armchair nanotubes are obtained by chemical vapour transport in the presence of iodine and C_{60} as a growth promoter. These nanotubes form bundles that are held together by the iodine atoms, which are situated in the trigonal voids between the nanotubes. Large amounts of Li atoms, up to 2.3 per MoS_2 molecule, could be intercalated into the partially empty trigonal voids between the hexagonally packed nanotubes.[128] This is to be compared with a loading of a single lithium atom per MoS_2 formula in the electrochemical intercalation of the bulk material. ESR measurements[129,130] of such samples revealed two kinds of peaks: a narrow peak attributed to Li clusters, and broad and nearly temperature-independent peaks. The latter feature was attributed to itinerant electrons from the lithium.

Chen and co-workers[131] studied various metal-dichalcogenide nanotubes as possible intercalation materials for both hydrogen and lithium. The intercalation was successful with TiS_2 nanotubes. Lithium intercalation in TiS_2 nanotubes was accomplished by stirring the nanotubes in a solution of n-butyllithium in hexane for 24 h. A high degree of Li loading, up to the stoichiometric composition of $LiTiS_2$, was obtained in this reaction. TiS_2 nanotubes have been used as cathode materials in rechargeable Mg^{2+} ion batteries.[132]

VO_x and related nanotubes have been examined as electrode materials for Li intercalation batteries. Vanadium oxide nanoscrolls were the first such nanostructures to be studied as cathode material in a rechargeable lithium battery.[133] Since the synthesis is generally performed with primary alkylamines as templating molecules, the embedded amine molecules can be exchanged by metal cations, *i.e.* alkaline, alkaline earth, and transition metal ions, with preservation of the tubular morphology. This property gives an opportunity to design tailor-made functional materials. When a 1M $LiClO_4$ solution of propylene carbonate was used for the electrochemical cell, the cathode capacity deteriorated fairly rapidly with the number of charge/discharge cycles. It was later possible to significantly improve the performance of the VO_x nanotubes-based cathode.[134] The electrode was tested by galvanostatic cycling in the potential range 1.8–3.5 V *vs.* Li/Li$^+$. The capacities were closely dependent on the type of lithium salt used in the electrolyte. Three salts were tested, $LiBF_4$, $LiPF_6$, and $LiN(CF_3SO_2)_2$. The imide salt,

LiN(CF$_3$SO$_2$)$_2$, gave the best result, with initial capacities of 200 mAh g^{-1}. The material cycled reversibly for at least 100 cycles. XRD results indicate that the tubular structure is preserved, even after prolonged cycling. In a related study, Whittingham and co-workers[135] studied nanotubes of the partially metal-exchanged manganese compound Mn$_{0.1}$V$_2$O$_5$ as electrode material for Li intercalation batteries.

Films of TiO$_2$ nanoparticles have been studied extensively for both photo-catalytic applications and for dye-sensitized photoelectrochemical cells for solar energy conversion. Since these applications require fast charge transfer across the nanoparticle interfaces, it was suggested that films made of TiO$_2$ nanotubes oriented perpendicular to the electrode surface could be beneficial for such applications. Recently, Adachi and co-workers have investigated the structure and functionality of films made of TiO$_2$ with anatase-like structure.[123,136] In the first study, the photocatalytic oxidation of I$^-$ into I$_3^-$ was investigated and compared to that of a film consisting of a quasi-spherical TiO$_2$ nanoparticles with the anatase structure. Here, an increased photocatalytic effect of the nanotubes was observed as compared to a film made of the quasi-spherical TiO$_2$ nanoparticles (Degussa-P-25), and high-performance ST-01 film of TiO$_2$ nanoparticles. This group has also studied films of TiO$_2$ nanotubes as photoanodes in dye-sensitized solar cells. Photocurrents as high as 16 mA cm^{-2} were obtained under simulated solar light (100 mW cm^{-2}) and overall solar into electricity conversion efficiency of close to 5%, which is approximately twice as large as obtained from the TiO$_2$ nanoparticles film. This study underscores the importance of the vectorial charge transfer across the photoanode, and the enhanced photocatalytic effect of the nanotubes, compared with quasi-spherical nanoparticles of the same formula.

Kohli *et al.*[137] describe highly-sensitive methods of electroanalysis based on Au nanotube membranes for biosensor applications. The analyte species can be detected by measuring a change in the trans-membrane current when the analyte is added to the nanotubule-based cell. The second method entails the use of a concentration change based on the nanotubule membrane. It is found that synthetic micropore and nanotube membranes can mimic the function of ligand-gated ion channels, *i.e.*, they can be switched from an off to an on state in response to the presence of a chemical stimulus. Using these methods, detection limits as low as 10 pM have been achieved. These methods may find applications in protein separation, DNA purification, amplification and sequencing, and drug discovery.

Chen *et al.*[138] demonstrated hydrogen loading of up to 2.5 wt% in open-ended TiS$_2$ nanotubes under 4 MPa. In an earlier study this group reported a reversible electrochemical intercalation of hydrogen in CaMoS$_2$ nanotube phase.[139] A charge–discharge capacity of 260 mAh g^{-1} was obtained at a current of 50 mA g^{-1}. Only 2% loss in capacity was observed after 30 charge/discharge cycles. Thermo-gravimetry/differential thermogravimetric analysis of BN nanomaterials such as nanotubes, nanocages and nanocapsules produced from LaB$_6$ and Pd/boron powder by using an arc melting method showed possibility of hydrogen storage of 1–3 wt%.[140] The conditions of H$_2$ gas storage in a B$_{36}$N$_{36}$ cluster, which was considered as a cap structure of B$_{99}$N$_{99}$ nanotubes, were predicted by first principle single-point energy calculations. H$_2$ molecules would be introduced from hexagonal rings of the cage structure. BN fullerene materials would store H_2 molecule more easily than carbon fullerene materials, and its stability for high temperature would be good. Chen *et al.*[141] have reviewed hydrogen adsorption/ storage in MoS_2, TiS_2, and BN nanotubes and addressed the hydrogen adsorption–desorption processes mechanisms for metal disulfide and BN nanotubes.

The catalytic properties of metallic gold, without interference from a catalyst support, by using nanotubes of gold deposited in polycarbonate membranes was studied for the oxidation of CO at gas–water interfaces.[142] These gold–nanotube membranes exhibit catalytic activity for the oxidation of CO by O_2 at room temperature, and this activity is enhanced by liquid water, promoted by increasing the pH of the solution, and increased using H_2O_2 as the oxidizing agent. The rate can also be increased by depositing KOH within these nanotubes. These rates are comparable with those found in heterogeneous catalysis studies with gold nanoparticles on oxide supports, which suggests that the high activity of these latter catalysts may be related to the promotional effect of hydroxyl groups.

Zettl and co-workers[143] have created insulated C_{60} nanowires by packing C_{60} molecules into the interior of insulating boron nitride nanotubes (BNNTs). For small-diameter BNNTs, the wire consists of a linear chain of C_{60} molecules; with increasing BNNT inner diameter, unusual C_{60} stacking configurations are obtained (including helical, hollow core, and incommensurate) that are unknown for bulk or thin-film forms of C_{60}. They also inserted 1D single crystals of potassium halides (with lengths up to several micrometres), including KI, KCl, and KBr, into boron nitride nanotubes at temperatures above 600 °C in a sealed tube.[144]

References

1. S. Iijima, *Nature*, 1991, **354**, 56.
2. H. W. Kroto, J. R. Heath, S. C. O'Brien, R. F. Curl and R. E. Smalley, *Nature*, 1985, **318**, 162.
3. E. Chianelli, T. Prestridge, T. Pecorano and J. P. DeNeufville, *Science*, 1979, **203**, 1105.
4. J. V. Sanders, *Chem. Scrip.* 1978–79, **14**, 141.
5. R. Tenne, L. Margulis, M. Genut and G. Hodes, *Nature*, 1992, **360**, 444.
6. L. Margulis, G. Salitra, R. Tenne and M. Talianker, *Nature*, 1993, **365**, 113.
7. M. Hershfinkel, L. A. Gheber, V. Volterra, J. L. Hutchison, L. Margulis and R. Tenne, *J. Am. Chem. Soc.*, 1994, **116**, 1914.
8. Y. Feldman, E. Wasserman, D. J. Srolovitch and R. Tenne, *Science*, 1995, **267**, 222.
9. R. Tenne, M. Homyonfer and Y. Feldman, *Chem. Mater.*, 1998, **10**, 3225.
10. R. Tenne, *PhysChemChemPhys*, 2002, **4**, 2095.
11. C. N. R. Rao and M. Nath, *Dalton Trans.*, 2003, 1.
12. G. R. Patzke, F. Krumeich and R. Nesper, *Angew. Chem. Int. Ed.*, 2002, **41**, 2446.
13. L. Pauling, *Proc. Nat. Acad. Sci. U.S.A.* 1930, **16**, 578.
14. W. Stöber, A. Fink and E. Bohn, *J. Colloid Interface Sci.*, 1968, **26**, 62.
15. M. Nakamura and Y. Matsui, *J. Am. Chem. Soc.*, 1995, **117**, 2651.

16. P. M. Ajayan, O. Stephane, Ph. Redlich and C. Colliex, *Nature*, 1995, **375**, 564.

17. B. C. Satishkumar, A. Govindaraj, E. M. Vogl, L. Basumallick and C. N. R. Rao, *J. Mater. Res.*, 1997, **12**, 604.

18. B. C. Satishkumar, A. Govindaraj, M. Nath and C. N. R. Rao, *J. Mater. Chem.*, 2000, **10**, 2115.

19. M. E. Spahr, P. Bitterli, R. Nesper, M. Müller, F. Krumeich and H. U. Nissen, *Angew. Chem. Int. Ed.*, 1998, **37**, 1263.

20. L. Pu, X. Bao, J. Zou and D. Feng, *Angew. Chem. Int. Ed.*, 2001, **40**, 1490.

21. Y. Miyamoto, A. Rubio, S. G. Louie and M. L. Cohen, *Phys. Rev. B*, 1994, **50**, 4976.

22. K. Kobayashi and N. Kurita, *Phys. Rev. Lett.*, 1993, **70**, 3542.

23. Z. W. Sieh, K. Cherrey, N. G. Chopra, X. Blasé, Y. Miyamoto, A. Rubio, M. L. Cohen, S. G. Louie, A. Zettl and P. Gronsky, *Phys. Rev. B*, 1994, **51**, 11229.

24. N. G. Chopra, J. Luyken, K. Cherry, V. H. Crespi, M. L. Cohen, S. G. Louie and A. Zettl, *Science*, 1995, **269**, 966.

25. P. Gleize, M. C. Schouler, P. Gadelle and M. Caillet, *J. Mater. Sci.*, 1994, **29**, 1575.

26. O. R. Lourie, C. R. Jones, B. M. Bertlett, P. C. Gibbons, R. S. Ruoff and W. E. Buhro, *Chem. Mater.*, 2000, **12**, 1808.

27. R. Ma, Y. Bando and T. Sato, *Chem. Phys. Lett.*, 2001, **337**, 61.

28. F. L. Deepak, C. P. Vinod, K. Mukhopadhyay, A. Govindaraj and C. N. R. Rao, *Chem. Phys. Lett.* 2002, **353**, 345.

29. J. Bao, C. Tie, Z. Xu, Q. Zhou, D. Shen and Q. Ma, *Adv. Mater.*, 2001, **13**, 1631.

30. B. Mayers and Y. Xia, *Adv. Mater.*, 2002, **14**, 279.

31. (a) A. Govindaraj and C. N. R. Rao, in *The Chemistry of Nanomaterials*, ed. C. N. R. Rao, A. Müller and A. K. Cheetham, Wiley-VCH Verlag GmbH & Co, Weinheim, 2004 Vol. 1; (b) J. C. Hulteen and C. R. Martin, *J. Mater. Chem.*, 1997, **7**, 1075.

32. M. Chhowalla and G. A. J. Amaratunga, *Nature*, 2000, **407**, 164.

33. P. A. Parilla, A. C. Dillon, K. M. Jones, G. Riker, D. L. Schulz, D. S. Ginley and M. J. Heben, *Nature*, 1999, **397**, 114.

34. P. A. Parilla, A. C. Dillon, B. A. Parkinson, K. M. Jones, J. Alleman, G. Riker, D. S. Ginley and M. J. Heben, *J. Phys. Chem. B*, 2004, **108**, 6197.

35. R. Tenne, *Chem. Eur. J.*, 2002, **8**, 5296.

36. T. Tsirlina, Y. Feldman, M. Homyonfer, J. Sloan, J. L. Hutchison, R. Tenne, *Fullerene Sci. Technol.*, 1998, **6**, 157.

37. M. Nath, A. Govindaraj and C. N. R. Rao, *Adv. Mater.*, 2001, **13**, 283.

38. M. Nath and C. N. R. Rao, *Chem. Commun.*, 2001, 2336.

39. M. Nath and C. N. R. Rao, *J. Am. Chem. Soc.*, 2001, **123**, 4841.

40. M. Nath and C. N. R. Rao, *Angew. Chem. Int. Ed.*, 2002, **41**, 3451.

41. A. P. Lin, C. Y. Mou, S. D. Liu, *Adv. Mater.*, 2000, **12**, 103.

42. J. Zhang, L. Sun, C. Liao and C. Yan, *Chem. Commun.*, 2002, 262.

43. M. Niederberger, H.-J. Muhr, F. Krumeich, F. Bieri, D. Günther and R. Nesper, *Chem. Mater.*, 2000, **12**, 1995.

44. C. N. R. Rao, A. Govindaraj, F. L. Deepak, N. A. Gunari, M. Nath, *Appl. Phys. Lett.*, 2001, **78**, 1853.

45. (a) T. Kasuga, M. Hiramatsu, A. Hason, T. Sekino and K. Niihara, *Langmuir*, 1998, **14**, 3160; (b) H. Imai, Y. Takei, K. Shimizu, M. Matsuda and H. Hirashima, *J. Mater. Chem.*, 1999, **9**, 2971.
46. C. M. Zelenski and P. K. Dorhout, *J. Am. Chem. Soc.*, 1998, **120**, 734.
47. S. Hou, C. C. Harrell, L. Trofin, P. Kohli and C. R. Martin, *J. Am. Chem. Soc.*, 2004, **126**, 5674.
48. F. Kokai, *Kokagaku*, 2003, **34**, 50.
49. J.-Q. Hu, X.-M. Meng, Y. Jiang, C.-S. Lee, S.-T. Lee, *Adv. Mater.*, 2003, **15**, 72.
50. A. Loiseau, F. Williame, N. Demonecy, G. Hug and H. Pascard, *Phys. Rev. Lett.*, 1996, **76**, 4737.
51. J. Goldberger, R. He, Y. Zhang, S. Lee, H. Yan, H.-J. Choi and P. Yang, *Nature*, 2003, **422**, 599.
52. (a) C. Tang, Y. Bando and D. Golberg, *J. Solid State Chem.*, 2004, **177**, 2670; (b) L. W. Yin, Y. Bando, D. Golberg and M. S. Li, *Appl. Phys. Lett.*, 2004, **85**, 3869.
53. Y. Ono, K. Nakashima, M. Sano, Y. Kanekiyo, K. Inoue, J. Hojo and S. Shinkai, *Chem. Commun.*, 1998, 1477.
54. P. Terech and R. G. Weiss, *Chem. Rev.*, 1997, **97**, 3133.
55. G. Gundiah, S. Mukhopadhyay, U. G. Tumkurkar, A. Govindaraj, U. Maitra and C. N. R. Rao, *J. Mater. Chem.*, 2003, **13**, 2118.
56. C. Mu, Y. Yu, R. Wang, K. Wu, D. Xu, G. Guo, *Adv. Mater.*, 2004, **16**, 1550.
57. S.-H. Zhang, Z.-X. Xie, Z.-Y. Jiang, X. Xu, J. Xiang, R.-B. Huang and L.-S. Zheng, *Chem. Commun.*, 2004, 1106.
58. N. Li, X. Li, X. Yin, W. Wang and S. Qiu, *Solid State Commun.*, 2004, **132**, 841.
59. L. Qu, G. Shi, X. W. and B. Fan, *Adv. Mater.*, 2004, **16**, 1200.
60. X-Y Liu, J-H. Zeng, S-Y. Zhang, R-B. Zheng, X-M. Liu and Y-T. Qian, *Chem. Phys. Lett.*, 2003, **374**, 348.
61. T. Kijima, T. Yoshimura, M. Uota, T. Ikeda, D. Fujikava, S. Mouri and S. Uoyama, *Angew. Chem. Int. Ed.*, 2004, **43**, 228.
62. N. Li, X. Li, X. Yin, W. Wang and S. Qiu, *Solid State Commun.*, 2004, **132**, 841.
63. Y. Sun and Y. Xia, *Adv. Mater.*, 2004, **16**, 264.
64. D. Ciuparu, R. F. Klie, Y. Zhu, L. Pfefferle, *J. Phys. Chem. B*, 2004, **108**, 3967.
65. H. Zhang, D. Yang, Y. Ji, X. Ma, J. X. and D. Que, *J. Phys. Chem. B*, 2004, **108**, 1179.
66. P. Hu, Y. Liu, L. Fu, L. Cao and D. Zhu, *Chem. Commun.*, 2004, 556.
67. P. V. Teredesai, F. L. Deepak, A. Govindaraj, C. N. R. Rao and A. K. Sood, *J. Nanosci. Nanotechnol.*, 2002, **2**, 495.
68. X.-P. Shen, A.-H. Yuan, F. Wang, J.-M. Hong and Z. Xu, *Solid State Commun.*, 2005, **133**, 19.
69. H. Cui, H. Liu, X. Li, J. Wang, F. Han, X. Zhang and R. I. Boughton, *J. Solid State Chem.*, 2004, **177**, 4001.
70. T. Kasuga, M. Hiramutsu, A. Hoson, T. Sekino and K. Niihara, *Adv. Mater.*, 1999, **11**, 1307.
71. C.-C. Tsai and H. Teng, *Chem. Mater.*, 2004, **16**, 4352.
72. Q. Chen, W. Zhou and L.-M. Peng, *Adv. Mater.*, 2002, **14**, 1208.

73. X. Sun and Y. Li, *Chem. Eur. J.*, 2003, **9**, 2229.

74. Y. Suzuki and S. Yoshikawa, *J. Mater. Res.*, 2004, **19**, 982.

75. X. Wang, J. Zhuang, J. Chen, K. Zhou and Y. Li, *Angew. Chem. Int. Ed.*, 2004, **43**, 2017.

76. R.-S. Chen, Y.-S. Huang, D.-S. Tsai, S. Chattopadhyay, C.-T. Wu, Z.-H. Lan and K.-H. Chen., *Chem. Mater.*, 2004, **16**, 2457.

77. G. Wu, L. Zhang, B. Cheng, T. Xie and X. Yuan, *J. Am. Chem. Soc.*, 2004, **126**, 5976.

78. R. Ma, Y. Bando and T. Sasaki, *J. Phys. Chem. B*, 2004, **108**, 2115.

79. Y. Liu, J. Dong and M. Liu, *Adv. Mater.*, 2004, **16**, 353.

80. L. Shi, Y. Gu, L. Chen, Z. Yang, J. M. and Y. Qian, *Carbon*, 2004, **43**, 195.

81. F. Krumeich, H.-J. Muhr, M. Niederberger, F. Bieri, B. Schnyder and R. Nesper, *J. Am. Chem. Soc.*, 1999, **121**, 8324.

82. H. J. Muhr, F. Krumeich, U. P. Schönholzer, F. Bieri, M. Niederberger, L. J. Gauckler and R. Nesper, *Adv. Mater.*, 2000, **12**, 231.

83. A. G. S. Filho, O. P. Ferreira, E. J. G. Santos, J. M. Filho and O. L. Alves, *Nano Lett.*, 2004, **4**, 2099.

84. W. Fan, S. Sun, X. Song, W. Zhang, H. Yu, X. Tan and G. Cao, *J. Solid State Chem.*, 2004, **177**, 2329.

85. F.-S. Cai, G.-Y. Zhang, J. Chen, X-L. Gou, H.-K. Liu and S.-X. Dou, *Angew. Chem. Int. Ed.*, 2004, **43**, 4212.

86. L. Guo, C. Liu, R. Wang, H. Xu, Z. W. and S. Yang, *J. Am. Chem. Soc.*, 2004, **126**, 4530.

87. T. Yamaguchi, S. Tashiro, M. Tominaga, M. Kawano, T. Ozeki and M. Fujita, *J. Am. Chem. Soc.*, 2004, **126**, 10818.

88. G. Du, Q. Chen, Y. Yu, S. Zhang, W. Zhou and L.-M. Peng, *J. Mater. Chem.*, 2004, **14**, 1437.

89. N. G. Chopra and A. Zettl, *Solid State Commun.*, 1998, **105**, 297.

90. E. Hernández, C. Goze, P. Bernier and A. Rubio, *Phys. Rev. Lett.*, 1998, **80**, 4502.

91. I. Kaplan-Ashiri, K. Gartsman, S. R. Cohen, G. Seifert and R. Tenne, *J. Mater. Res.* 2004, **19**, 454.

92. R. Tenne and A. Zettl, *Topics in Applied Physics*, (*Carbon Nanotubes*) ed. M. S. Dresselhaus, G. Dresselhaus and Ph. Avouries, Springer-Verlag, Heidelberg, 2001, Vol. 80, pp. 81–112.

93. V. V. Pokropivnyi, *Powder Metall. Metal Ceram.*, 2001, **40** (9–10), 485.

94. V. V. Pokropivnyi, *Powder Metall. Metal Ceram.*, 2001, **40**, (11–12), 582.

95. L. Vaccarini, C. Goze, L. Hrnand, E. Hernández, P. Bernier and A. Rubio, *Carbon*, 2000, **38**, 1681.

96. M. L. Cohen, *Solid State Commun.*, 1994, **92**, 45; A. Rubio, J. L. Corkill and M. L. Cohen, *Phys. Rev. B*, 1994, **49**, 5081.

97. Y. Miyamoto, A. Rubio, S. G. Louie and M. L. Cohen, *Phys. Rev. B*, 1994, **50**, 18360.

98. F. Jensen and H. Toftlund, *Chem. Phys. Lett.*, 1993, **201**, 95.

99. G. Seifert, H. Terrones, M. Terrones, G. Jungnickel and T. Frauenheim, *Phys. Rev. Lett.*, 2000, **85**, 146.

100. G. Seifert, H. Terrones, M. Terrones, G. Jungnickel and T. Frauenheim, *Solid State Commun.*, 2000, 114, 245.

101. (a) V. V. Ivanovskaya, A. N. Enjashin, A. A. Sofronov, Makurin, N. Yu, N. I. Medvedeva and A. L. Ivanovskii, *J. Mol. Struct. (Theochem)*, 2003, 625, 9; (b) V. V. Ivanovskaya, A. N. Enyashin, N. I. Medvedeva and A. L. Ivanovskii, *Phys. Status Solidi*, 2003, 238, R1.

102. G. Seifert, H. Terrones, M. Terrones and T. Frauenheim, *Solid State Commun.*, 2002, 115, 635.

103. G. L. Frey, S. Elani, M. Homyonfer, Y. Feldman and R. Tenne, *Phys. Rev. B*, 1998, 57, 6666.

104. L. Scheffer, R. Rosentsveig, A. Margolin, R. Popovitz-Biro, G. Seifert, S. R. Cohen and R. Tenne, *Phys. Chem. Chem. Phys.*, 2002, 4, 2095.

105. P. Santiago, J. A. Ascencio, D. Mendoza, M. Pérez-Alverez, A. Espinosa, C. Reza-Sangermán, P. Schabes-Retchkiman, G. A. Camacho-Bragado and M. José-Yacamán, *Appl. Phys. A*, 2004, 78, 513.

106. M. Remskar, A. Mrzel, R. Sanjines, H. Cohen and F. Levy, *Adv. Mater.*, 2003, 15, 237.

107. V. Nemanic, V. Zumer, B. Zajec, J. Pahor, M. Remskar, A. Mrzel, P. Panjan and D. Mihailovic, *Appl. Phys. Lett.*, 2003, 82, 4573.

108. M.-S. Wu, J.-T. Lee, Y.-Y. Wang and C.-C. Wan, *J. Phys. Chem. B*, 2004, 108, 16331.

109. Y. Z. Jin, W. K. Hsu, Y. L. Chueh, L. J. Chou, Y. Q. Zhu, K. Brigatti, H. W. Kroto and D. R. M. Walton, *Angew. Chem. Int. Ed.*, 2004, 43, 5670.

110. C. Su, H.-T. Liu and J.-M. Li, *Nanotechnology*, 2002, 13, 746.

111. P. Zhang and V. H. Crespi, *Phys. Rev. Lett.*, 2002, 89, 056403.

112. A. Quandt, A. Y. Liu and I. Boustani, *Phys. Rev. B*, 2001, 64, 125422.

113. A. L. Chernozatonskii, *JETP Lett.*, 2001, 74, 335.

114. V. G. Bamburov, V. V. Ivanovskaya, A. N. Enyashin, I. R. Shein, N. I. Medvedeva, Makurin, N. Y. and A. L. Ivanovskii, *Dokl. Phys. Chem.*, 2003, 388, 624.

115. S. Guerini and P. Piquini, *Microelectron. J.*, 2003, 34, 495.

116. M. Kociak, A. Yu. Kasumov, S. Guéron, B. Reulet, I. I. Khodos, Yu. B. Gorbatov, V. T. Volkov, L. H. Vaccarini and H. Bouchiat, *Phys. Rev. Lett.*, 2001, 86, 2416.

117. Z. K. Tang, L. Y. Zhang, N. Wang, X. X. Zhang, G. H. Wen, G. D. Li, J. N. Wang, C. T. Chan and P. Sheng, *Science*, 2001, 292, 2462.

118. M. Nath, S. Kar, A. K. Raychaudhuri and C. N. R. Rao, *Chem. Phys. Lett.*, 2003, 368, 690.

119. V. V. Ivanovskaya, A. N. Enyashin, A. A. Sofronov, Makurin, N. Yu, N. I. Medvedeva and A. L. Ivanovskii, *Solid State Commun.*, 2003, 126, 489.

120. L. K. Elbaun, D. M. Newns, H. Zeng, V. Derycke, J. Z. Sun and R. Sandstrom, *Nature*, 2004, 431, 672.

121. (a) J.-F. Xu, Czerwa, S. Webster, D. L. Carroll, J. Ballato and R. Nesper, *Appl. Phys. Lett.*, 2002, 81, 1711; (b) J. Cao, J. Choi, J. L. Musfeldt, S. Lutta and M. S. Whittingham, *Chem. Mater.*, 2004, 16, 731.

122. V. V. Ivanovskaya, A. N. Enyashin and A. L. Ivanovskii, *Mendeleev Commun.*, 2003, 5.
123. M. Adachi, Y. Murata, I. Okada and S. Yoshikawa, *J. Electrochem. Soc.*, 2003, **150**, G488.
124. L. Rapoport, Yu. Bilik, Y. Feldman, M. Homyonfer, S. R. Cohen and R. Tenne, *Nature*, 1997, **387**, 791.
125. L. Rapoport, N. Fleischer and R. Tenne, *Adv. Mater.*, 2003, **15**, 651.
126. Y. Q. Zhu, T. Sekine, K. S. Brigatti, S. Firth, R. Tenne, R. Rosentsveig, H. W. Kroto and D. R. M. Walton, *J. Am. Chem. Soc.*, 2003, **125**, 1329.
127. A. Zak, Y. Feldman, H. Cohen, V. Lyakhovitskaya, G. Leitus, R. Popovitz-Biro, S. Reich and R. Tenne, *J. Am. Chem. Soc.*, 2002, **124**, 4747.
128. R. Dominko, D. Arcon, A. Mrzel, A. Zorko, P. Cevc, P. Venturini. and M. Gaberscek, *Adv. Mater.*, 2002, **14**, 1531.
129. D. Mihailovic, Z. Jaglicic, D. Arcon, A. Mrzel, A. Zorko, M. Remskar, V. V. Kabanov, R. Dominko, M. Gaberscek, C. J. Gomez-Garcý, J. M. Martýnez-Agudo and E. Coronado, *Phys. Rev. Lett.*, 2003, **90**, 146401.
130. D. Arcon, A. Zorko, P. Cevc, A. Mrzel, M. Remskar, R. Dominko, M. Gaberscek and D. Mihailovic, *Phys. Rev. B*, 2003, **67**, 125423.
131. J. Chen, Z.-L. Tao, and S.-L. Li, *Angew. Chem. Int. Ed.*, 2003, **42**, 2147.
132. Z.-L. Tao, L.-N. Xu, X.-L. Gou, J. Chen and H.-T. Yuan, *Chem. Commun.*, 2004, 2080.
133. M. E. Spahr, P. Stoschitzki-Bitterli, R. Nesper, O. Haas and P. Novak, *J. Electrochem. Soc.*, 1999, **146**, 2780.
134. S. Nordlinder, K. Edström and T. Gustafsson, *Electrochem. Solid State Lett.*, 2001, **4**, A129.
135. A. Dobley, K. Ngala, T. Shoufeng, P. Y. Zavalij and M. S. Whittingham, *Chem. Mater.*, 2001, **13**, 4382.
136. M. Adachi, Y. Murata, M. Harada, and S. Yoshikawa, *Chem. Lett.*, 2000, 942.
137. P. Kohli, M. Wirtz and C. R. Martin, *Electroanalysis*, 2004, **16**, 9.
138. J. Chen, S.-L. Li, Z.-L. Tao, Y.-T. Shen and C.-X. Cui, *J. Am. Chem. Soc.*, 2003, **125**, 5284.
139. J. Chen, N. Kuriyama, H. Yuan, H. T. Takeshita and T. Sakai, *J. Am. Chem. Soc.* 2001, **123**, 11813.
140. T. Oku, M. Kuno and I. Narita, *J. Phys. Chem. Solids*, 2004, **65**, 549.
141. J. Chen and F. Wu, *Appl. Phys. A*, 2004, **78**, 989.
142. M. A. Sanchez-Castillo, C. Couto, W. B. Kim and J. A. Dumesic, *Angew. Chem. Int. Ed.*, 2004, **43**, 1140.
143. W. Mickelson, S. Aloni, Wei-Qiang Han, J. Cumings and A. Zettl, *Science*, 2003, **300**, 467.
144. W.-Q. Han, C. W. Chang and A. Zettl, *Nano Lett.*, 2004, **4**, 1355.

CHAPTER 3

Inorganic Nanowires

1 Introduction

Following the discovery of carbon nanotubes by Iijima,[1] there has been great interest in the synthesis and characterization of other 1D structures, which include nanowires, nanorods and nanobelts. Inorganic nanowires can act as active components in devices, as revealed by recent investigations. In the last 4–5 years, nanowires of various inorganic materials have been synthesized and characterized. Thus, nanowires of elements, oxides, nitrides, carbides and chalcogenides have been generated by employing several strategies. One of the crucial factors in nanowire synthesis is the control of composition, size and crystallinity. Among the methods employed, some are based on vapour phase techniques and others are solution techniques. Compared to some of the physical methods, such as patterning techniques, chemical methods are more versatile for the synthesis of the nanowires. Thus, techniques involving chemical vapour deposition (CVD), precursor decomposition, as well as solvothermal, hydrothermal and carbothermal methods have been widely employed. Several physical methods, especially microscopic techniques such as scanning electron microscopy (SEM), transmission electron microscopy (TEM), scanning tunnelling microscopy (STM) and atomic force microscopy (AFM) are used to characterize nanowires. There are a few surveys on nanowires and related materials in the literature.[2–5] In this chapter, we present a comprehensive, up-to-date review of the various families of inorganic nanowires, wherein we discuss their synthesis along with their properties. Wherever possible, we have also indicated potential applications. We have not covered molecular nanowires, polymer wires and wires of composites due to limitations of scope and space. The term nanorod is often used to describe nanowires with short aspect ratios.

2 Synthetic Strategies

One of the important aspects of the 1D structures relates to their crystallization,[6] wherein the evolution of a solid from a vapour, a liquid, or a solid phase involves nucleation and growth. As the concentration of the building units (atoms, ions, or molecules) of a solid becomes sufficiently high, they aggregate into small nuclei or clusters through homogeneous nucleation. The clusters serve as seeds for further growth to form larger clusters. Several synthetic strategies have been developed for 1D nanowires with different levels of control over the growth parameters. They include: (i) use of the anisotropic crystallographic structure of the solid to facilitate 1D nanowire growth; (ii) introduction of a solid–liquid interface, (iii) use of templates (with 1D morphologies) to direct the formation of nanowires, (iv) supersaturation control to modify the growth habit of a seed; (v) use of capping agents to kinetically control the growth rates of the various facets of a seed and (vi) self-assembly of zero-dimensional (0D) nanostructures. They are conveniently categorized into (a) growth in the vapour phase and (b) solution-based growth, and we discuss these in the next section. Various media have been employed in solution-based growth, including supercritical fluids.[7]

Vapour Phase Growth

Vapour phase growth is extensively used to produce nanowires. Starting with the simple evaporation technique in an appropriate atmosphere to produce elemental or oxide nanowires, vapour–liquid–solid, vapour–solid and other processes are also used.

Vapour–Liquid–Solid Growth

The growth of nanowires *via* a gas phase reaction involving the vapour–liquid–solid (VLS) process has been widely studied. Wagner,[6] during his studies of growth of large single-crystalline whiskers, proposed in the 1960s, a mechanism for the growth *via* gas-phase reaction involving the so-called vapour–liquid–solid process. He studied the growth of mm-sized Si whiskers in the presence of Au particles. According to this mechanism, the anisotropic crystal growth is promoted by the presence of the liquid alloy/solid interface. This mechanism has been widely applied to explain the growth of various nanowires, including those of Si and Ge. The growth of Ge nanowires using Au clusters as a solvent at high temperatures is explained based on the Ge–Au phase diagram (Figure 3.1). Ge and Au form a liquid alloy when the temperature is higher than the eutectic point (363 °C) (Figure 3.1a-I). The liquid surface has a large accommodation coefficient and is, therefore, a preferred deposition site for the incoming Ge vapour. After the liquid alloy becomes supersaturated with Ge, precipitation of the Ge nanowire occurs at the solid–liquid interface as shown in a-II–III of Figure 3.1. Until recently, the only evidence for this mechanism was the presence of alloy droplets at the tips of the nanowires. Real-time observations of Ge nanowire growth in an *in situ* high-temperature TEM demonstrate the validity of the VLS growth mechanism.[8] The observations suggest that there are three growth stages: metal alloying, crystal nucleation and axial growth (Figure 3.2). Figure 3.2(a–f) shows a sequence of TEM images during the *in situ* growth of a Ge nanowire. Three stages, I–III, are clearly identified. (I) Alloying process, (Figure 3.2a–c): The maximum temperature that could be attained in the system was 900 °C, upto which the Au clusters remain in the solid state in the absence of Ge vapour. With increasing amount of Ge vapour condensation and dissolution, Ge and Au form an alloy and liquefy. The volume of

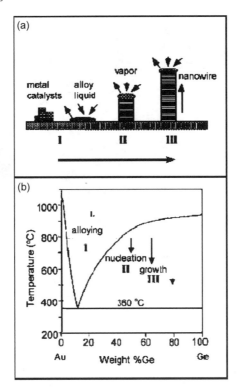

Figure 3.1 (a) *Schematic illustration of the three stages in the vapour–liquid–solid (VLS) mechanism: (I) alloying, (II) nucleation and (III) axial growth. The three stages are projected onto the coventional Au-Ge phase diagram: The compositional and phase evolution during the nanowire growth process is shown in* (b) (Reproduced from ref. 8)

the alloy droplet increases and the elemental contrast decreases while the alloy composition crosses sequentially, from left to right, a biphasic region (solid Au and Au/Ge liquid alloy) and a single-phase region (liquid). An isothermal line in the Au–Ge phase diagram (Figure 3.1b) shows the alloying process. (II) Nucleation (Figures 3.2d–e): As the concentration of Ge increases in the Au–Ge alloy droplet, nucleation of the nanowire begins. Knowing the alloy volume change, it is estimated that the nucleation generally occurs at a Ge weight percentage of 50–60%. (III) Axial growth (Figure 3.2d–f): Once the Ge nanocrystal nucleates at the liquid–solid interface, further condensation/dissolution of the Ge vapour into the system increases the amount of Ge precipitation from the alloy. The incoming Ge vapours diffuse and condense at the solid–liquid interface, thus suppressing secondary nucleation events. The interface is then pushed forward (or backward) to form nanowires (Figure 3.2f). This study confirms the validity of the VLS growth mechanism at the nanometre scale.

Since the diameter of the nanowires is determined by the diameter of the catalyst particles, this method provides an efficient means to obtain uniform-sized nanowires. Knowing the phase diagram of the reacting species, the growth

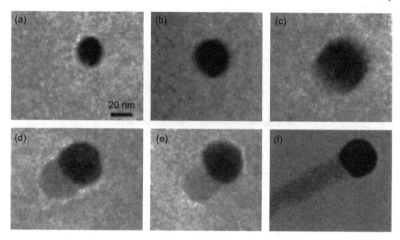

Figure 3.2 *In situ TEM images recorded during nanowire growth*: (a) *Au nanoclusters in the solid state at 500 °C;* (b) *alloying initiated at 800 °C, where Au exists mostly in the solid state;* (c) *liquid Au/Ge alloy;* (d) *nucleation of the Ge nanocrystal on the alloy surface;* (e) *elongation of the Ge nanocrystal with further Ge condensation, eventually forming a wire* (f)
(Reproduced from ref. 8)

temperature can be set in between the eutectic point and the melting point of the material. Physical methods such as laser ablation or thermal evaporation as well as chemical methods such as CVD are used to generate the reactant species in vapour form, required for nanowire growth. Catalyst particles are sputtered onto the substrates or metal nanoparticles prepared by solution-based routes used as the catalysts. An advantage of this route is that patterned deposition of catalyst particles yields patterned nanowires. Using this growth mechanism, nanowires of materials, including elements, oxides, carbides, phosphides, *etc.*, have been successfully obtained, as detailed in the following sections.

Oxide-assisted Growth

Lee and co-workers[9,10] have proposed the oxide-assisted growth mechanism for nanowires. No metal catalyst is required for the synthesis of nanowires by this means. These workers found that the growth of Si nanowires is enhanced when SiO_2-containing Si powder targets are used. Limited quantities of Si nanowires were obtained even with a target made of pure Si powder (99.995%). Here, the growth of the Si nanowires is assisted by the Si oxide, where the Si_xO $(x > 1)$ vapour generated by thermal evaporation or laser ablation plays a key role. Nucleation of the nanoparticles is assumed to occur on the substrate as:

$$Si_xO \longrightarrow Si_{x-1} + SiO \ (x > 1), \text{ and}$$
$$2SiO \longrightarrow Si + SiO_2$$

The above decomposition reactions cause the precipitation of Si nanoparticles, which act as the nuclei of the silicon nanowires covered by silicon oxide.

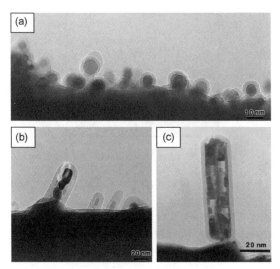

Figure 3.3 *TEM micrographs of* (a) *Si nanowire nuclei formed on the Mo grid and* (b), (c) *initial growth stages of the nanowires*
(Reproduced from ref. 9)

Precipitation, nucleation and growth of the nanowires occur near the cold finger, suggesting that the temperature gradient provides the external driving force for the formation and growth of the nanowires.

Figure 3.3(a–c) show the TEM images of the formation of nanowire nuclei at the initial stages. Figure 3.3(a) shows Si nanoparticles covered by an amorphous silicon oxide layer. The tip of the Si crystalline core contains a high concentration of defects. Figure 3.4 shows a schematic of the nanowire growth by this mechanism. The growth of the silicon nanowires is determined by three factors: (1) catalytic effect of the Si_xO $(x > 1)$ layer on the nanowire tips, (2) retardation of the lateral growth of nanowires by the SiO_2 component in the shells formed by the decomposition of SiO, (3) stacking faults along the nanowire growth direction of $\langle 112 \rangle$, which contains easy-moving $1/6[112]$ and nonmoving $1/3[111]$ partial dislocations, and micro-twins present at the tip areas. Only the nuclei that have their $\langle 112 \rangle$ direction parallel to the growth direction grow fast (Figure 3.4b).

Vapour–solid Growth

The vapour–solid (VS) process for whisker growth holds for the growth of nanowires as well.[6] In this process, evaporation, chemical reduction or gaseous reaction generates the vapour. The vapour is subsequently transported and condensed onto a substrate. The VS method has been used to prepare whiskers of oxides as well as metals with micrometre diameters. It is, therefore, possible to synthesize 1D nanostructures using the VS process if one can control the nucleation and subsequent growth process. Using the VS method, nanowires of several metal oxides have been obtained.

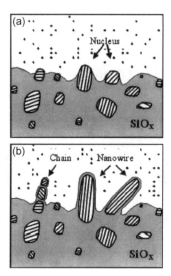

Figure 3.4 *Schematic describing the nucleation and growth mechanism of Si nanowires. Parallel lines indicate the [112] orientation. (a) Si oxide vapour is deposited first and forms the matrix within which the Si nanoparticles are precipitated. (b) Nanoparticles in a preferred orientation grow fast and form nanowires. Nanoparticles with nonpreferred orientations may form chains of nanoparticles (Reproduced from ref. 9)*

Carbothermal Reactions

Nanowires of oxides, nitrides and carbides can be synthesized by carbothermal reactions. For example, carbon (activated carbon or carbon nanotubes) in mixture with an oxide produces sub-oxidic or metal vapour species that reacts with C, O_2, N_2 or NH_3 to produce the desired nanowires. Thus, heating a mixture of Ga_2O_3 and carbon in N_2 or NH_3 produces GaN nanowires. Carbothermal reactions generally involve the following steps:

$$\text{metal oxide} + C \longrightarrow \text{metal suboxide} + CO$$

$$\text{metal suboxide} + O_2 \longrightarrow \text{metal oxide nanowires}$$

$$\text{metal suboxide} + NH_3 \longrightarrow \text{metal nitride nanowires} + CO + H_2$$

$$\text{metal suboxide} + N_2 \longrightarrow \text{metal nitride nanowires} + CO$$

$$\text{metal suboxide} + C \longrightarrow \text{metal carbide nanowires} + CO$$

Depending on the desired product, the suboxide heated in the presence of O_2, NH_3, N_2 or C yields oxide, nitride or carbide nanowires.

Solution-based Growth of Nanowires

This synthetic strategy makes use of the anisotropic growth dictated by the crystallographic structure of the solid material, confined and directed by templates, kinetically controlled by supersaturation, or by the use of an appropriate capping agent.

Highly Anisotropic Crystal Structures

Solid materials such as poly(sulphur nitride), $(SN)_x$, grow into 1D nanostructures, the habit being determined by the anisotropic bonding in the structure.[11,12] Other materials such as selenium,[13,14] tellurium[15] and molybdenum chalcogenides[16] are easily obtained as nanowires due to anisotropic bonding, which dictates that the crystallization occurs along the *c*-axis, favouring the stronger covalent bonds over the weak van der Waals forces between the chains.

Template-based Synthesis

Template-directed synthesis is a convenient, versatile method for generating 1D nanostructures. The template serves as a scaffold against which other materials with similar morphologies are synthesized. The *in situ* generated material is shaped into a nanostructure with a morphology complementary to that of the template. Templates could be nanoscale channels within mesoporous materials, porous alumina or polycarbonate membranes. The nanoscale channels are filled using the solution, the sol–gel or the electrochemical method. The nanowires so produced are released from the templates by removing the host matrix.[17] Unlike polymer membranes fabricated by track etching, the anodic alumina membrane (AAM) containing hexagonally packed 2D arrays of cylindrical pores of uniform size is prepared using anodization of an aluminium foil in an acidic medium (Figure 3.5). Several materials have been fabricated into nanowires using AAMs in the templating process. Examples of inorganic materials include Au, Ag, Pt, TiO_2, MnO_2,

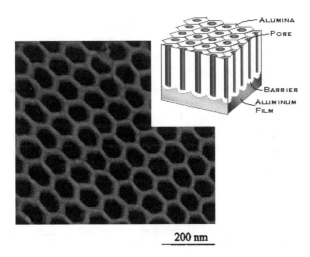

Figure 3.5 *TEM micrograph of an anodic alumina membrane (AAM)*
(Reproduced from ref. 17c)

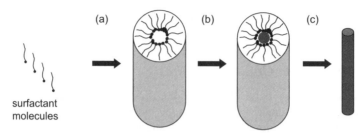

Figure 3.6 *Schematic nanowire formation by templating against mesostructures that are self-assembled from surfactant molecules; (a) formation of cylindrical micelle, (b) formation of the desired material in the aqueous phase encapsulated by the cylindrical micelle, (c) removal of the surfactant molecule with an appropriate solvent (or by calcination) to obtain individual nanowires*
(Reproduced from ref. 18a)

ZnO, SnO$_2$, In$_2$O$_3$, CdS, CdSe, CdTe, electronically conducting polymers such as polypyrole, poly(3-methylthiophene) and polyaniline as well as carbon nanotubules. However it is often difficult to obtain single-crystalline materials by this method.

Mesophase structures self-assembled from surfactants (Figure 3.6) provide another class of useful, versatile templates for generating 1D nanostructures in large quantities. At the critical micellar concentration (c.m.c.) surfactant molecules spontaneously organize into rod-shaped micelles.[18] These anisotropic structures can be used as soft templates to promote the formation of nanorods when coupled with an appropriate chemical or electrochemical reaction. The surfactant needs to be selectively removed to collect the nanorods/nanowires. Based on this principle, nanowires of CuS, CuSe, CdS, CdSe, ZnS and ZnSe have been grown, by using surfactants such as Na-AOT or Triton X-100.[19,20]

Nanowires themselves are used as templates to generate the nanowires of other materials. The template may be attached to the nanowire, forming coaxial nanocables,[21] or it might react with the nanowires, forming a new material.[22] Dissolution of the original nanowires may also lead to nanotubes of the coated materials. The sol–gel coating method is a generic route to synthesize co-axial nanocables that may contain electrically conductive metal cores and insulating sheaths.

Several metal nanowires of 1–1.4 nm diameter have been prepared by filling SWNTs, opened by acid treatment.[23] Nanowires of Au, Pt, Pd and Ag have been synthesized by employing sealed-tube reactions as well as solution methods. In addition, thin layers of metals have been incorporated into the intertubular space of the SWNT bundles.

Solution–Liquid–Solid Process

Buhro and co-workers[24] have developed a low-temperature solution–liquid–solid (SLS) method for the synthesis of crystalline nanowires of group III–V semiconductors.[25] In a typical procedure, a metal (*e.g.* In, Sn, Bi) with a low melting point is used as a catalyst, and the desired material generated through

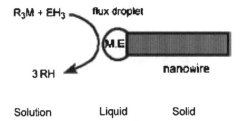

Figure 3.7 *Schematic of the growth of nanowires through the solution–liquid–solid (SLS) mechanism, which is similar to the vapour–liquid–solid (VLS) process* (Reproduced from ref. 24)

the decomposition of organometallic precursors. Nanowhiskers of InP, InAs and GaAs have been prepared by low-temperature ($\leq 203\,^{\circ}$C) solution phase reactions. The schematic illustration in Figure 3.7 clearly shows the growth of nanowires or whiskers through the SLS method. The products obtained are generally single-crystalline. Holmes *et al.*[26] have used the supercritical fluid–liquid–solid (SFLS) method to synthesize bulk quantities of defect-free silicon and germanium nanowires.

By using selective capping capacities of mixed surfactants, it is possible to extend the synthesis of group II–IV semiconductor nanocrystals to that of nanorods.[27]

Solvothermal Synthesis

Solvothermal methodology is employed extensively as a solution route to produce semiconductor nanowires and nanorods. In this process, a solvent is mixed with certain metal precursors and crystal growth regulating or templating agents such as amines. The solution mixture is placed in an autoclave to carry out the crystal growth and the assembly process. The methodology is versatile and has enabled the synthesis of crystalline nanowires of semiconductors and other materials.

Growth Control and Integration

A significant challenge in the chemical synthesis of nanowires is how to rationally control the nanostructure assemblies so that their size, dimensionality, interfaces and their 2D and 3D superstructures can be tailor-made towards desired functionality. Many physical and thermodynamic properties are diameter-dependent. Several groups of workers have synthesized uniform-sized nanowires by the VLS process using clusters with narrow size distributions.

Controlling the growth orientation is important for the applications of nanowires. By applying the conventional epitaxial crystal growth technique to the VLS process, a vapour–liquid–solid epitaxy technique has been developed for the controlled synthesis of nanowire arrays. Nanowires generally have preferred growth directions. For example, zinc oxide nanowires prefer to grow along their *c*-axis, *i.e.* along the $\langle 001 \rangle$ direction.[28,29] Silicon nanowires grow along the $\langle 111 \rangle$ direction

PDMS Micromold

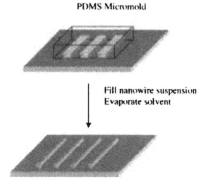

Fill nanowire suspension
Evaporate solvent

Figure 3.8 *Schematic of the microfluidic-assisted nanowire integration process for*
nanowire surface patterning
(Reproduced from ref. 31b)

when grown by the VLS growth process, but may grow along the $\langle 112 \rangle$ or $\langle 110 \rangle$ directions by the oxide-assisted growth mechanism.

Clearly from the VLS nanowire growth mechanism, the initial positions of Au clusters or Au thin films control the positions of the nanowires. By creating the desired patterns of Au using a lithographic technique, it is possible to grow ZnO nanowires of the same designed pattern since they grow vertically from the region coated with Au and form the designed patterns of the nanowire arrays.[28,29] Networks of nanowires with the precise placement of individual nanowires on substrates with the desired configuration is also achieved by the surface patterning strategy.[28,29]

Integration of nanowire building blocks into complex functional networks in a controlled manner remains a challenge. In the direct one-step growth process, nanowires grown by the VLS method are patterened on substrates by selectively depositing catalyst particles. Another way is to place the nanowire building blocks together into the functional structure to develop a hierarchical assembly. By using a simple dubbed microfluidic-assisted nanowire integration process, wherein the nanowire solution/suspension is filled in the microchannels formed between a poly(dimethylsiloxane) (PDMS) micromould and a flat Si substrate, followed by the evaporation of the solvent, nanowire surface patterning and alignment have been achieved.[30,31a,b] Figure 3.8 gives a schematic illustration of the microfluidic-assisted nanowire integration process. The Langmuir–Blodgett technique has also been used to obtain aligned, high-density nanowire assemblies.[32]

3 Elemental Nanowires

Silicon

Silicon nanowires (SiNWs) have been prepared by various methods, which include physical evaporation of the metal and CVD. The methods employ SiO_x and other precursors as silicon sources. The first report of the synthesis of SiNWs by thermal evaporation was by Yu *et al.*,[33] who sublimed a hot-pressed Si powder target mixed with Fe at 1200 °C in flowing Ar gas at ~100 Torr. Using this simple method, they could obtain SiNWs with a diameter of ~15 nm and length varying from a few tens to hundreds of microns (Figure 3.9a). The inset in the figure shows a selected-area electron diffraction (SAED) pattern, which is similar to that of bulk silicon. The nanowires were sheathed by an amorphous oxide layer of about 2 nm, which could be etched out by treatment with a dilute HF solution. A TEM image of the nanowires after this treatment is shown in Figure 3.9(b). By varying the ambient pressure between 150 to 600 Torr,[34] the diameter of the nanowires was controlled. The average size of the nanowires increases with the increasing gas pressure. By using thermal evaporation over Fe-patterned Si substrates, nanowires can be positioned.[35] The silicon substrates are patterned with a 5 nm thick Fe film by electron beam evaporation and lithography and the SiNWs selectively grown onto them. By heating pure Si powder at 1373 K under Ar flow onto a quartz substrate coated with $Fe(NO_3)_3$, it has been possible to obtain Si and SiO_x ($x = 1$ to 2) nanostructures.[36] The products include fist-capped SiO_x fibres (Si core), tree-like SiO_x nanofibres and tadpole-like SiO_x nanofibres.

Carbon-assisted synthesis of single-crystal SiNWs has been accomplished with Si powder as well as solid Si substrates.[37] Here carbon is coated lightly on Si and the material heated to ~1300 °C. The carbon reacts with the surface oxide to produce the suboxide species.

The VLS method has been successful in producing SiNWs in large quantities by a low temperature route.[38] SiNWs with uniform diameters of ~6 nm were

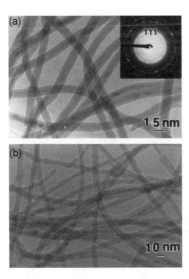

Figure 3.9 (a) *TEM image of the SiNWs with an average diameter of around 15 nm. Inset: the SAED.* (b) *TEM of the SiNWs after etching the outer oxide layer in dilute HF*
(Reproduced from ref. 33)

synthesized using Ga as the molten solvent at below 400 °C in a hydrogen plasma. Defect-free SiNWs with diameters in the range 4–5 nm and lengths of several microns were synthesized using a supercritical fluid solution-phase approach wherein alkanethiol-coated Au nanocrystals (2.5 nm in diameter) were used as seeds to direct the 1D crystallization of Si in a solvent heated and pressurized above its critical point.[26] The reaction pressure controlled the orientation of the nanowires. Single-crystal SiNWs with diameters approaching molecular dimensions are obtained by using gold nanocluster-catalysed 1D growth.[39a] Branched and hyper-branched SiNWs are obtained by a multi-step nanocluster catalysed VLS method.[39b]

Application of a voltage between a Si substrate and an Au STM tip[40] produces SiNWs. The most common technique, however, is laser ablation, which affords high-purity, crystalline nanowires in high yields.[41] These have diameters ranging from 3 to 43 nm with lengths extending up to a few hundred microns. TEM studies show them to possess a high density of structural defects, which may play a role in the formation of the SiNWs and in the determination of the morphology.[42] The diameters of the nanowires synthesized using laser ablation change with the ambient gas.[43]

Laser ablation has been combined with the VLS method to synthesize semi-conductor nanowires.[44] In this process, laser ablation is employed to prepare nanometric catalyst clusters that define the size of the Si/Ge nanowires produced by the VLS growth. In Figure 3.10(a) we show a TEM image of SiNWs obtained by the ablation of a $Si_{0.9}Fe_{0.1}$ target at 1200 °C, with diameters of ~10 nm and lengths above 1 μm. The presence of the catalyst particles at the ends of nanowires suggests that they grow by the VLS mechanism. An oxide layer, as evidenced from the TEM image in Figure 3.10(b), sheaths the nanowires. The inset shows that the nanowires are single-crystalline. The HREM image in Figure 3.10(c) reveals that the nanowires grow along the ⟨111⟩ direction. The use of targets of Si mixed with SiO_2 appears to enhance the formation and growth of SiNWs obtained by laser ablation.[10,45] SiO_2 plays a more important role than the metal in the laser ablation synthesis of SiNWs. To describe the formation of SiNWs by laser ablation, a cluster–solid mechanism has been proposed.[46] SiNWs (diameter ≥5 nm, length ~μm) have been fabricated with metal- and SiO_2-catalysis assisted by laser ablation.[47] In the catalytic growth of single-crystalline SiNWs by pure metal catalysts (Fe, Ru, and Pr), (111) is the most stable plane and the wire growth axis is along ⟨111⟩. The growth mechanism follows a VLS process, and the synthesized SiNWs typically have metal-tips composed of $FeSi_2$, $RuSi_3$, $PrSi_4$ and such solid solutions.

SiNWs have been grown on Si(111) by the VLS process using silane as the Si source and Au as the mediating solvent.[48] The wires so obtained were single crystalline, exhibiting growth defects such as bends and kinks. Using well-defined Au nanoclusters as catalysts for 1D growth *via* the VLS mechanism, SiNWs have been synthesized using SiH_4 as the Si source.[49] The diameters of the nanowires obtained are similar to those of the catalytic Au clusters. Amorphous SiNWs (10–50 nm diameters) have been obtained with Au Pd co-deposited Si oxide substrates by thermal CVD using SiH_4 gas at 800 °C.[50] SiNWs are produced by the Ti-catalysed decomposition of SiH_4 in different atmospheres such as H_2 and N_2.[51]

Figure 3.10   (a) *TEM image of silicon nanowires by the laser ablation of a $Si_{0.9}Fe_{0.1}$ target. Scale bar, 100 nm. (b) TEM image of a single silicon nanowire, showing the crystalline core (dark) and the amorphous SiO_x sheath (light). Scale bar 10 nm. Inset: the SAED pattern. (c) HREM image of the crystalline Si core and amorphous SiO_x sheath. The (111) planes (black arrows) with a spacing of 0.31 nm are oriented perpendicular to the growth direction (white arrow)* (Reproduced from ref. 44)

Thermal evaporation of a mixture of Si and SiO_2 yields SiNWs.[52] The nanowires consist of a polycrystalline Si core with a high density of defects and a silicon oxide shell. Highly oriented, long SiNWs are obtained in good yields on flat silicon substrates by the thermal evaporation of SiO.[53] The SEM images in Figure 3.11 reveal the aligned nature of the nanowires, with the length of the individual nanowires extending up to 1.5–2 mm. SiNWs have also been synthesized by the thermal evaporation of SiO powders without any metal catalyst.[54] Substrate temperature is crucial for controlling the diameter of the nanowires as well as the morphologies resulting from thermal evaporation of SiO powders mixed with 0–1% Fe.[55] Ultrafine SiNWs of diameters between 1 and 5 nm, sheathed with a SiO_2 outer layer of 10–20 nm, were synthesized by oxide-assisted growth *via* the disproportionation of thermally evaporated SiO using a zeolite template.[56] The zeolite restricts the growth of the nanowires laterally.

Dimensionally ordered SiNWs are formed within mesoporous silica using a supercritical fluid solution-phase technique.[57] The mesoporous silica matrix provides a means of producing a high density of stable, well ordered arrays of SiNWs. Ordered SiNWs arrays have been prepared on Si wafers without the use of a template in an aqueous HF solution containing silver nitrate near room temperature.[58]

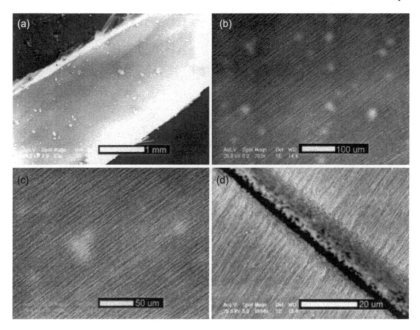

Figure 3.11 (a)–(d) *SEM images of oriented SiNWs at different magnifications* (Reproduced from ref. 53)

Synthesis of $NiSi_2/Si$ and $CoSi_2/Si$ nanowires has been demonstrated on the surface of bare SiNWs using metal vapour vacuum arc implantation.[59] Nanowires of $ScSi_2$, $ErSi_2$, $DySi_2$ and $GdSi_2$ have also been grown on Si (001) substrates, with widths and heights in the ranges 3–11 and 0.2–3 nm respectively.[60] Detailed study of the structural and electronic properties of Gd disilicide nanowires on Si(100) have been made using STM and STS.[61] Free-standing $DySi_2$ nanowires have been formed on Si(001) by self-assembly.[62]

The "grow-in-place" approach to Si nanowire devices uses a silicon precursor gas (*e.g.* SiH_4) to produce self-assembled, electrically contacted, crystalline Si nanowires without any intervening silicon material formation or collection/positioning steps.[63] The approach uses the VLS mechanism and lithographically fabricated, permanent, nanochannel growth templates to control the size, shape, orientation, and positioning of the nanowires and ribbons. These horizontal templates are an integral component of the final devices and provide contacts, interconnects, and passivation/encapsulation. The approach results in self-assembly of the Si nanowires (SiNWs) and nanoribbons (SiNRs) into interconnected devices without any "pick-and-place" or printing steps, thereby avoiding the most serious problems encountered in process control, assembly, contacting, and integration of SiNWs and SiNRs for IC applications.

Various physical methods have been used to characterize SiNWs. SiNWs, when excited with green light, emit red light due to the recombination of the electron–hole pairs across the band gap. Yu *et al.*[33] obtained SiNWs that emit stable blue light unrelated to quantum confinement, which they attributed to the presence of the amorphous silicon oxide over-coating layer. Li *et al.*[56] obtained a strong emission at ~720 nm from nanowires with diameters of <5 nm. Zhang *et al.*[43] obtained different photoluminescence (PL) emissions centred at 624 nm (1.99 eV) and 783 (1.58 eV) depending on the synthetic conditions used, which they attributed to quantum size effects in the thin SiNWs. Raman spectra of SiNWs match those predicted by the quantum confinement model for Si microcrystals.[64] However, the sizes predicted do not match those observed in TEM, possibly because the SiNWs are composed of smaller Si grains. If the size of the grains is taken into account, better agreement is obtained.

Doped SiNWs of n- and p-types have been prepared by introducing B or P dopants during the growth of SiNWs by laser ablation.[65] It is possible to heavily dope SiNWs and approach the metallic regime. Doping of SiNWs by Li has been carried out by an electrochemical insertion method at room temperature.[66] The crystalline structure of the SiNWs, investigated by HREM, was gradually destroyed with increasing Li$^+$ ion dose. Ma *et al.*[67] have performed STM and STS measurements on B-doped and undoped SiNWs. The STM images (Figure 3.12a–d) showed the presence of nanoparticle chains and nanowires in the B-doped SiNWs sample, while STS measurements showed an enhancement in the electrical conductance due to boron doping. B- and P-doped SiNWs were used as building blocks to assemble three types of semiconducting nanodevices.[68] Passive diode structures consisting of crossed p- and n-type nanowires exhibit rectifying transport similar to planar p-junctions. Active bipolar transistors, consisting of heavily and lightly n-doped nanowires crossing a common p-type wire base, exhibit common base

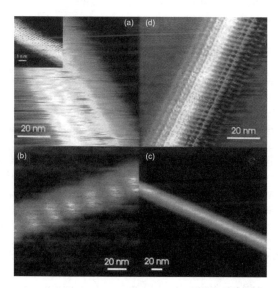

Figure 3.12 *STM images of individual SiNWs: (a) undoped SiNW. Inset: image of an oxide-removed, H-terminated SiNW; (b) a B-doped nanoparticle chain, (c) a B-doped nanowire, and (d) boron-induced reconstruction of SiNW*
(Reproduced from ref. 67)

and emitter current gains as large as 0.94 and 0.16, respectively. Doped nanowires have been used to assemble complementary inverter-like structures.

Catalytic growth of metal–semiconductor junctions between carbon nanotubes and SiNWs has been reported.[69] The junctions exhibit reproducible rectifying behaviour and could act as building blocks for nanoelectronics. Room-temperature Coulomb blockade effects and the influence of a capacitively coupled gate on the transport properties of conducting silicon wires have been studied.[70]

Transport measurements have been carried out on 15–35 nm diameter SiNWs grown using SiH_4 CVD *via* Au/Zn particle-nucleated VLS growth at 440 °C.[71] Using these SiNWs, several types of devices were fabricated, including crossed nanowire devices, 4- and 6- terminal devices, and 3-terminal (gate) devices.[72] The resistivity could be varied from $>10^5 \, \Omega$ cm to $\sim 10^{-3} \, \Omega$ cm based on the nature of the electrical contact (Schottky or Ohmic) and the doping levels.

Intramolecular junctions (IMJs) in SiNWs have been formed from a single growth process.[73] STM shows IMJs formed by fusing two straight wire segments at an angle of 30°, which repeats itself in a regular pattern across the nanowire.

Chemical sensitivity of SiNW bundles has been studied.[74] Upon exposure to NH_3 gas and water vapour, the electrical resistance of the HF-etched SiNWs relative to the non-etched SiNWs decreases, even at room temperature. This phenomenon serves as the basis for a sensor. Sensitive sequence-specific DNA sensors have been fabricated based on SiNWs.[75]

Germanium

Germanium nanowires (GeNWs) with diameters in the 10–100 nm range have been synthesized by the VLS method, using Au clusters as catalysts in a sealed-tube chemical vapour transport system.[76] Melting and recrystallization processes of individual nanowires have been observed by recording the TEM images while heating the nanowires. The growth and nucleation of individual nanowires were monitored within a high-temperature TEM when Ge was evaporated into monodisperse Au clusters, to demonstrate the validity of the VLS growth mechanism at the nanometre scale,[8] as discussed in an earlier section. A mixture of Ge + GeI_4, when sublimed on to Au-coated Si substrates, produces single-crystalline GeNWs with diameters less than 30 nm.[77]

Single-crystalline GeNWs are obtained in high yields by CVD of GeH_4 at 275 °C with Au nanocrystals as seed particles.[78] The SEM image in Figure 3.13(a) shows the nanowires to have diameters of ~ 25 nm and lengths up to tens of μm. The HREM image and the electron diffraction pattern in Figure 3.13(b) show the nanowires to be single crystalline. The nanowires form by the VLS growth mechanism, as evidenced by the presence of catalyst particles at the ends of the nanowires. High yields of GeNWs and core shell structures are obtained by CVD of $Ge(C_5H_5)_2$.[79]

GeNWs, of 10–150 nm diameter and lengths of several microns, were grown in cyclohexane heated and pressurized above its critical point.[80] Alkanethiol-protected Au nanocrystals 2.5–6.5 nm in diameter were used to seed the formation of the wires, which occurs through a solution–liquid–solid mechanism.

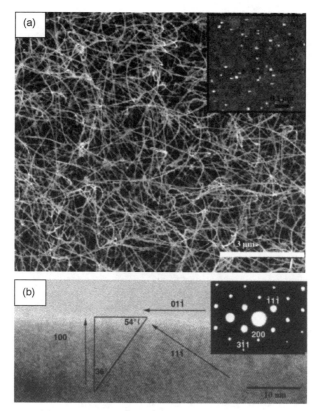

Figure 3.13 (a) *SEM image of GeNWs synthesized by CVD at 275 °C on a SiO$_2$/Si substrate. Inset: an AFM image of Au nanoclusters on the substrate recorded prior to CVD.* (b) *HREM of a single GeNW. Inset: the SAED pattern* (Reproduced from ref. 78)

A supercritical fluid solution-phase method has also been demonstrated for the synthesis of GeNWs within the pores of an ordered mesoporous material.[81] Diphenylgermane was decomposed in hexane at 773 K and 375 bar in the presence of mesoporous silica. Reduction of GeCl$_4$ and phenyl-GeCl$_3$ by Na metal in an alkane solvent at elevated temperature and pressure produces GeNWs with diameters in the range 7–30 nm and lengths up to 10 μm.[82]

High-vacuum electron beam evaporation has been used to synthesize Ge cone-arrays on N$^+$-type Si(100) and Si$_3$N$_4$ using Ti as catalyst.[83,84] The surface morphology of Ti nanocrystal catalyst and Ge cone-arrays was investigated. GeNWs, consisting of a crystalline Ge core and an amorphous GeO$_2$ sheath, have been produced by the laser ablation of a mixture of Ge and GeO$_2$.[85] The crystalline Ge core lies in the axial[211] direction and is terminated by the {111} facets on the surface. Photoluminescence and Raman scattering measurements have been reported on Ge wires formed by self-assembly on Si(113) substrate.[86]

Thermal evaporation of Ge powder at 950 °C onto Au nanoparticles at 500 °C produces GeNWs.[87] The diameters of the nanowires depend on the diameters of

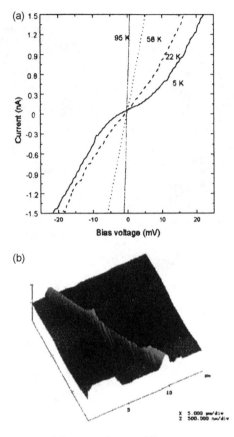

Figure 3.14 (a) *I–V curves of Ge nanowires at different temperatures.* (b) *AFM image of a Ge nanowire device*
(Reproduced from ref. 87)

the catalyst nanoparticles used. Temperature-dependent *I–V* characteristics of a single GeNW with a diameter of 120 nm is shown in Figure 3.14(a). An AFM image is shown in Figure 3.14(b). Transport measurements indicate that the wires are heavily doped during the growth process.

Boron

Films of aligned boron nanowires (BNWs) have been synthesized by radio-frequency (RF) magnetron sputtering of a mixture of boron and B_2O_3 powders in Ar.[88] Figure 3.15 shows SEM images of the BNW arrays peeled off the substrates. These have diameters of ~40–50 nm and grow perpendicular to the substrate. The tips of the nanowires are flat rather than hemispherical. TEM studies reveal that the conventional growth mechanisms are not suitable in this case. A vapour cluster–solid mechanism is proposed for the growth of amorphous BNWs.[89] Magnetron sputtering using a B target in an Ar atmosphere also produces ordered

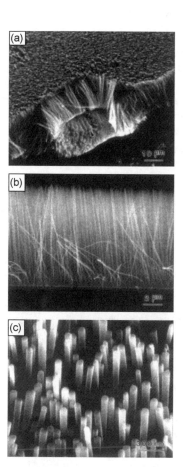

Figure 3.15 *SEM images of the boron nanowire arrays grown on Si substrates. (a) A low-magnification image, showing that the nanowire arrays grew uniformly on the substrate over large areas. The arrowhead shows the root part of the nanowire arrays, which was exposed by peeling operations. (b) Cross-sectional image, showing that the nanowire arrays grew perpendicularly to the substrate surface. (c) High-magnification SEM image, showing that most of the B nanowire tips have a platform-shaped morphology with a diameter of 60–80 nm*
(Reproduced from ref. 88)

BNW arrays.[90] Feather-like BNWs arranged in arrays, with multiple Y- or T-nanojunctions, are produced by using RF magnetron sputtering in Ar.[91] The target used was a mixture of B and B_2O_3 powders.

BNWs are obtained by the laser ablation of B targets at high temperatures.[92] The nanowires have diameters ranging from 30 to 60 nm with lengths of several tens of microns. Laser ablation of a B pellet in a furnace produces B nanobelts that are rectangular in cross section with a width-to-thickness ratio of about 5, several tens of nm to about 150 nm in width and several μm to mm in length.[93] These are well crystallized in a tetragonal structure and have a 2–4 nm thick amorphous sheath. The effect of various factors such as types of catalysts, additives and laser beam intensity on the morphology and size of BNWs prepared by laser ablation have been examined.[94] BNWs have been aligned on Au-coated Si substrates.[95]

B_2H_6 in Ar passed over NiB at 1100 °C yielded crystalline BNWs,[96] 20 to 200 nm in diameter and several microns long. The nanowires were semi-conducting and have properties akin to those of elemental boron.

Superconducting MgB_2 nanowires ($T_c \sim 33$ K) with diameters between 50 and 400 nm have been prepared by the reaction of BNWs with Mg vapour.[97]

In, Sn, Pb, Sb and Bi

The growth of In on a Si(001) $2 \times n$ nanostructured surface has been investigated by *in situ* STM.[98] The deposited In atoms predominantly occupy the normal 2×1 dimer-row structure, and develop into an uniform array of In nanowires. Long-chain amines have been used as templates for the synthesis of In nanowhiskers from InC_p ($C_p = C_5H_5{}^{-1}$).[99] These workers also synthesized nanowires of In_3Sn.

β-Sn nanowires surrounded by graphitic material, ≤ 100 nm in diameter and ≤ 2 μm long are produced by passing a current between graphite rods immersed in a molten mixture of LiCl and $SnCl_2$ under Ar at 600 °C.[100]

Pb nanowire arrays have been fabricated in an AAM, by anodization of a pure Al foil and subsequent electrodeposition of Pb.[101] The nanowires are single-crystalline with an average diameter of 40 nm. The nanowire arrays embedded in the AAM can only transmit polarized light vertical to the wires. Large-scale synthesis of single-crystal Pb nanowires has been achieved by refluxing lead acetate with poly(vinylpyrrolidone) (PVP) in ethylene glycol.[102] A paramagnetic Meissner effect is observed in Pb NWs[103] and superconducting arrays of the nanowires have been fabricated.[104]

Single-crystalline Sb nanowire arrays are obtained by pulsed electrodeposition in AAM.[105] Figure 3.16(a) shows a field-emission SEM image of an array after the AAM was partially etched. Figure 3.16(b) shows the degree of filling of the template. The nanowires have diameters of ~ 40 nm (Figure 3.16c). As revealed by the XRD pattern in Figure 3.16(d), the nanowires grow along the [11$\bar{2}$0] direction.

Bismuth nanowires (BiNWs) can be extruded at room temperature from the surfaces of freshly grown composite thin films consisting of Bi and chrome-nitride.[106] The nanowires range from 30 to 200 nm in diameter and are up to several mm long. Highly oriented hexagonal arrays of parallel Ni and Bi nanowires with diameters ~ 50 nm and lengths of up to 50 μm have been synthesized by electro-deposition in AAMs.[107] Reduction of sodium bismuthate with ethylene glycol in the presence of PVP also yields BiNWs.[108] $Bi_{1.7}V_8O_{16}$ and NH_3 react at 450 °C to yield BiNWs,[109] as also does the decomposition of $Bi[N(SiMe_3)_2]_3$ in the presence of a PVP related polymer [poly(1-hexadecene)$_{0.67}$-*co*-(1-vinylpyrro-lidinone)$_{0.33}$].[110]

Single-crystal BiNWs are formed inside carbon SWNTs by capillary filling.[111] Bismuth is introduced as a gas, solution or solid, the solution phase process being the most efficient method. Size-dependent electrical resistivity of BiNWs has been reported.[112]

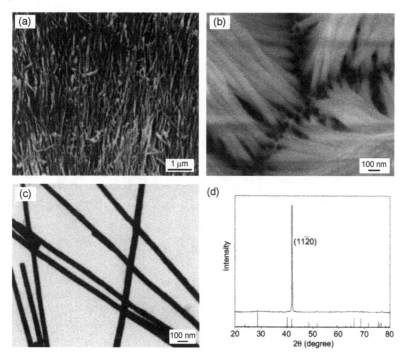

Figure 3.16 (a) *Field emission SEM image, showing the general morphology of the Sb nanowire array.* (b) *Field emission SEM image, showing the degree of filling of the template and the height variation of the nanowires.* (c) *TEM image of the Sn nanowires and* (d) *XRD pattern of the Sb nanowire array; the sole diffraction peak indicates the same orientation of all the nanowires* (Reproduced from ref. 105)

Se and Te

Laser ablation of Se powder produces Se nanorods of different sizes.[113] By controlling the experimental conditions, Se nanorods with diameters ranging from 20 nm to several hundred nm and lengths up to 10 μm have been obtained. Hexagonal Se nanowires were synthesized by simple thermal evaporation of Se powder with the assistance of Si powder as source material (catalyst) at 950 °C.[114]

Making use of a soft, solution-based method, nanowires of trigonal Se with controllable diameters, 10–800 nm, and lengths up to hundreds of microns have been prepared.[14] Selenous acid, on refluxing with hydrazine at a suitable temperature, forms Se nanoparticles that act as seeds for the growth of the nanowires. Figure 3.17(a) and 3.17(b) show Se nanowires obtained by this method. The reaction, when carried out in ethylene glycol, also yields nanowires, as evidenced from Figure 3.17(c) and 3.17(d). Optical properties of the nanowires as well as their photoconductivity have been studied. In aqueous solution, Se molecules produced from the decomposition of selenodigluthathiones continually stack on previously formed α-monoclinic Se nanoparticles along the $\langle 001 \rangle$ direction, gradually producing α-monoclinic Se nanowires.[115]

Refluxed at ~110 °C   Refluxed at ~130 °C

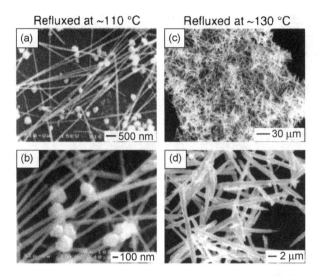

Figure 3.17   (a), (b) *SEM images of t-Se nanowires that are ~100 nm in diameter, prepared by refluxing the solution at 110 °C. Some a-Se colloidal nanoparticles
are still present in the sample.* (c), (d) *SEM images of t-Se nanowires with
diameters of ~800 nm, prepared by refluxing at 130 °C. The solvent used was
ethylene glycol instead of water*
(Reproduced from ref. 14)

Selenium nanowires are produced by the reduction of selenate (SeO_4^{2-}) by the
protein cytochrome C_3.[116] Nanoparticles of Se precipitate from an aqueous solution
at room temperature, followed by spontaneous self-assembling into nanowires.
Trigonal Se nanowires have been converted into single-crystalline nanowires of
Ag_2Se by reacting with aqueous $AgNO_3$ solution at room temperature.[22]

Selenium nanorods have been prepared at room temperature using Se powder
and $NaBH_4$ in an aqueous medium.[117] Se initially reacts with $NaBH_4$ to give
NaHSe, which in turn can be used as an aqueous Se^{2-} source. The NaHSe solution,
within a few hours, disproportionates to give a red dispersion of α-Se particles. On
ageing this solution for one week, t-Se nanorods with dimensions of ~100 nm and
several microns in length can be obtained. Under solvothermal conditions, this
chemical procedure yields nanobelts and other nanostructures. By a similar procedure, starting from Te powder, Te nanowires, nanobelts and other nanostructures
have been obtained (Figure 3.18).[118] Decomposition of $TeCl_4$ in the presence of
trioctylphosphine oxide (TOPO) in polydecene solution at 250–300 °C gave
TeNWs.[119] TeNWs have also been prepared by microwave-assisted synthesis.[120]

Single-crystal Te nanorods (14 nm diameter and 300 nm long) have been prepared through the reduction of $[TeS_4]^{2-}$ by SO_3^{2-} using the surfactant sodium
dodecyl benzenesulfonate.[121] Adjusting the concentration of the reactants permits
control over the diameters of the nanorod. The nucleation and growth processes are
understood in terms of the solid–solution–solid transformation and surfactant-
assisted growth mechanism. A solution-phase approach has been used to synthesize
nanorods of Se/Te alloys.[122]

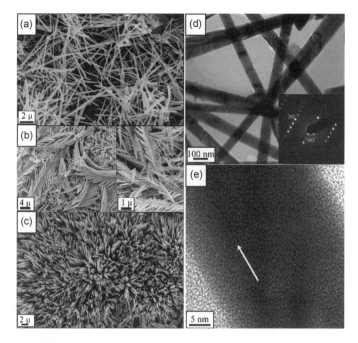

Figure 3.18 *SEM images of* (a) *the* t-*Te nanorods obtained by reacting 0.03 g of Te with 0.05 g of NaBH₄ in 20 mL water* (b) *feather-like structures obtained with high reactant concentrations (0.3 g of Te powder in 20 mL of water) and of* (c) t-*Te nanoflowers* (d) *TEM image of the* t-*Te nanorods obtained by using the reaction conditions as in* (a). *Inset: electron diffraction pattern of a nanorod, showing its single crystalline nature.* (e) *HREM image of a nanorod, showing the [001] lattice planes of hexagonal Te*
(Reproduced from ref. 118)

Gold

Gold nanorods with high aspect ratios of up to 200 have been synthesized in high yields by a seed-mediated growth process.[123] Figure 3.19 shows a TEM image of the nanorods. The first step involves the formation of Au nanoparticles by the citrate route, followed by the addition of cetyltrimethylammonium bromide (CTAB), HAuCl₄, ascorbic acid and NaOH. The spontaneous reduction of the chloroaurate ions present in the Langmuir monolayers of 4-hexadecylaniline leads to the formation of highly oriented, flat Au sheets and ribbon-like nanocrystals bound to the monolayer.[124] This one-step process constrains the growth of the Au nanocrystals to within the plane of the Langmuir monolayers. Recently, liquid–liquid interfaces have been used to make nanocrystalline films of Au and other metals by taking the metal organic precursors in the organic layer and the reducing agent in the aqueous layer.[125] By thermolysis of HAuCl₄ impregnated in alumina membranes gold nanorods are obtained,[126] the acetone present in the solution giving C-Au composite nanotubes.

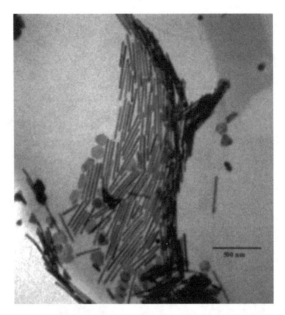

Figure 3.19   *TEM image of gold nanorods after one round of purification*
(Reproduced from ref. 123)

Single-crystalline Au nanowire arrays are electrochemically fabricated within AAMs with pore diameters of 35–100 nm.[127] AAMs containing 200 nm diameter pores have also been replicated electrochemically with Au and Ag to make free-standing nanowires several microns long.[128] The *I–V* characteristics of the nanowires show current rectifying behaviour. An electric-field assisted assembly has been described to position individual nanowires suspended in a dielectric medium between two electrodes defined lithographically on a silica substrate.[129] This approach has facilitated rapid electrical characterization of nanowires of 350 and 70 nm diameter, having room-temperature resistivities of ~2.9 and 4.5×10^{-6} Ω cm. Porous polycarbonate membranes can be employed for the synthesis of Au nanowires as well.[130] AFM has enabled the *I–V* characteristics to be obtained. Au nanowires and nanotubes have been prepared by electrodeless deposition of Au onto the pore walls of a porous polymeric membrane.[131]

Nanowires synthesized using AAMs have been assembled using DNA.[132] Oligonucleotides were adsorbed as monolayer coatings on these wires through Au-thiol linkage. Duplexes are formed between the strands on the nanowires and on Au-coated glass slides bound the two surfaces together. Tris(hydromethyl)-phosphine-capped Au nanoparticles that bind to DNA have been used to synthesize Au nanowires.[133] The particles bound to DNA are immobilized on silicon, followed by electroplating, to obtain nanowires of 30–40 nm diameter and low conductivity (compared to bulk gold). Biological molecules such as histidine-rich peptides have also been used as templates to synthesize Au nanowires.[134] Self-assembly of dodecanethiol-capped Au nanoparticles in a matrix of dipalmitoylphosphatidylcholine at the air–water interface yield continuous Au nanowires resembling a molecular electronic circuit board.[135]

Carbon nanotubes are effectively used as templates for the self-assembly and thermal processing of Au nanowires.[136] Nanowires of several metals such as Au, Ag, Pt and Pd have been produced in the capillaries of single-walled carbon nanotubes (SWNTs) by means of sealed tube reactions.[23] The Au nanowires had diameters between 1–1.4 nm and lengths of 15–70 nm. Figure 3.20 shows TEM images of the Au nanowires inside SWNTs obtained by this route. These are single-crystalline with a spacing of ~0.23 nm, corresponding to the (111) planes. They melt around 350 °C, giving nanoparticles as products.

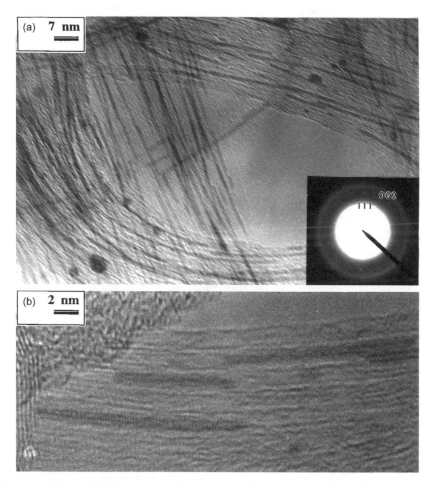

Figure 3.20 (a) *TEM image of Au nanowires inside carbon SWNTs obtained by a sealed tube reaction. Inset: corresponding SAED.* (b) *HREM of the Au nanowires, showing them to be single-crystalline*
(Reproduced from ref. 23)

Helical, multi-shell Au nanowires are formed under UHV conditions by an electron beam thinning technique.[137] By irradiating thin organic films containing Au or Ag nanoparticles with a series of ultrashort laser pulses, wire arrays containing more than 30 nearly parallel wires, several μm long, are obtained.[138]

On excitation with an intense pulsed laser, Au nanorods change shape.[139] The final products depend on the energy of the laser pulse as well as on the width. The shape transformations are followed by changes in the plasmon absorption and TEM. Visualization of 2D confined suspensions of Au nanorods has been reported[140] and their diffusion coefficients determined by particle tracking measurements. The diffusion coefficient is affected by the surface chemistry of the nanowires, the substrate and by the dimensions of the nanowires. Spin coating of thiol-passivated Au nanoparticles onto silicon produces nanostructured cellular networks.[141] On annealing between 500 and 600 K, the cellular morphology of the nanocrystalline foam is preserved. A nano-replication method for the preparation of an artificial template for Au nanowire self-assembly has been demonstrated.[142]

A pseudo-1D network of AuNWs separated by 1–3 nm and interconnected by thiols has been fabricated by lipid-templated self-assembly.[143] Thiolcarboxylic acids assemble gold nanorods, giving rise to uniaxial plasmon coupling (Figure 3.21).[144] Templateless assembly of Au nanowire networks has been achieved by agitating nanoparticles in a toluene–water mixture.[145]

Electronic transport properties of Au nanowires have been investigated experimentally and also based on theoretical analysis.[146] Resistivities of polycrystalline Au nanowires have been measured at some locations connected through small grain boundaries.[147] These nanowires were obtained by high-resolution microcontact printing of a self-assembled monolayer as a resist, followed by selective etching.

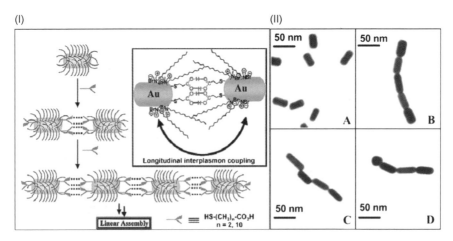

Figure 3.21   (I) *Schematic representation of the linear self-assembly of gold nanorods. Inset: intermolecular hydrogen bonding between 3-mercaptopropionic acid (MPA) on adjacent nanorods. (II) TEM images of (A) Au nanorods in the absence of MPA and (B–D) linearly self-assembled Au nanorods in the presence of MPA*
(Reproduced from ref. 144)

The electron relaxation dynamics and the thermal cooling of Au nanorods and truncated tetrahedra in air and water have been studied, after excitation with femtosecond laser pulses.[148] The possibility of using Au nanowires as building blocks for self-assembling logic and memory circuits has been discussed.[149] A method of connecting nanowires consisting of ligand stabilized Au_{55} clusters with metal arrays prepared by using metal evaporation through a mask of monodisperse latex beads has been developed.[150] The yield strength and yield mechanisms of AuNWs have been investigated by atomistic simulations.[151,152]

Silver

Using seed-mediated growth in micellar media, silver nanorods and nanowires have been prepared.[153] In the first step, Ag nanoparticles of ~4 nm diameter were prepared by the citrate route. Silver nitrate was then reduced in the presence of the Ag seeds, CTAB, NaOH and ascorbic acid. The rods, wires and spheres were separated by centrifugation. Figure 3.22 shows a TEM image of the Ag nanowires (AgNWs) obtained by this method. These are between 1 and 4 μm long with an aspect ratio 50–350. AgNWs with sizes of ~35 × 300 nm have been synthesized in a lamellar liquid crystalline alignment of oleate vesicles by UV irradiation under ambient conditions.[154] Passivation of oleate amphiphiles on the surface of Ag nanoparticles is utilized for the nucleation and oriented growth of AgNWs. A self-seeded process has been used to obtain AgNWs, in which $AgNO_3$ and poly(vinylpyrrolidone) (PVP) solutions were simultaneously injected into refluxing ethylene glycol through a two-channel syringe pump.[155] Ag nanoparticles formed initially serve as seeds for the subsequent nanowire growth. A polymer–surfactant-assisted-sandwiched reduction route, in which the sandwich organometallic compound acts as a reducing agent, permits the formation of elemental Ag along a certain axis in the xy-plane.[156] The curl and extension of poly(ethylene glycol) in a

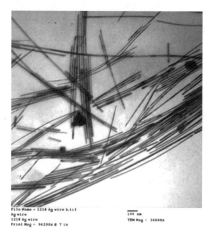

Figure 3.22 *TEM images of Ag nanowires obtained using the procedure outlined in ref. 153* (Reproduced from ref. 153)

mixed solvent provide microemulsions for confining the 1D growth. A solution-phase approach has been used to synthesize large-scale quantities of AgNWs with diameters in the 30–40 nm range and lengths of ~50 μm.[157] Transport property measurements at room temperature indicate that the nanowires are electrically continuous with a conductivity of 0.8×10^5 S cm^{-1}.

A reduction method has been used to synthesize long, straight and continuous AgNWs by using nanocrystalline AgBr and a developer (containing *N*-methyl-paminophenol, citric acid, AgNO$_3$ solution and water) as precursors, in the presence of gelatin.[158] UV irradiation of AgNO$_3$ yields Ag nanorods and dendritic supramolecular nanostructures at room temperature using poly(vinyl alcohol) as the protecting agent.[159] Ag wires of 0.4 nm diameter and micrometre length were grown inside the pores of self-assembled calix[4]hydroquinone nanotubes by electro-/photochemical redox reaction.[160] The wires are stable in aqueous environments and occur as coherently oriented 3D arrays of ultrahigh density. Double-hydrophobic block copolymers enable the synthesis of AgNWs in aqueous solutions without the use of UV irradiation or electrochemistry.[161] Crystalline AgNW arrays have been fabricated by electrodeposition from a reverse hexagonal liquid-crystalline phase containing 1D aqueous channels.[162] A high electric field applied during the electrodeposition helps to align the liquid-crystalline phase.

Single-walled carbon nanotubes have been filled with AgBr–AgCl without causing significant chemical or thermal damage. Exposure to light or electronic beam results in the reduction the silver halide to give continuous metallic silver nanowires within the tubules.[163] Figure 3.23(a) and 3.23(b) show a structural representation and a HREM image of an empty unfilled SWNT, respectively. Figure 3.23(c) shows a HREM image of the SWNT filled with a KCl–UCl$_4$ eutectic mixture. Figure 3.23(d) gives a HREM image of a large diameter (~3.8 nm) SWNT filled with 17 layers of silver metal. The initial filling medium in this case was AgCl. A HREM image of the product of SWNT filling with AgBr within a bundle of (10,10) SWNTs is shown in Figure 3.23(e). The paths of empty and filled SWNTs are indicated by dark and light arrows respectively. Zeolites have also been used to synthesize AgNWs.[164] The Ag$^+$ ions are exchanged with zeolites and treated with a solution of AgNO$_3$ at 260 °C to further introduce Ag$^+$ ions into the zeolite pores. On exposure to the electron beam, AgNWs are formed. Highly ordered AgNW arrays are obtained by the pulsed electrodeposition in self-ordered porous alumina templates.[165] Figure 3.24 shows SEM images of silver-filled alumina membranes. Electrochemical plating into monodomain porous alumina templates also yields AgNWs with high aspect ratios.[166] The nanowires have lengths greater than 30 μm with diameters in the 180–400 nm range.

Plasmon resonance of AgNWs with a nonregular cross-section on the 20–50 nm diameter range has been investigated.[167] Atomic arrangements and quantum conductance of AgNWs generated by mechanical elongation have been analysed.[168] The structural behaviour of the nanowires has been employed to interpret quantum conductance data. Quantum transport properties of ultrathin silver nanowires have also been theoretically investigated.[169] A method to synthesize Ag/SiO$_2$ coaxial nanocables, 2–100 nm thick and up to ~50 μm long, has been developed.[21]

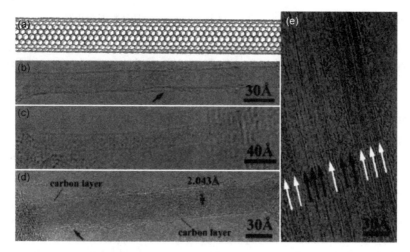

Figure 3.23 (a) *Schematic representation of a SWNT.* (b) *HREM image showing an empty SWNT. The defect region is shown by the arrow.* (c) *SWNT filled with a KCl–UCl₄ eutectic mixture.* (d) *Wide capillary SWNT filled with Ag metal formed by capillary insertion of AgCl followed by photolytic decomposition. The indicated d-spacing corresponds to the (020) lattice planes of Ag metal.* (e) *Incorporation of AgBr into an SWNT bundle. Filled SWNTs are indicated by light arrows and unfilled by dark arrows* (Reproduced from ref. 163)

These nanostructures were prepared by coating bicrystalline AgNWs with sheaths of silica through a sol–gel process.

Thin films containing interwoven bundles of single-crystalline AgNWs of 35 nm diameter have been synthesized by chemical reduction of AgNO₃ in poly-(methacrylic acid) solutions.[170] Reduction of AgCl by glucose at 180 °C yields uniform AgNWs.[171] Mesoporous silica has been used to obtain highly ordered nanowires of Ag, Ni and other nanowires.[172] The mesostructures are unusual with coaxially multilayered helical and stacked donut structures and exhibit hierarchical

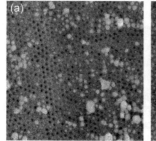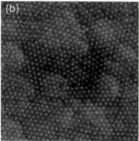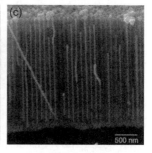

Figure 3.24 *SEM images of silver-filled alumina membranes:* (a) *Top view of an unthinned sample,* (b) *top view of the same sample ~200 nm underneath the initial surface and* (c) *side view of a fracture* (Reproduced from ref. 165)

organization and novel surface properties. AgNW arrays prepared in AAMs by electrodeposition exhibit plasmon resonance depending on the direction of the incident light.[173] Arrays of heterojunctions of AgNWs and CNTs have been prepared using alumina membranes[174] while close-packed monolayers of aligned AgNWs have been prepared by the Langmuir–Blodgett technique.[175]

Iron and Cobalt

Large arrays of oriented ferromagnetic iron nanorods have been fabricated by aqueous chemical growth, without the use of a template or surfactant and without applying electric or magnetic fields.[176] The method involves the growth of β-iron oxyhydroxide nanorods from an aqueous ferric chloride solution onto substrates, followed by reduction in a hydrogen atmosphere at mild temperatures. Fe nanowires (FeNWs) have been grown in a mesoporous silica matrix by the decomposition of $Fe(CO)_5$ by UV radiation.[177] Magnetic susceptibility measurements indicate superparamagnetic properties of these samples. Local spin-density-functional theory has been applied to describe the structural and magnetic properties of FeNWs consisting of chains of single atoms.[178] An unsupported isolated wire appears to be unstable with respect to both dimerization and bending.

Oxidation-resistant FeNWs encapsulated in aligned carbon nanotube bundles have been synthesized by a simple route[179] based on the pyrolysis of a ferrocene–hydrocarbon mixture, affording aligned CNTs that contain a good proportion of encapsulated FeNWs. TEM observations reveal these FeNWs to be single-crystalline. Magnetization measurements show Barkhausen jumps, similar to those observed with amorphous iron nanowires. A typical hysteresis curve, with a M_s of 24 emu g^{-1}, exhibiting such features is shown in Figure 3.25(a). The jumps with 5 emu g^{-1} steps in magnetization arise from the magnetization reversal in the encapsulated iron nanowires or depinning of large domains on increase of the magnetization field. On repetition of the magnetization measurement on the sample, the shape of the hysteresis curve shows slight changes (Figure 3.25b). Magnetization reversal in 2D arrays of parallel ferromagnetic FeNWs embedded in AAMs has been examined.[180] By combining bulk magnetization measurements with field-dependent measurements, the macroscopic hysteresis loop has been decomposed in terms of the irreversible magnetization response of individual nanowires. Self-assembled FeNWs are obtained by the thermal decomposition of $La_{0.5}Sr_{0.5}FeO_3$ under reducing conditions.[181a] The magnetic structure (texture) of ordered FeNWs, prepared by electrodeposition in alumina membranes,[181b] becomes weaker on increasing the diameter of the nanowires, the small diameter samples showing high coercive fields when the field is applied along the wire axis.

Ultrahigh-density arrays of nanopores with high aspect ratios have been obtained by using the self-assembled morphology of asymmetric block copolymers.[182] Electrodeposition enables the formation of vertical arrays of Co nanowires (CoNWs) with densities in excess of 1.9×10^{11} wires cm^{-2}. Polyaniline nanotubules supported with alumina membranes have been used as templates for the synthesis of CoNWs.[183] SEM images of the polyaniline nanotubules and polyaniline nanotubules filled with CoNWs are shown in Figure 3.26. The alumina layer is

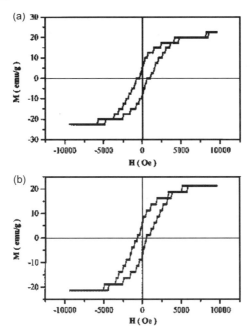

Figure 3.25 (a) *Hysteresis loop, showing multiple steps due to the magnetization reversal in interacting iron nanowires encapsulated in carbon nanotube bundles.* (b) *Hysteresis loop obtained with the same sample as in* (a) *on repeating the experiment*
(Reproduced from ref. 179)

removed by dissolution in NaOH solution. Figure 3.27 shows two hysteresis loops of the cobalt nanowire arrays with the membrane support, when the field is applied parallel and perpendicular to the nanowires. The array exhibits uniaxial magnetic anisotropy with the easy axis parallel to the nanowires: the magnetization being

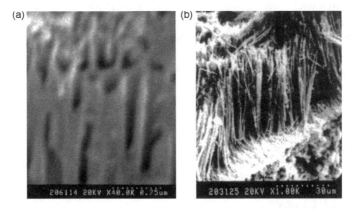

Figure 3.26 *SEM images of* (a) *polyaniline nanotubules and* (b) *polyaniline nanotubules filled with Co nanowires obtained after dissolution of alumina*
(Reproduced from ref. 183)

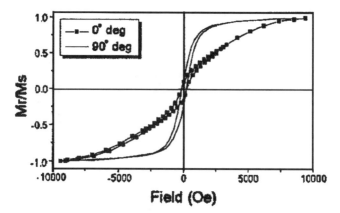

Figure 3.27 *Hysteresis loop of a supported array of Co nanowires varied for various angles between the applied magnetic field and the membrane plane* (Reproduced from ref. 183)

perpendicular to the membrane. The coercivities obtained for the nanowires are also enhanced compared to bulk Co. Electrodeposition into the pores of poly-carbonate membranes also yields CoNW arrays.[184] A magnetic field applied during deposition, parallel or perpendicular to the membrane plane, controls the wire growth. Magnetic properties of an ordered array of concentric composite nano-structures of ZrO_2 nanotubules and CoNWs containing continuous and uniform CoNWs in the ZrO_2 nanotubules have been reported.[185] The ZrO_2 envelope protects the metal nanowires from oxidation and corrosion.

Using STM, the growth of Co onto Cu(110) surfaces can be altered by first terminating the Cu(110) surface with an atomic N-induced (2×3) structure.[186] Growth of Co on such surfaces results in ordered arrays of nanowires. Magneto-transport measurements on single CoNWs fabricated by electrodeposition in nanoporous membranes and contacted by an electron beam lithographic technique show inhomogeneity of the magnetization reversal processes along the wire.[187] Magnetoresistance (MR) of single CoNWs, prepared by electron beam lithography, shows anisotropy due to the reversal process.[188] Monte Carlo simulations suggest magnetization distribution during the reversal process, revealing different mechanisms, depending on the width of the wire. Field emission properties of vertically aligned CoNW arrays seem to promising for potential applications.[189]

$Fe_{1-x}Co_x$ $(0 \leq x \leq 1)$ nanowires have been self-assembled by electrodeposition in porous alumina films and magnetic studies performed on them.[190] Ordered ferromagnetic–nonmagnetic alloy (Co-Cu, Co-Ag, Fe-Ag) nanowire arrays embedded in the nanochannels of anodic alumina membranes have also been fabricated by electrodeposition and their magnetic properties investigated.[191]

Nickel and Copper

Ni nanowires (NiNWs), ~4 nm diameter, have been formed in CNTs with an inner tubule diameter of 20 nm.[192] Carbon was initially deposited into the channels of

Figure 3.28 *Bright-field TEM images at different areas for carbon tubes/nickel nanowire composites prepared by MOCVD*
(Reproduced from ref. 192)

porous AAM followed by MOCVD of nickelocene. The templates were dissolved in NaOH to obtain NiNWs in the CNTs. Figure 3.28 shows TEM images of the CNT/Ni NW composites. The magnetic behaviour of periodic arrays of NiNWs embedded in an ordered alumina pore matrix has been characterized.[193] Reducing the diameter of the nanowires from 55 to 30 nm, while keeping the interwire distance constant, leads to increasing coercive fields (from 600 to 1200 Oe) and remanence (from 30 to 100%). Ni and Co wires of diameters 35–500 nm have been electrodeposited into the cylindrical pores of track-etched polymer membranes and their magnetic properties studied.[194] The study shows intrinsic differences between the magnetization reversal mechanism in the two systems. Static and dynamic aspects of magnetization reversal in nanowire arrays of Fe, Co and Ni produced by electrodeposition in porous AAM templates have been examined along with the magnetization properties.[195]

Ni nanorods have been synthesized by a solution-based route by using hexadecylamine as the shape-controlling agent.[196] Increasing the concentration of the amine favours the formation of monodisperse nanorods.

NiNW arrays with a diamond-shaped cross-section are produced in nanoporous single mica crystal membranes by electrodeposition.[197] The wires have diameters of 120 nm and lengths of 5 μm. Magnetization anisotropy is found in the wires due to the quasi-1D shape and the diamond cross section. Magnetic nanocontacts are obtained by a templating method involving the electrodeposition of Ni within the pores of track-etched polymer membranes.[198] At room temperature, the electrical conductance shows quantization steps in units of e^2/h, as expected of ferromagnetic metals without spin degeneracy.

Polarization-dependent scattering of light has been reported from homogeneous, multi-segment Ag, Au and NiNWs.[199] Inelastic light scattering has enabled a study of the dynamic properties of uniform 2D arrays of NiNWs.[200] Polycrystalline Ni wires, with diameters of ~100 nm and resistance up to 20 Ω, have been prepared by the controlled etching of microscopic wires, and the magnetoresistance properties studied.[201] Highly ordered composite Ni-Cu nanowire arrays have been electro-deposited into the pores of anodic alumina templates and characterized by various techniques.[202]

Adelung et al.[203] report that the deposition of Cu onto layered crystals (VSe_2) induces the formation of self-assembled nanowire networks, which are stable under ambient conditions. Cu deposition induces the formation of nanotunnels on the layered-crystal surface. High aspect ratio Cu nanowires encapsulated in poly-(dimethylsiloxane) are obtained under mild conditions, by the solvent-free reduction of solid CuCl by $(Me_3Si)_4Si$ vapour.[204] The nanowires grow by the VS mechanism. Cu metal has been deposited onto DNA, forming nanowire-like structures that are ~3 nm high.[205] DNA is first aligned on a Si substrate followed by treatment with $Cu(NO_3)_2$, when Cu^{2+} ions become associated with the DNA. On reduction by ascorbic acid, a metallic Cu sheath is formed around the DNA. Ultrahigh density arrays of Cu and Ni nanowires have been synthesized inside ordered porous inorganic materials through a Pd-catalysed electrodeless deposition process.[206] Pd nanoparticles were introduced into the pores of mesoporous SBA-15 followed by the electroless deposition of Ni and Cu to yield nanowires inside the channels of the mesoporous material.

The seedless growth of free-standing single-crystalline Cu nanowires has been achieved by a CVD process below 300 °C, using Cu(I) complexes such as $Cu(etac)[P(OEt)_3]_2$ (etac = ethyl 3-oxobutanoate, Et = ethyl) as a precursor at low temperatures.[207] A template synthesis of large-scale Y-junction Cu nanowires has been reported.[208] A Y-shaped nanochannel anodic alumina template is first prepared by the anodization of aluminium in which Cu is electrodeposited to form the Y-junction nanowires. The SEM images in Figure 3.29 reveal Y-junction Cu nanowires fabricated by this method. Thermal expansion of CuNW arrays in alumina membranes has been measured. The nanowires possess larger lattice parameters and small thermal expansion coefficients compared to bulk Cu.[209]

Other Metals and Alloys

Zinc nanobelts are produced by the carbothermal reduction of ZnS using graphite at 1000 °C under Ar.[210] The nanobelts have widths of 40–200 nm and a thickness of ~20 nm. Nanowires of Zn as well as of Cd, Pb and Co have been obtained by the nebulized spray pyrolysis of precursor compounds (Figure 3.30).[211] This pro-cedure promises to be a continuous process. Nanowires are obtained from a cold

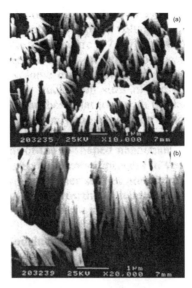

Figure 3.29 *SEM images of Y-junction copper nanowires:* (a) *top view and* (b) *cross-sectional view*
(Reproduced from ref. 208)

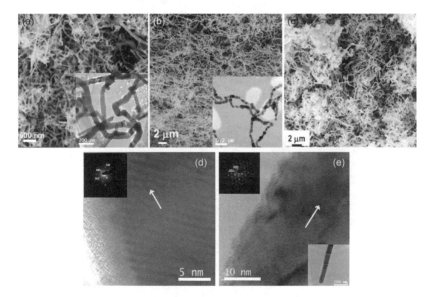

Figure 3.30 *Metal nanowires obtained by the nebulized spray pyrolysis route:* (a) *SEM image of Zn nanowires. Inset: the TEM image.* (b) *SEM image of Co nanowires. Inset: the TEM image of Co necklaces.* (c) *SEM image of Cd nanowires.* (d) *HREM image of Zn, showing the (101) lattice planes. Inset: the SAED pattern of Zn.* (e) *HREM image of Cd nanowires, showing the (100) planes. Inset: top left corner shows SAED pattern of Cd and bottom right corner is the TEM image of a single Cd nanowire*
(Reproduced from ref. 211)

field-emission tip in the presence of a precursor, typically an organometallic compound.[212] Electron emission from the newly grown nanowire tip continues the growth and gives rise to nanowires that are tens of microns long. Nanowires of W, Co and Fe have been grown by this method, along with composite wires. Pyrolysis and carbothermal reduction of lamellar composites of tungsten oxide with CTAB takes place *in vacuo* between 500 and 850 °C, and yields metallic W nanowires with diameters of 20–80 nm and lengths of 5–30 μm.[213] Figure 3.31(a) shows a TEM image of the nanowires. The HREM image of an individual nanowire in Figure 3.31(b) along with the SAED in Figure 3.31(c) reveals the growth direction as ⟨110⟩. EDAX analysis of the core and sheath parts of the W nanowire in Figure 3.31(d) shows that the nanowires are sheathed by an amorphous carbon coating. Treatment of lamellar surfactant/inorganic composite precursors hydrothermally yields crystalline metal nanowires[214] and the method has been used to synthesize nanowires of W, Co, Ni and Cu.

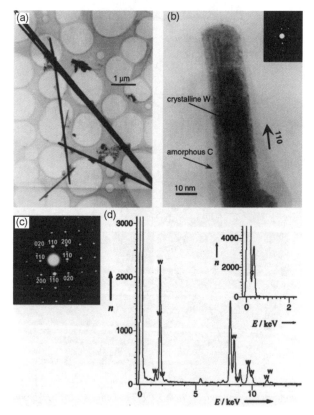

Figure 3.31   (a) *TEM image showing the morphology of W nanowires.* (b) *HREM image of an individual nanowire. Inset: the SAED pattern.* (c) *Diffraction indexes of the SAED pattern in* (b). (d) *Compositions of the core and sheath (inset) parts of the W nanowire shown in* (b), *obtained by EDAX analysis* (Reproduced from ref. 213)

Metallic Mo nanowires 15 to 1000 nm in diameter and 0.5 mm long have been prepared by a two-step procedure.[215] Molybdenum oxide nanowires are electrodeposited selectively at step edges of graphite and then reduced in H_2 at 500 °C to yield Mo nanowires. By a similar strategy, nanowires of metals such as Cu, Ni, Au and Pd have also been fabricated.[216] Ruthenium nanowires, prepared by electrochemical deposition in porous polycarbonate membranes, are metallic, but not superconducting, down to 0.3 K.[217]

Metal thin films composed of ordered arrays of Pd metal nanowires have been fabricated electrochemically within mesoporous silica channels.[218] Large arrays of Pd nanowires with diameters of 75 nm have been produced by a simple technique.[219] Holographic laser interference exposure of a photoresist and anisotropic etching are used to pattern the surface of InP(001) substrates into V-shaped grooves of 200 nm period. Subsequently, the metal is evaporated at an angle onto these substrates, resulting in arrays of parallel nanowires. This method has also been applied for the synthesis of Au, Ni and Ta nanowires. Pd nanowires are formed inside MWNTs during the synthesis of MWNTs on electroplated Pd nanocrystals using a microwave plasma-enhanced CVD system in a mixture of $CH_4 + H_2$.[220] The CNTs are burnt off in an oxygen plasma to obtain nanowires of Pd. Palladium nanowires have also been obtained using mesoporous silica templates.[221]

Platinum nanowires are readily grown inside CNTs.[222] Nanowires of Pt and Pd have been produced in the capillaries of SWNTs by a sealed tube reaction.[23] In Figure 3.32(a), we show the TEM images of Pt nanowires, obtained by the sealed tube reaction, with a diameter of ~1 nm and a length of ~90 nm. Pt nanowires obtained by the solution route are shown in Figure 3.32(b). Pt nanowires have also been synthesized in mesoporous silica by the photoreduction of H_2PtCl_6.[223]

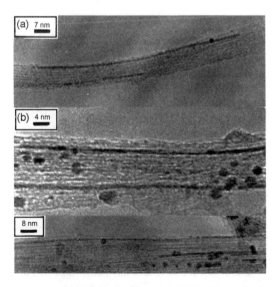

Figure 3.32   *TEM image of Pt nanowires inside SWNTs obtained* (a) *by the sealed-tube reaction and* (b) *by the solution method*
(Reproduced from ref. 23)

Single-crystal nanowires of Pt can be synthesized in large quantities by controlling the reaction rate of the polyol process through the introduction of a trace amount of Fe^{2+} (or Fe^{3+}) species.[224]

FeCo alloy nanowire arrays, prepared by using an AAM template, have been annealed in the 300 to 600 °C range and their structures and magnetic properties studied.[225] Annealed nanowires of $Fe_{69}Co_{31}$ exhibit good coercivity and remanence magnetization.[226] Nanowire arrays of CoNi alloys have also been fabricated, using porous alumina templates, to study their magnetic properties.[227] Ordered ferromagnetic Co-Cu, Co-Ag and Fe-Ag nanowire arrays embedded in nanochannels of AAMs have been fabricated by electrodeposition and their magnetic properties studied as a function of annealing temperature.[228,229]

Necklace-shaped Pt-Rh, Pt-Pd, Pt and Rh nanowires in hybrid organic-inorganic mesoporous materials have been synthesized.[230] AgNWs coated with Ag/Pd alloy sheaths have excellent hydrogen solubility.[231] They can be used as substrates for reversible absorption and desorption of hydrogen.

4 Metal Oxide Nanowires

MgO

MgO nanowires have been obtained at relatively low temperatures (800–900 °C) by a vapour–solid process, by using MgB_2 powder as the precursor.[232] MgB_2 decomposes into Mg and MgB_4 under a constant flow of Ar and the Mg vapour reacts with the traces of O_2 present in the reaction system to form MgO vapour. The vapour, on supersaturation, condenses onto a substrate placed on top of an alumina boat. The different products formed include whiskers, tapered nanowires, whose diameters varied from ~200 to 30 nm over a length of ~50 μm, and nanowires of uniform diameters (15–20 nm) and up to ~30 μm long. Figure 3.33(a) shows a SEM image of the cone-shaped microwhiskers. The SEM image in Figure 3.33(b) reveals tapered nanowires with a much smaller cross section at their base (~300 nm *versus* 5 μm in lateral dimension), their tips growing until a dimension of ~20–30 nm is reached. Figure 3.33(c) shows a TEM image of the middle portion of a tapered nanowire. Within this segment, the diameter of the nanowire reduces from ~200 to ~50 nm over ~20 μm. Figure 3.33(d) and 3.33(e) show SEM and TEM images of nanowires with a uniform diameter of ~15 nm. The SAED pattern of the nanowires in Figure 3.33(f) indicates they are single crystalline with the fcc structure and a growth direction of ⟨001⟩.

A large-scale synthesis of crystalline MgO nanobelts was achieved through a simple evaporation method starting from Mg powder heated in an alumina crucible at 650 °C for 1 h in a constant flow of Ar/O_2 (4:1); the growth of the nanobelts occurs by the vapour–solid mechanism.[233a] A TEM image of two straight MgO nanobelts is shown in Figure 3.34(a). An individual MgO nanobelt has a uniform width along its entire length. The ripple-like contrast observed in the TEM image is due to the strain resulting from the bending of the belt. The HREM image in Figure 3.34(b) shows lattice fringes, and indicates the single crystalline nature of the nanobelt. In this image, the spacing of ~0.208 nm between the arrowheads corresponds to the distance between (200) planes and the arrow indicates the growth direction of the

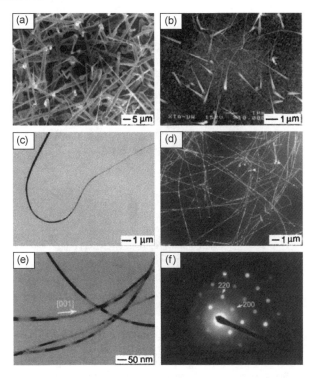

Figure 3.33 *SEM image of (a) MgO microwhiskers, (b) tapered MgO nanowires, (c) TEM image of a single tapered MgO nanowire, (d) and (e) SEM and TEM images of the uniform diameter MgO nanowires, and (f) ED pattern of the nanowires, showing their single-crystalline nature*
(Reproduced from ref. 232)

nanobelt. The SAED pattern of the nanobelt in the inset indicates that the nanobelt grows along ⟨001⟩ and is enclosed by ±(100) and ±(010) crystallographic facets.

MgO nanowires and related nanostructures have also been produced by carbothermal synthesis, starting from polycrystalline MgO or Mg with or without the use of metal catalysts.[233b,234] This study has been carried out with different sources of carbon, all of them yielding interesting nanostructures such as nanosheets, nanobelts, nanotrees and aligned nanowires. Figure 3.35(a) shows a SEM image revealing the various MgO nanostructures formed, including nanosheets and nanobelts by the reaction of activated carbon with MgO powder in the molar ratio 1:1 under flowing Ar (50 sccm) at 1300 °C for 5 h. The diameter of these nanostructures varies from 20 to 200 nm. The nanosheets are about 8 μm wide and tens of micrometres long (inset in Figure 3.35a). The nanobelts, however, are 50–200 nm wide and up to several micrometres long. The TEM image of a nanostructure given in Figure 3.35(b) shows an interesting feature with jagged-edges. The inset in Figure 3.35(b) shows the SAED pattern of this jagged-edged nanostructure, with the reflections corresponding to (222) and (420) planes of cubic MgO, revealing its single crystalline nature. Mg powder reacts with activated carbon (1:1) under

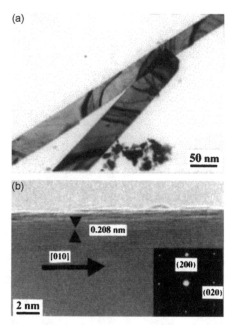

Figure 3.34 (a) *and* (b) *TEM and HREM of the MgO nanobelts. The inset in* (b) *is the SAED of the nanobelts, revealing their single-crystalline nature. The arrow in* (b) *indicates the ⟨010⟩ growth direction of the nanobelt. The spacing between arrowheads in* (b) *corresponds to the (200) planes with an interlayer spacing of 0.208 nm*
(Reproduced from ref. 233a)

flowing Ar (50 sccm) at 850 °C to give nanowire-networks with complex junctions along with star-shaped nanostructures. The TEM image in Figure 3.35(c) shows such nanostructures with T-junction, having rectangular cross-section morphology and a peculiar dot-pattern. The diameter of the nanostructures varies from 100 to 300 nm and their length extends to tens of micrometres. The slight shrinkage in cross section along the direction of branch length may be due to the gradual decrease in vapour supply during growth. The reaction of Mg with graphite powder (1:1) under flowing Ar (50 sccm) at 850 °C yields interesting nano-structures. Figure 3.35(d) shows one such nanostructure, a nanotree like morpho-logy with nanotube branches. These nanotubes are about 100 nm in diameter and their length extends to a few microns. The reactions of Mg and MgO with activated carbon on gold-coated Si substrate yielded uniform nanowires of MgO. Aligned MgO nanowires can be obtained by carbon-assisted synthesis over Au-coated Si substrates. Figure 3.35(e) shows the SEM image of aligned MgO nanowires with gold catalyst particles at their tips, prepared by the reaction of MgO with activated carbon (1:0.5) on gold-coated Si substrate at 800 °C for 5 h under Ar (50 sccm). The diameter of the nanowires is about 100 nm and their length extends to several tens of microns. The inset in Figure 3.35(e) shows the high yields of mainly long uniform nanobelts of MgO grown on an uncoated silicon substrate by the reaction of MgO powder with graphite at 1300 °C (3 °C min^{-1}) for 5 h in Ar (50 sccm).

Figure 3.35 (a) *SEM image of MgO nanostructures prepared by the reaction of activated carbon with MgO. Inset: SEM image of the nanosheets and nanobelts. (b) TEM image of a jagged-edged nanostructure obtained by this procedure. Inset: corresponding SAED pattern. (c) TEM image of T-junctions prepared by the reaction of Mg with activated carbon. (d) TEM image of a MgO nanotree with protruding nanotube fingers, obtained by the reaction of Mg with graphite powder. (e) SEM image of aligned MgO nanowires obtained by the reaction of MgO with activated carbon on gold-coated Si substrate. Inset: SEM image of nanobelts prepared by the reaction of MgO and graphite on an uncoated silicon substrate. (f) TEM image of the MgO nanowires obtained from the reaction of Mg with graphite on gold-coated Si substrate. Inset: the corresponding SAED pattern*
(Reproduced from ref. 233b)

Figure 3.35(f) shows the TEM image of the MgO nanowires with tapered tips obtained by the reaction of Mg with graphite (1:0.5) at 800 °C for 5 h under Ar (50 sccm) on a gold-coated Si substrate. The spots in SAED pattern (inset in Figure 3.35(f)) correspond to the (220) Bragg planes of cubic MgO, revealing their single-crystalline nature. A VS mechanism of 1D growth seems to be operative in these reactions when carried out in bulk, but a VLS mechanism applies when the Au-coated Si substrates are used.

The PL spectrum of MgO nanobelts shows a strong, broad band mainly located in the green region, with its maximum intensity centred at 508 nm and a weak ultralight emission band around 383 nm. This may be assigned to the existence of a large number of oxygen vacancies or surface states. A weak red-light emission band at 721 nm is assigned to the relaxation luminescence of defect centres excited by mechanical stress during fracture and crack propagation.[233]

Al_2O_3, Ga_2O_3 and In_2O_3

An innovative method for producing single-crystalline α-Al_2O_3 fibres, using readily available raw materials, at relatively low temperatures has been developed.[235] The method is based on the VLS growth mechanism and consists of heating Al and SiO_2 in an alumina crucible in the presence of 0.1–10% Fe_2O_3, under flowing Ar at 1300–1500 °C for 2–4 h. Alumina nanopillar arrays have been grown by chemical etching of an ordered porous alumina film.[236] The diameter, height and inter-pillar spacing of the nanopillar arrays can be controlled. Such controllable self-organization process offers a simple, non-lithographic means to fabricate 2D periodic nanostructures. High yields of alumina nanowires are obtained by etching porous alumina membranes in an aqueous NaOH solution at room temperature.[237]

Nanostructures of Al_2O_3 that include nanowires and other network-like structures have been prepared by a carbothermal process involving the heating of Al and graphite powders in a zirconia boat under flowing Ar (50 sccm) at temperatures of 1300 °C for 6 h.[238] In Figure 3.36(a) and 3.36(b) we show SEM images of the alumina nanostructures prepared by the carbothermal process. The nanostructures consist of nanowires with high aspect ratios. Certain other network structures, with a radial outgrowth originating from a single stem-like structure, are also seen. The nanowires are single-crystalline with an interlayer spacing of 0.253 nm, corresponding to the (104) lattice planes. We show a typical HREM image of a nanowire in Figure 3.37. The nanowire appears to grow in a direction that makes an acute angle of 35° with the (104) planes. The growth mechanism of the nanowires can be understood on the basis of a VS mechanism. The relevant reactions being

$$Al(s) + C(s) + O_2(g) \longrightarrow AlO_x(v) + CO(g)$$

$$AlO_x(v) + O_2(g) \longrightarrow Al_2O_3(s)$$

α-Al_2O_3 nanowires and nanobelts have been obtained starting from Al pieces and SiO_2 nanoparticles at 1200 °C under flowing Ar.[239] The nanowires possess diameters of 20–70 nm and are 15–25 μm long, and the nanobelts are several micrometres long, with a width of 0.1–1 μm and thickness of 10–50 nm. A typical SEM image of the nanobelts is shown in Figure 3.38(a). The TEM image in Figure 3.38(b) reveals a thin nanobelt (about 20 nm) with two parallel flat faces. The corresponding HREM image in Figure 3.38(c) and the ED pattern in the inset

Figure 3.36 (a) *and* (b) *SEM images of nanowires and other nanostructures of* Al_2O_3
(Reproduced from ref. 238)

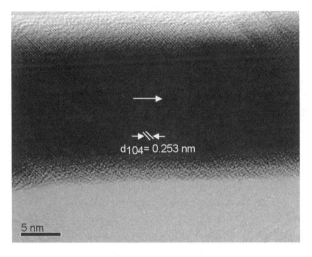

Figure 3.37 *HREM image of an* Al_2O_3 *nanowire, showing a lattice spacing of 0.253* nm,
*corresponding to the (104) planes. The arrow denotes the growth direction of
the nanowire, making an angle of* ~35° *with the normal to the [104] plane*
(Reproduced from ref. 238)

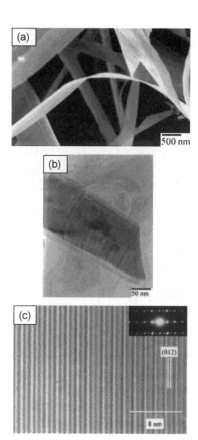

Figure 3.38   (a) *SEM image of α-Al₂O₃ nanobelts.* (b) *Typical TEM image of a very thin nanobelt with two parallel flat faces.* (c) *HREM image and SAED pattern (inset), showing the single-crystalline nature of the nanobelts* (Reproduced from ref. 239)

reveal the nanobelts to be single-crystalline. The lattice fringes are 0.347 nm apart–characteristic of the (012) planes of the α-Al$_2$O$_3$. The growth of the nanowires is controlled by a VLS process, whereas the growth of the nanobelts may be due to a VS process.

Nanowires and other nanostructures of Al$_2$O$_3$ show an emission centred at 380 nm. The PL spectra of the nanowires and nanobelts of α-Al$_2$O$_3$ show a peak centred around 393 nm.[238,239] Optical transitions of oxygen-related defects and F$^+$ (oxygen vacancy with one electron) centres could be responsible for the observed emission.

Ga$_2$O$_3$ nanowires were obtained by the evaporation from a bulk gallium target in flowing Ar and H$_2$ at ~300 °C for about 24 h, by the VS mechanism.[240] The nanowires so obtained had diameters in the range of 60 nm with lengths up to 100 μm. Heating a mixture of GaAs and pre-evaporated Au at 1240 °C in an Ar/O$_2$ atmosphere yields Ga$_2$O$_3$ nanowires with diameters of 20–50 nm and lengths of several micrometres.[241] The PL spectrum of the nanowires show bands centred

Figure 3.39 *SEM images of* (a) *a cluster of paint-brush-like 1D structures of gallium oxide and* (b) *an individual gallium oxide nanopaint-brush*
(Reproduced from ref. 243)

at 475 and 330 nm at room temperature. The presence of Au at the tips of the nanowires indicates a possible VLS growth mechanism. Thermal annealing of milled GaN powder at 930 °C under nitrogen results in bundles of Ga_2O_3 nanowires with high aspect ratios.[242] Bulk synthesis of highly crystalline Ga_2O_3 nanowires and nanopaint-brushes has been carried out by using molten gallium and microwave plasma containing a mixture of oxygen and hydrogen.[243] Multiple nucleation and growth of gallium oxide nanostructures occur directly out of the molten gallium. Figure 3.39(a) and 3.39(b) shows SEM images of the clusters and individual paint-brush nanostructures with tips of 10–100 nm diameter. A plausible mechanism is shown in Figure 3.40, wherein the growth of the paint-brushes is explained on the basis of the initial nucleation and nuclei movement on the molten gallium surface. Agglomeration of neighbouring nanometre-scale nuclei at an intermediate stage leads to the paint-brush morphology.

Ga_2O_3 powder reacts with activated charcoal, carbon nanotubes or activated carbon around 1100 °C in flowing Ar to give nanowires, nanorods and novel nanostructures of Ga_2O_3 such as nanobelts and nanosheets.[244] The diameter and the proportion of the nanowires depend on the flow rate of Ar through the reaction zone. Nanosheets and nanowires of Ga_2O_3 obtained by this method are shown in Figure 3.41(a). The diameter of the nanowires is between 300 and 400 nm and the length extends to tens of microns. The nanosheets are rectangular with a width of around 5 μm and are up to tens of microns long. The reaction with activated carbon

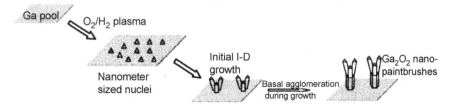

Figure 3.40 *Schematic to explain the plausible growth mechanism of nanopaint-brushes of gallium oxide*
(Reproduced from ref. 243)

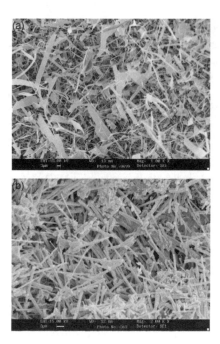

Figure 3.41 (a) *SEM image of Ga₂O₃ nanosheets and nanowires prepared with activated charcoal; (b) SEM image of Ga₂O₃ nanorods prepared with activated carbon* (Reproduced from ref. 244)

affords β-Ga₂O₃ nanorods (Figure 3.41b). These have diameters of 500 to 1000 nm and lengths of 10–15 μm. On increasing the flow rate of Ar from 40 to 60 sccm, there was a distinct change in morphology. Nanosheets and nanowires were formed at 40 sccm, while at 60 sccm nanorods were obtained. The reaction of Ga₂O₃ powder with multi-walled carbon nanotubes at an Ar flow rate of 40 sccm yielded a mixture of nanosheets and nanobelts. The nanobelts have a much smaller width (Figure 3.42a). The nanobelts thus obtained are typically 150–200 nm wide, with lengths extending to tens of microns. The SAED pattern in the inset of Figure 3.42(a) shows reflections corresponding to the (104), ($\bar{2}$11) and ($\bar{2}$02) planes of β-Ga₂O₃. On further increasing the flow rate of Ar to 80 sccm, the product consisted entirely of nanowires of considerably smaller diameter than that obtained at lower flow-rates. The low magnification TEM image in Figure 3.42(b) shows nanowires with diameters of around 70 nm. A high-resolution electron microscopic (HREM) image of a nanowire obtained at an Ar flow-rate of 60 sccm is shown in Figure 3.42(c). The image shows a lattice spacing of 0.47 nm, corresponding to the ($\bar{1}$02) planes of β-Ga₂O₃. The growth of the nanowires is explained on the basis of a VS mechanism.

The PL spectrum of the nanostructures of gallium oxide show two broad peaks at 324 and 405 nm. The intensity of the peak at 324 nm is considerably lower than that of the 405 nm peak.[244] The mechanism of PL in Ga₂O₃ is likely to involve the recombination of an electron on the donor and a hole on the acceptor formed

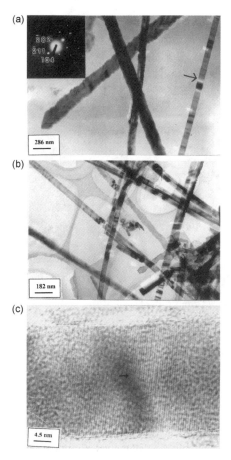

Figure 3.42 *Low magnification TEM images of Ga_2O_3 nanowires obtained by the use of multi-walled carbon nanotubes (a) at a flow rate of Ar maintained at 40 sccm. The arrow shows a nanobelt. Inset: SAED pattern of the sample. (b) with the flow rate of Ar maintained at 80 sccm. (c) HREM image of a Ga_2O_3 nanowire obtained on maintaining the flow of Ar at 60 sccm. The arrow indicates the growth direction that makes an angle of ~6° with the normal to the ($\bar{1}02$) planes*
(Reproduced from ref. 244)

by the gallium vacancies. The acceptor would be formed by a gallium–oxygen vacancy pair. After excitation of the acceptor, a hole on the acceptor and an electron on a donor would be created. These combine radiatively to emit a blue photon. By increasing the temperature, the blue emission can be quenched either by electron detrapping from a donor to the conduction band or by hole detrapping from an acceptor to the valence band. The holes and electrons recombine *via* a self-trapped exciton to emit a UV photon. Well-aligned Ga_2O_3 nanowires have been obtained from the low-temperature decomposition of gallium acetylacetonate.[245]

An efficient route for the synthesis of semiconducting In_2O_3 fibres based on thermal evaporation–oxidation has been reported.[246] The method involves rapid

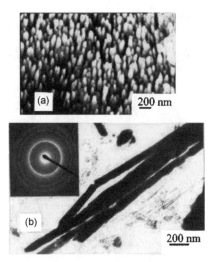

Figure 3.43   (a) *SEM image of In₂O₃ nanowire arrays embedded in AAM, and* (b) *TEM micrograph of the nanowires*
(Reproduced from ref. 247)

heating of a InP substrate coated with a thin Au layer under Ar/O$_2$ at 1000 °C for about 1.5 h. Investigations of the PL spectra have revealed that the nanofibres show light emission in the blue-green region due to oxygen vacancies.[246] Ordered In$_2$O$_3$ nanowire arrays can be assembled into the hexagonally ordered nanochannels of AAMs by electrodeposition and oxidization.[247] Indium nanowires are electrodeposited into the nanoholes by a three-probe dc method in a solution containing 8.5 g L^{-1} InCl$_3$ and 25 g L^{-1} Na$_3$C$_6$H$_5$O$_7$.2H$_2$O solution at room temperature. After electrodeposition, the assembly systems were annealed in air at different temperatures to form ordered In$_2$O$_3$ nanowire arrays. Figure 3.43(a) shows a SEM image (top view) of In$_2$O$_3$ nanowire arrays grown in an AAM template after cleaning in an ultrasonic bath for 3 min, followed by etching of the alumina in aqueous NaOH. The In$_2$O$_3$ nanowires are nearly of the same height and are uniformly embedded in the AAM. Figure 3.43(b) gives a TEM image of the nanowires after removal of the AAM. The surface of the nanowires is not smooth and the diameter is not uniform. The ED pattern in the inset indicates the nanowires to be polycrystalline. The spectra of the nanowires embedded in an AAM also show the blue-green emission. The PL intensity and peak position depend on the annealing temperature.[247] In$_2$O$_3$ nanowires, nanoboquets and nanotrees (Figure 3.44) have been obtained by carbon-assisted synthesis, starting from the oxide or In metal.[248a] An auto VLS mechanism operates here. Copious quantities of single crystalline and optically transparent Sn-doped In$_2$O$_3$ (ITO) nanowires have been grown on gold-sputtered Si substrates by carbon-assisted synthesis starting with a powdered mixture of the metal nitrates or with a citric acid gel formed by the metal nitrates.[248b] The SEM image in Figure 3.45(a) reveals high yield of ITO nanowires obtained by this procedure. TEM images of the ITO nanowires (Figures 3.45b and 3.45c) show these have diameters between 40 and 100 nm and lengths extending to

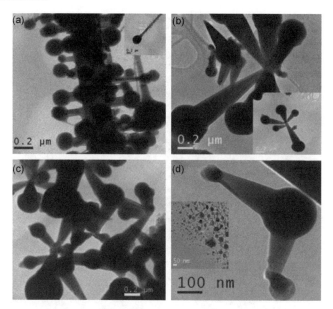

Figure 3.44 *TEM images of In$_2$O$_3$ nanostructures*: (a) *nanotrees*, (b,c) *nanobouquets and* (d) *a highway-type branch structure. Inset in* (d) *shows short nanowires as pegs or nails*
(Reproduced from ref. 248a)

tens of microns. Figure 3.45(c) shows the presence of an Au catalyst particle at the end of the wire. The SAED pattern in Figure 3.45(c) shows diffraction spots corresponding to (222) and (622) planes of cubic In$_2$O$_3$. The TEM image in Figure 3.45(d) corresponds to that of a nanotube, obtained along with nanowires, with SAED pattern in the inset showing it to be single-crystalline, the diffraction spots corresponding to the (622), (400) and (440) Bragg planes of cubic In$_2$O$_3$.

Laser ablation of an indium target with the use of a Au film or Au clusters results in good yields of In$_2$O$_3$ nanowires with a broad diameter distribution.[249] During the laser ablation, the chamber is maintained at 770 °C and 220 Torr, under a constant flow of Ar mixed with 0.02% O$_2$. Field-effect transistors (FET) have been fabricated based on In$_2$O$_3$ nanowires.[249] In$_2$O$_3$ nanowires are n-type semiconductors, and the on/off ratio of the FET can reach up to 1000. Figure 3.46(a) shows the *I–V* curves recorded on an In$_2$O$_3$ nanowire of 10 nm diameter with the gate voltage varying from 5 to ~25 V, indicating n-type semiconductor behaviour. Figure 3.46(b) shows *I–V*$_g$ data of the device at $V_{ds} = 10$ mV. The on/off ratio is ~1000. The inset in the figure shows a SEM image of the nanowire between the Ti/Au source and drain electrodes. In$_2$O$_3$ nanowires can be used as sensors for NO$_2$ and ethanol.[250,251]

SnO$_2$

Single crystalline SnO$_2$ nanoribbons have been synthesized by a simple thermal evaporation of SnO or SnO$_2$ powders in an alumina crucible heated to high

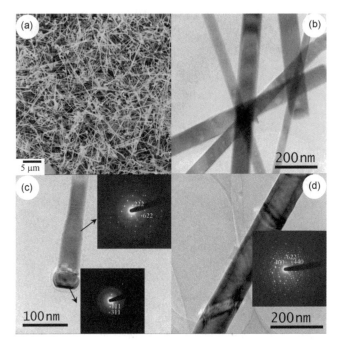

Figure 3.45 (a) *SEM image of ITO nanowires obtained by the reaction of citric acid gel formed with the metal nitrates over Au-patterned Si substrates at 900° C (3 °C min⁻¹) in a flow of 100 sccm of Ar. (b) and (c) TEM images of nanowires. Insets in (c) show SAED patterns of the nanowire and the Au catalyst particle. (d) TEM image of a nanotube. Inset: the SAED pattern of the nanotube*
(Reproduced from ref. 248b).

temperatures.[252] Thermal evaporation was carried out at 1000 °C for SnO powders and 1350 °C for SnO_2 powders for 2 h under 300 Torr with an Ar gas flow rate of 50 sccm. The nanoribbons are structurally perfect and uniform, with widths of 30–200 nm, width-to-thickness ratio of ~5–10, and lengths varying between several hundred micrometres and a few millimetres. A similar thermal evaporation technique, using SnO/Sn and SnO based mixtures at 1050–1150 °C, resulted in nanowires and sandwiched nanoribbons of SnO_2.[253] Novel SnO diskettes have been prepared by evaporating SnO or SnO_2 powders at elevated temperature.[254] Other novel nanostructures obtained by the thermal evaporation of SnO powder include junctions and networks of SnO nanoribbons.[255] Figure 3.47(a) shows a typical SEM image of the junctions and networks of SnO nanoribbons, a magnified image of which is shown in Figure 3.47(b). The nanoribbons are straight, and form a crossed network. The tip of a nanoribbon has a large head of a Sn metal particle that serves as a catalyst for the growth.

SnO_2 nanowires are produced in bulk quantities at 680 °C by the thermal evaporation of SnO powder are single-crystalline in nature with diameters in the 10–190 nm range and lengths of up to tens of micrometres,[256] the growth occurring by a self-catalytic VLS process. Nanoribbons of SnO_2 with a ribbon-like

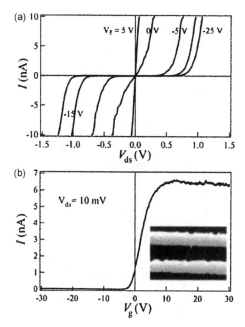

Figure 3.46 *FET fabricated based on In_2O_3 nanowires*: (a) *I–V curves recorded on an In_2O_3 nanowire of 10 nm diameter*, (b) *I–V_g data of the same device at V_{ds} = 10 mV. Inset: SEM image of the nanowire between the source and drain electrodes*
(Reproduced from ref. 249)

morphology have been prepared by the rapid oxidation of elemental tin mixed with $Fe(NO_3)_3$ powder at 1080 °C in Ar/H_2.[257] The PL spectrum of the as-synthesized SnO_2 nanoribbons show two strong peaks, at 392 and 439 nm, and two more at 486 and 496 nm, which may arise due to defects.[257] Single crystalline rutile SnO_2 nanobelts have been synthesized by a thermal evaporation of Sn powders in an alumina boat heated to 800 °C for 2 h in flowing Ar.[258] The SnO_2 nanobelts grows through the VS mechanism. A representative high magnification SEM image of several curved SnO_2 belt-like nanostructures with thickness of about 10–30 nm is shown in Figure 3.48(a). The corresponding TEM and SAED patterns of the nanobelts are shown in Figure 3.48(b). The nanobelts are both straight as well as curved. The width and thickness is uniform, typically in the range of 60 to 250 and 10 to 30 nm, respectively. Figure 3.48(c) shows the TEM image of a single SnO_2 nanobelt with a sharp tip end, the absence of the particle at the tip suggestive of a VS growth mechanism.

SnO_2 nanorods with the rutile structure have been prepared by annealing fine SnO_2 powder in a NaCl flux.[259] The starting SnO_2 powder, the NaCl flux, and surfactant NP9 were mixed and heated at 800 °C for 2.5 h to obtain nanorods of diameter ~20–40 nm and lengths of up to 1 μm. The growth of the nanorods can be explained by an Ostwald ripening mechanism, *i.e.* the dissolving of fine particles and the deposition of components on larger particles. A similar method has been

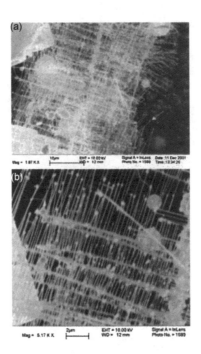

Figure 3.47   (a) *and* (b) *SEM images of the junction networks of SnO nanoribbons, showing*
the orthogonal growth of the branches
(Reproduced from ref. 255)

reported for the preparation of SnO_2 nanorods by annealing precursor powders in
which sodium chloride, sodium carbonate and stannic chloride are homogeneously
mixed along with surfactants and calcined at 750–820 °C for different lengths of
time.[260] Uniform SnO_2 nanowire arrays have been fabricated on a large scale in
AAM templates, by electrodeposition and thermal oxidation.[261] The rate of oxidation and reduction at the surface of a SnO_2 nanowire has been modified by
manipulating the electron density.[262]

SiO$_2$ and GeO$_2$

Amorphous silica nanowires have been prepared by using Au nanoparticle catalysts
on a Si substrate at ~850 °C and a mixture of SiH_4 and He by the VLS growth.[263]
The diameter and the lengths of the nanowires so obtained were around 20 nm and
10 μm, respectively. The PL spectrum of α-SiO_2 nanowires shows two broad bands,
at 2.8 and 3.0 eV, the blue light emission arising due to the presence of oxygen
vacancies.[263]

Highly aligned closely-packed SiO_2 nanowire bunches are obtained in high
yields by using molten Ga as the catalyst and the silicon wafer as the source of Si at
1150 °C in flowing Ar for 5 h, *via* the VLS process.[264] Various silica morphologies
such as carrot-shaped rods (CSRs) whose walls are composed of highly aligned

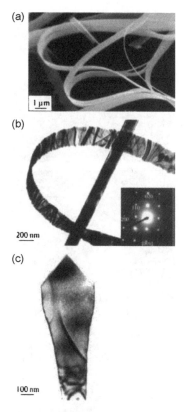

Figure 3.48 (a) *High magnification SEM image of curved SnO₂ nanobelts*, (b) *TEM images of SnO₂ nanobelts, the inset showing the SAED pattern along the [001] axis, and* (c) *TEM image of a single SnO₂ nanobelt with a sharp tip end* (Reproduced from ref. 258)

silica nanowires, with diameters of 15–30 nm and lengths of 10–40 μm, as well as comet-like structures composed of highly oriented silica nanowires, with diameters of 50–100 nm and lengths of 10–50 μm, were thus formed. In Figure 3.49(a), we show the CSRs with diameters of 10–50 μm and lengths up to ~1 mm. Several tens of CSRs radially grow upward, forming a sisal-like structure (Figure 3.49b). Each CSR terminates at its top end in a large spherical ball with a diameter comparable to that of the connected rod. Energy dispersive X-ray spectroscopy (EDX) analysis reveals the balls to be composed of Ga, covered by a thin oxide layer composed of Ga, O, and a small amount of Si. The growth mechanism of CSRs with stair-like inner structures (Figure 3.49c) can be understood by the steps shown in Figure 3.50. CSRs with a series of upside down bell-like cavities (Figure 3.49d) are formed, if the nanowires grow from a region below a band around the lower hemisphere surface of the Ga ball.

Silicon oxide nanowires have been prepared by the CVD process as well, using molten Ga placed on a Si wafer heated at ~920–940 °C in H₂ for 1–3 h.[265] Figure 3.51(a) shows a picture of the so-obtained product. The ball is white

Figure 3.49   *SEM images of the SiO$_2$ nanowires grown on a silicon wafer: (a) low- magnification image, showing carrot-shaped rods (CSR) growing in groups in the
silicon wafer, (b) high magnification SEM image of the boxed area in (a),
showing several tens of CSR forming a sisal-like structure; (c) and (d) are two
cross sections viewed along the 3–3 direction, displaying two types of inner
structures of the CSRs: (c) CSR with a continuous central hole with stair-like
structure, (d) discontinuous upside down, bell-like cavities. Arrows in (c) and
(d) are the growth directions*
(Reproduced from ref. 264)

and ~10 mm in diameter. It is broken into two halves to show the inside, as seen in
the photograph. Molten Ga occupies the centre of the ball and highly ordered and
millimetre-long wires extend from the Ga ball. The SEM image in Figure 3.51(b)
reveals aligned nanowires. The TEM image in Figure 3.51(c) shows the
general morphology of the SiO$_x$ nanowires. The nanowires are amorphous and
homogeneous without elemental Si cores, as revealed by the HREM image in
Figure 3.51(d).

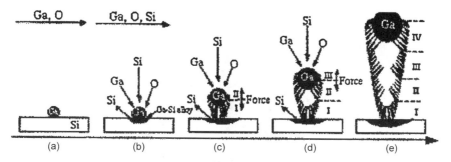

Figure 3.50 *Proposed growth model for CSRs with stair-like inner structures*
(Reproduced from ref. 264)

Well-aligned nanocrystalline Si/SiO$_x$ composite nanowires have been synthesized on catalyst-free substrates using SiCl$_4$/H$_2$ in a hot-filament chemical vapour deposition (HFCVD) reactor.[266] The average diameter of the nanowires was about 80 nm with lengths of about 3.5 μm. The PL spectrum of nano-crystalline Si/SiO$_x$ composite nanowires shows a broad emission band in the 420–585 nm region.[266] Various silica nanostructures consisting of nanowire bundles and brush-like arrays (Figures 3.52a–d) have been prepared by heating equimolar mixtures of Si/SiO$_2$ in argon/nitrogen atmospheres at ~1400 °C for 12 h.[267] The TEM images reveal typical cage structures composed of aligned silica nanofibres. The nanofibres grow in bundles, while paralleling a structure that has a cylindrical symmetry. The cage is 0.3–1 μm wide. Figure 3.52(e) shows an unusual Chinese lantern structure

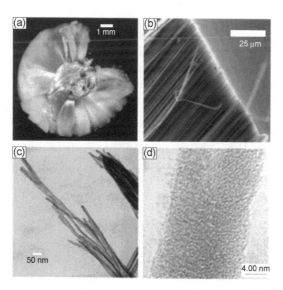

Figure 3.51 (a) *Picture of the SiO$_x$ product*, (b) *SEM image of oriented silicon oxide nanowires*, (c) *TEM image of the SiO$_x$ nanowires*, (d) *HREM image of SiO$_x$ nanowires*
(Reproduced from ref. 265)

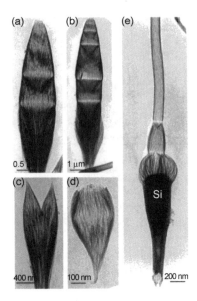

Figure 3.52 *TEM images of bundled arrays and cages of silica*
(Reproduced from ref. 267)

composed of Si and SiO$_x$, where an SiO$_x$ tube extends from the top of a silica wire bundle. One-dimensional nanocomposites consisting of a self-organized chain of gold silicide nanocrystals pea-podded in a silicon oxide nanowire (Au$_2$Si@ SiONW) are obtained by a process involving the deposition of a Si film on a Si (100) substrate by ion-beam sputtering, followed by dc sputtering of Au onto the Si film.[268] A sharp peak around 683 nm in the PL spectrum observed in the Au$_2$Si@SiONW samples is assigned to a radiative recombination process related to interband transitions in the gold silicide nanosphere. Thermal annealing of the gold/ amorphous-silicon bilayers resulted in straight and smooth pea-podded nanowires.

Laser ablation of a composite target, consisting of a mixture of Si powder and 20 wt% silica together with 8 wt% Fe, yields amorphous silica nanowires of ~15 nm diameter and up to 100 μm long.[269] The growth of silicon oxide nanowires is likely to be by the VLS growth mechanism. Shear-aligned surfactant mesophases can be used as templates to produce silica nanowires.[270] Crystalline SiO$_2$ nanowires (α-cristabolite form) have been synthesized by the reaction of fumed silica with activated charcoal around 1300 °C.[271]

Physical evaporation of Ge powder mixed with 8 wt% Fe powder in flowing Ar at 820 °C results in GeO$_2$ nanowires. The diameter and length of the nanowires are in the ranges 15–80 nm and a few micrometres, respectively.[272] Carbothermal reduction of GeO$_2$ with active carbon at 840 °C for 3.5 h in flowing N$_2$ results in GeO$_2$ nanowires of hexagonal structure.[273] The diameter and the length of the nanowires were 50–120 nm and hundreds of micrometres, respectively. The most striking property of the nanowires is that they emit bright blue light. A broad PL peak is observed at 485 nm under excitation at 221 nm at room temperature. This is

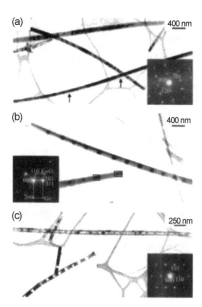

Figure 3.53 *TEM images and SAED patterns (inset) of the α-GeO₂ nanowires. (a) Common nanowires, (b) dot-patterned nanowires, (c) hole-patterned nanowires. The image in (c) is composed of two independent images, located at the upper and lower left*
(Reproduced from ref. 275)

attributed to radiative recombination between an electron on $V_o{}^x$ and a hole on $(V_{Ge}, V_o)^x$ in the GeONWs.[273]

GeO₂ nanorods have been synthesized by a carbon-nanotube confined reaction wherein Ge powders were heated at 850 °C in flowing Ar along with carbon nanotubes and a Si–SiO₂ mixture.[274] The synthesis and nanostructuring of patterned α-GeO₂ wires were carried out by thermal oxidation of metallic Ge and Fe(NO₃)₃ powder at 1300 °C in a mixture of flowing Ar mixed with 5% H₂.[275] A large percentage (70%) of the as-synthesized wires had a diameter of ∼150 nm and a length of several tens of micrometres. A significant proportion of the nanowires possessed a common wire morphology with a uniform structure (Figure 3.53a), while some of them have distinct patterned structures. These structures consist either of regular arrays of Ge dots as in Figure 3.53(b), or of penetrating holes, as in Figure 3.53(c), along their axial directions (*i.e.* dot-patterned and hole-patterned nanowires, respectively). The SAED patterns in the insets suggest that the dots and the matrix wire consist of single crystalline Ge and GeO₂, respectively. The growth of the nanowires seems to be through the VS mechanism since no catalyst particle is observed at the tip of the nanowires.

TiO₂

Micro and nano structures of metal oxides have been obtained by a sol–gel synthesis inside porous template membranes.[276] Oxides synthesized by this method include TiO$_2$ (anatase), V$_2$O$_5$, MnO$_2$, Co$_3$O$_4$, ZnO, WO$_3$, and SiO$_2$. Both porous alumina and track-etched polycarbonate membranes have been employed. TiO$_2$ nanostructures are likely to have potential applications because of their unique properties. For instance, TiO$_2$ nanostructures with high surface area show increased rates of decomposition of salicylic acid in sunlight.[276] Ordered TiO$_2$ nanowire arrays have been fabricated by anodic oxidative hydrolysis of TiCl$_3$ within a hexagonal close-packed nanochannel alumina.[277] Single-crystalline anatase TiO$_2$ nanowires with diameters of 15 nm and lengths of 6 μm are obtained after annealing at 500 °C. Ordered TiO$_2$ nanowire arrays have also been obtained by the sol–gel method by the use of AAMs.[278,279] A TEM image of TiO$_2$ nanowires obtained by a cathodically induced sol–gel method from an aqueous solution of a Ti precursor is shown in Figures 3.54(a) and 3.54(b).[280] The nanowires have relatively straight morphologies and smooth surfaces. Figure 3.54(c) shows the SAED pattern of a nanowire, with the diffraction spots corresponding to the (004), (200), and (103) planes. The anatase nanowires have a uniform diameter of around 50 nm. Room-temperature PL measurements show a broad band with three

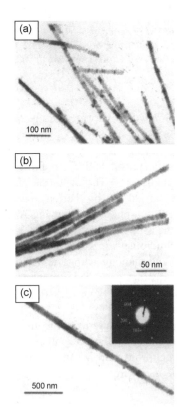

Figure 3.54   *TEM images of* (a) *20 nm TiO$_2$ nanowires,* (b) *10 nm nanowires,* (c) *40 nm nanowire (corresponding ED pattern in the inset)*
(Reproduced from ref. 280)

peaks at 425, 465 and 525 nm, arising from self-trapped excitons, F and F^+ centres respectively.

A novel spin-on process has been employed to fabricate titania nanowires.[281] In this method, nanowires are generated along the grooves of a patterned azobenzene-functionalized polymer, which acts as the template *via* gel cracking. This was possible especially when the concentration of the sol precursor was low. Gelation of the viscous sol deposited on the flat polymer substrate led to the formation of nanowires. By combining sol–gel processing with electrophoretic deposition, binary metal oxides (TiO_2, SiO_2) and complex oxides such as $BaTiO_3$, $Sr_2Nb_2O_7$ and $Pb(Zr_{0.52}Ti_{0.48}O_3)$ have been obtained as unidirectionally aligned nanorods, 45–200 nm in diameter and 10 μm long.[282]

Single crystalline anatase nanowires have been successfully prepared hydrothermal means starting from TiO_2 nanoparticles.[283] The nanowires have diameters in the range 30–45 nm and lengths of several micrometres. Titania nanowires so prepared emit blue-green photoluminescence, peaking at 487 nm.[283] Hydrothermal treatment of the hydrolysate of $TiCl_4$ with caustic soda yields anatase nanorods.[284] A hydrothermal procedure has also yielded titania nanoribbons that are several nanometres thick and 30–200 nm wide.[285] Figure 3.55(a) shows a low magnification TEM image of TiO_2 with a ribbon-like structure. The width of the nanoribbon varies from 30 to 200 nm and the geometry is uniform. In Figure 3.55(b), an individual nanoribbon with a rolled end is shown. From the rolled region one can see that the ribbon is thin (~5 nm). Figures 3.55(c) and 3.55(d) shows typical HREM images of the titania nanoribbons growing along the ⟨001⟩ direction. Two sets of lattice fringes are observed in the lattice resolved image. The fringes parallel to the ribbon axis [(001) plane] correspond to an interplanar spacing distance of about 0.75 nm. This set of fringes may correspond to the structural features of cis-skewed chains, characteristic of the anatase structure in the ⟨001⟩ direction. The fringes with a spacing of 0.35 nm, skewed in the direction of the ribbon axis, may correspond to the (101) plane of anatase. Titania whiskers are obtained in a sheet form, with a length of about 1 μm and a width of 60 nm, on sonicating TiO_2 particles in NaOH solution.[286]

The hydrothermal reaction of TiO_2 nanoparticles with a KOH solution results in $K_2Ti_6O_{13}$ nanowires.[287] The diameters and lengths are in the 10 nm and 1–2 μm ranges, respectively. Crystalline $K_2Ti_6O_{13}$ particles, first formed inside the anatase particle matrix, grow out along the ⟨010⟩ direction into 1D nanowires. These titanate nanowires are wide-bandgap semiconductors ($E_g \sim 3.45$ eV). TiO_2 (B) nanowires have been prepared by thermal treatment of layered hydrogen titanates.[288]

MnO_2 and Mn_3O_4

α-, β-, γ-, and δ-MnO_2 single-crystal nanorods and nanowires, with diameters of 5–80 nm and lengths varying in the 5–10 μm range, can be prepared hydrothermally, involving neither a catalyst nor a template.[289–291] α-MnO_2 nanowires are also obtained by electrodeposition[292] while β-MnO_2 have been produced in mesoporous silica.[293]

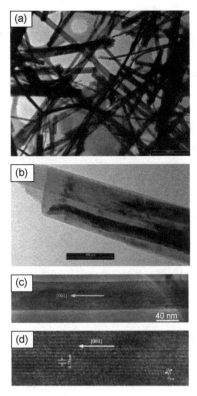

Figure 3.55   (a) *Low magnification TEM image of titania nanoribbons*, (b) *TEM image of one straight nanoribbon with a rolled end, revealing the small thickness.* (c) *and* (d) *HREM image of a single titania nanoribbon growing along [001], showing well-defined structure; the arrows show the growth direction* (Reproduced from ref. 285)

Single-crystal Mn_3O_4 nanowires have been obtained from the solid state reaction of $MnCl_2.4H_2O$, Na_2CO_3, NaCl and nonylphenyl ether (NP-9).[294] In Figure 3.56(a) shows the TEM image of the Mn_3O_4 nanowires, 40–80 nm in diameter and up to 150 μm long. A typical SAED pattern shown as an inset in the figure confirms the nanowire to be a single crystal of tetragonal Mn_3O_4. The HREM image in Figure 3.56(b) also reveals the nanowire to be single crystalline, with an interplanar spacing of about 0.248 nm, corresponding to the (211) planes. TEM images of the tips of the as-synthesized nanowires reveal no spherical droplets, and it is proposed that the nanowires grow by Ostwald ripening.

Cu_xO

A vapour phase method involving the VS process has been employed to synthesize single-crystalline CuO nanowires supported on Cu surfaces.[295] The procedure involves thermal oxidation of the substrate in air at 400–700 °C. The nanowires have controllable diameters in the 30–100 nm range, with lengths of up to 15 μm.

Figure 3.56 (a) *TEM image of the Mn₃O₄ nanowires. Inset: SAED pattern of a nanowire.*
(b) *HREM image of a 40 nm Mn₃O₄ nanowire*
(Reproduced from ref. 294)

Figure 3.57(a) and 3.57(b) shows SEM images of a copper grid after heating in air at 500 °C for 4 h. Although the entire surface of the grid are covered by a high density of nanowires, those protruding from the edges appear to be much straighter, longer and more uniform in diameter as shown in Figure 3.57(b). SEM images of a copper wire (0.1 mm diameter) heated in air at 500 °C for 4 h are shown in Figure 3.57(c) and 3.57(d). The wire was completely covered by a dense array of uniform CuO nanowires. Because of the surface curvature of the substrate, each nanowire grows in the direction essentially perpendicular to the support. Thermal dehydration of $Cu(OH)_2$ nanowires, prepared by the reaction of $CuSO_4$ with ammoniacal KOH, yields CuO nanowires.[296]

Hex-pod-like Cu_2O whiskers are obtained under hydrothermal conditions using $Cu(CH_3COO)_2$ as the precursor.[297] CuO nanorods with an average ~8 nm diameter and lengths of up to 400 nm have been prepared by the reaction of $CuCl_2.2H_2O$ and NaOH in the presence of a polyethylene glycol.[298] A modification of the method involving the reduction by hydrazine hydrate results in Cu_2O nanowires with high aspect ratios.[299] In Figure 3.58(a), a typical TEM image of Cu_2O nanowires prepared by the reduction route is shown. The nanowires are 10–20 nm range in diameter, but smaller diameter nanowires were also found. A typical HREM image of a nanowire (8 nm in diameter) is shown in Figure 3.58(b). The lattice fringes in the image illustrate that the nanowire is single crystalline, with an interplanar spacing of 0.2465 nm, corresponding to the (111) plane of cubic Cu_2O, which is also the growth plane.

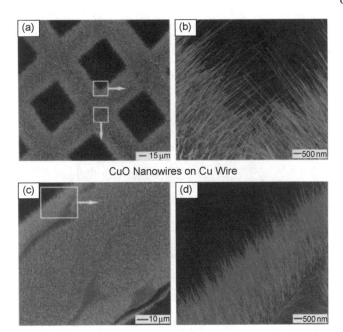

Figure 3.57 (a) *and* (b) *SEM images of CuO nanowires prepared by directly heating copper TEM grids in air at 500 °C for 4 h;* (c) *and* (d) *SEM images of CuO nanowires formed on the surface of a copper wire (0.1 mm in diameter) by heating at 500 °C for 4 h*
(Reproduced from ref. 295)

ZnO

Several methods have been employed to synthesize nanowires of ZnO. They have been grown by a vapour-phase transport process *via* the VLS mechanism using a gold catalyst.[28] Zn vapour is generated by carbothermal or hydrogen reduction of ZnO and the size of the nanowires is controlled by the thickness of the Au film. ZnO NW arrays are obtained by patterning the Au catalyst. Figure 3.59 shows a SEM image of patterned networks of ZnO nanowires at two different magnifications. The nanowires are single-crystalline with the wurtzite structure and grow along the *c*-axis. Evaporation and condensation of different Zn vapour sources in O_2 yields tetrapods and dendrites by adjusting the reaction temperature and O_2 pressure.[300] ZnO NWs have been obtained by a carbothermal route involving the reaction of Zn oxalate or ZnO powders mixed with active carbon or MWNTs.[238] The nanowires obtained are single-crystalline with the wurtzite structure and have diameters of 300–400 nm, with lengths extending to tens of microns. Figure 3.60(a) shows a SEM image of nanowires obtained by the solid-state reaction of zinc oxalate with MWNTs at 900 °C. The HREM image of a single nanowire in Figure 3.60(b) reveals that the nanowires grow along the ⟨001⟩ direction.

A CVD method has been used to grow aligned ZnO nanorods at low temperatures.[301] Vapourization of zinc acetylacetonate hydrate around 135 °C, carried by

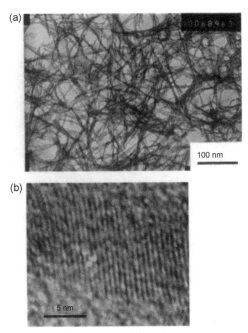

Figure 3.58 (a) *TEM images of the Cu$_2$O nanowires*, (b) *HREM image of individual Cu$_2$O nanowires*
(Reproduced from ref. 299)

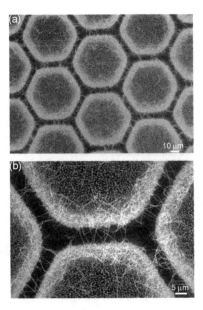

Figure 3.59 (a) *and* (b) *SEM images of ZnO nanowire networks at two different magnifications*
(Reproduced from ref. 28)

Figure 3.60   (a) *SEM image of ZnO nanowires obtained by the reaction of Zn oxalate with MWNTs and* (b) *HREM of a single nanowire. Inset: electron diffraction pattern, indicating that the nanowires are single crystalline*
(Reproduced from ref. 238)

N_2/O_2 onto a Si/SiO_2 substrate at 500 °C, afforded nanorods with diameters in the range 60–80 nm. When the carrier gas was N_2, Zn/ZnO coaxial nanocables and ZnO nanotubes were obtained.[302] Aligned ZnO nanorods are obtained by a similar catalyst-free method.[303] An amorphous SiO_x layer was seen at the interface of the nanorods and the Si (100) substrate.

Well-aligned ZnO NWs have been synthesized using a NiO catalyst on an alumina substrate by a metal vapour deposition method at 450 °C.[304] The nanowires had diameters of ~55 nm, with lengths of up to 2–6 μm. Vertically aligned ZnO nanorods were obtained on Al_2O_3 (001) at 400 °C, by a metalorganic vapour-phase epitaxial growth method employing diethylzinc.[305]

Ordered polycrystalline ZnO NW arrays embedded in AAMs have been fabricated by generating alumina templates with nanochannels, electrodepositing Zn in them, and then oxidizing the Zn nanowire arrays.[306] A one-step electrochemical deposition technique was used to produce a ZnO NW array using nanoporous AAMs.[307] Directed growth of ordered arrays of small diameter ZnO nanowires has been accomplished by simple nanofabrication techniques.[308] High-density ZnO rods can be grown on pregrown 1D nanostructures *via* thermal CVD of Zn at 500 °C.[309] Deformation-free nanohelixes of ZnO have been grown on a Zn-terminated (0001) surface through self-catalysis.[310]

Solution-based routes are also successful in the synthesis of ZnO nanowires and nanorods. ZnO nanorods were prepared using poly(vinyl pyrrolidone) as the capping agent.[311] Perpendicularly oriented ZnO nanorods have been grown on thin ZnO templates from aqueous solutions of zinc acetate and hexamethylenete-traamine.[312a] Later, a template-less and surfactant-free aqueous method using zinc nitrate and hexamethylenetetraamine was used to obtain arrays of nanorods and nanowires.[312b] A microemulsion-mediated hydrothermal process has also been employed to synthesize ZnO NWs.[313]

Nanobelts of zinc oxide have been obtained by the thermal evaporation of the metal oxide powder under flowing Ar at 1400 °C in the absence of a catalyst.[314] The nanobelts are single-crystalline having width-to-thickness ratios of 5 to 10, and lengths of a few millimetres. Hierarchical ZnO nanostructures such as nanonails on nanowires and nanorods on nanobelts have been synthesized by thermal vapour evaporation and condensation.[315]

PL studies on ZnO NWs at 10 K have been reported in detail.[316] Size-dependent surface PL has been observed in ZnO NWs.[317] Field-emission measurements with vertically aligned ZnO NWs grown by the vapour deposition method at 550 °C show the turn-on voltage to be $\sim$0.6 V μ^{-1}m at a current density of 0.1 μA cm^{-2}.[318] The emission current density from the nanowires reached 1 mA cm^{-2} at a bias field of 11 V μm^{-1}, which could give sufficient brightness as a field emitter in a flat panel display. An extremely low operating electric field has been achieved on ZnO nanowire emitters grown on carbon with a current density of 1 mA cm^{-2} at 0.7 V μm^{-1}.[319] Transient PL spectroscopy and time-resolved second-harmonic generation have been utilized to probe ultrafast carrier dynamics in single ZnO nanowires.[320]

ZnO NWs have been used to create sensitive switches by exploiting the photo-conducting properties of individual nanowires.[321] The conductivity of ZnO is sensitive to UV light exposure. The light-induced conductivity increase allows reversible switching of the nanowires between the off and on states, an optical gating phenomena analogous to electric gating. The effect of UV illumination on the electronic conductivity of ZnO nanowires is sensitive to oxygen pressures.[322] Using near-field scanning optical microscopy, the second- and third-harmonic generations (SHG and THG) are imaged on single ZnONWs.[323] The absolute magnitudes of the two independent $\lambda^{(2)}$ elements of a single nanowire have been determined, and the SHG/THG emission as a function of incident polarization is attributed to the hexagonal nanowire geometry and $\lambda^{(2)}$ tensor symmetry.

UV lasing has been demonstrated from single ZnO NWs and nanowire arrays at room temperature.[29,324] The line-widths, wavelengths and power dependence of the nanowire emission demonstrate that the nanowires act as an active optical cavity. The individual nanolasers could serve as miniaturized light sources for microanalysis, information storage and optical communication. Nanowires with diameters of 20–150 nm form natural laser cavities. Under optical excitation, sur-face-emitter lasing action was observed at 385 nm, with an emission linewidth less than 0.3 nm. Figure 3.61 shows the lasing action observed in the ZnO NWs during the evolution of the emission spectra with increasing pump power. At low excita-tion intensities, the spectrum consists of a single broad peak with full-width at half-maximum at $\sim$17 nm. As the pump power increases, the emission peak narrows due to the preferential amplifications of frequencies close to the maximum of the gain

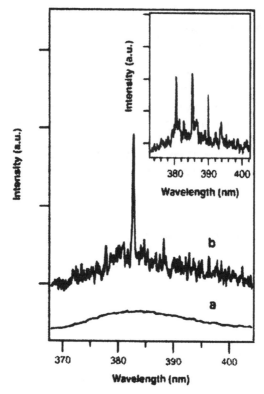

Figure 3.61 *Emission spectra from the ZnO nanowire arrays below (line a) and above (line b) the lasing threshold*
(Reproduced from ref. 29)

spectrum. When the excitation intensity exceeds a threshold of $\sim$40 kW cm^{-2}, sharp peaks emerge in the emission spectra with line-widths of <0.3 nm.

ZnO nanowires have a hydrogen storage capacity of 0.83 wt% under a pressure of about 3.03 MPa.[325] ZnO nanowires can act as ethanol sensors.[326] ZnO NWs can, probably, also act as sensors for other gases such as H$_2$.

Vanadium and Tungsten Oxides

V$_2$O$_5$ fibres have been obtained from a V$_2$O$_5$ sol, which was prepared by ion-exchanging ammonium (meta) vanadate.[327] Individual V$_2$O$_5$ fibres deposited on prefabricated electrodes reveal nonlinear *I–V* characteristics with high resistance.[327] Increased conductivity is achieved by the deposition of AuPd electrodes on top of the fibres. The conductivity of an individual fibre is estimated to be $\sim$0.5 S cm^{-1} at room temperature. Compared to the fibre networks, the individual fibres are characterized by a smaller hopping activation energy, possibly reflecting the absence of transport barriers at the inter-fibre contacts. Single crystalline nanorods of V$_2$O$_5$, 20–55 nm in diameter and lengths up to a few microns long, have been prepared by templating against carbon nanotubes.[328] The method involves coating acid-treated carbon nanotubes with vanadic acid (HVO_3) followed by calcination at 450 °C, and burning in air at 500 °C results in the nanorods.

VO_2 nanobelts have been prepared by hydrothermal reduction of ammonium metavanadate with formic acid; the reaction of the metavanadate with HNO_3 under the same conditions yield V_2O_5 nanobelts.[329] VO_2 nanowires can also be made by ethylene glycol reduction of V_2O_5.[330]

WO_x nanowires have been prepared on metal tungsten tips, by electrochemical etching of tungsten wires and heating in Ar at 700 °C for 10 min.[331] Figure 3.62(a), shows a SEM image of such nanowires. The nanowires protruding from the tip have diameters in the 10–30 nm range, and are up to 300 nm long. WO_x NWs can also be grown on flat tungsten substrates, by H_2 reduction followed by heating in Ar at 700 °C (Figure 3.62b). The growth of the nanowires on the flat W substrates seems to occur due to an oxygen leak in the system. Through the controlled removal of the surfactant from the presynthesized mesolamellar tungsten oxide precursor at elevated temperatures, tungsten oxide nanowires with diameters ranging from 10–50 nm and lengths up to several micrometres have been obtained.[332] Oriented potassium-doped tungsten oxide nanowires are obtained by heating W metal and KI or KBr in air.[333]

Other Binary Oxides

Single crystalline nanorods of oxides such as WO_3, MoO_3 and Sb_2O_5 as well as metallic MoO_2, RuO_2 and IrO_2, 10–200 nm in diameter and up to a few microns long, have been prepared by templating against carbon nanotubes.[328] The method involves coating acid-treated carbon nanotubes with an alkoxide followed by calcination at 450 °C and burning in air at 700 °C. Figure 3.63(a) and 3.63(b) shows HREM images of the carbon nanotubes coated with MoO_3 and IrO_2, respectively, obtained on calcination at 450 °C for 12 h. The images reveal the uniform oxidic coating throughout the length of the nanotube. Interaction of the oxidic coating with the surface of the carbon nanotube is due to the presence of surface acidic sites, resulting from the acid treatment. On removal of the nanotube template, the resulting oxidic species showed interesting nanostructures. Nanorods of MoO_3 thus obtained are shown in the Figure 3.64(a). The nanorods are 80–150 nm in diameter and 5–15 μm long. By subjecting the MoO_3-coated carbon nanotubes to repeated washing (before calcination), it was possible to obtain thinner nanorods of the oxide, after removal of the template. A TEM image of a MoO_3 nanorod is shown in Figure 3.64(b). The SAED pattern, revealing the single-crystalline nature of the nanorod, is shown in the inset of the figure. Similar nanorods were obtained with IrO_2, which is a metallic oxide. A typical TEM image of the intermediate structure formed during the process is shown in Figure 3.65(a). HREM images showed the polycrystalline nature for the intermediate nanostructure. Apparently, during template removal, the oxide coatings on neighbouring carbon nanotubes coalesce to form the rod-like nanostructures. The nanorods are 30–60 nm in diameter and up to 1 μm long Figure 3.65(b). The inset in this figure shows the SAED pattern of a single nanorod, with the spots arising from the (100) and (001) planes of tetragonal IrO_2.

Figure 3.62 *SEM images of* (a) *WO_x nanowires at a W tip and* (b) *WO_x nanowires grown from tungsten plates*
(Reproduced from ref. 331)

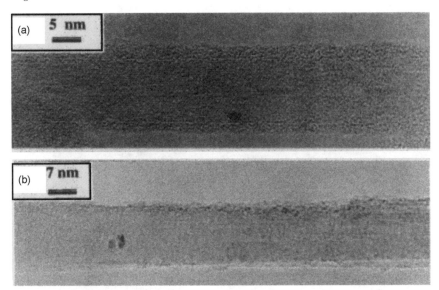

Figure 3.63 *HREM images of carbon nanotubes coated with* (a) *MoO_3 and* (b) *IrO_2, obtained on calcination of the oxide-coated carbon nanotubes* (Reproduced from ref. 328)

Four-armed α-MoO_3 building blocks have been prepared *via* a TiO_2-capping method under hydrothermal conditions.[334] Because of the external bonding capacity and self-assembly of these novel nano-units, other geometrically complex α-MoO_3 nanostructures such as armed and/or holed fork-like rods, tridents, and paint-brushes may also be fabricated under similar conditions. MoO_2 nanowire arrays have been prepared on Si substrates by thermal evaporation of Mo in a flow of argon.[335]

Thermally stable hollw α-Fe_2O_3 nanowires have been prepared by vacuum pyrolysis of β-FeOOH nanowires.[336] Large arrays of Fe_2O_3 nanowires are obtained by the oxidation of iron.[337] Fe_2O_3 nanowires can be encapsulated in carbon microtubes by the pyrolysis of an ethanol–ferrocene mixture.[338] Fe_3O_4 nanowires have been prepared hydrothermally with an applied external magnetic field.[339] An epitaxial shell of Fe_3O_4 has been deposited onto single crystal MgO nanowires by pulsed laser deposition.[340]; the resulting wires show 1.2% magnetoresistance at room temperature.

Ternary and Quarternary Oxides

Single-crystalline $BaTiO_3$ nanowires have been obtained by a solution-phase method.[341,342] Decomposition of barium titanium isopropoxide in the presence of coordinating ligands resulted in nanowires of $BaTiO_3$ (5 to 70 nm in diameter and up to tens of micrometres long). Ferroelectric properties of individual nanowires have been studied. Nonvolatile electric polarization is reproducibly induced and

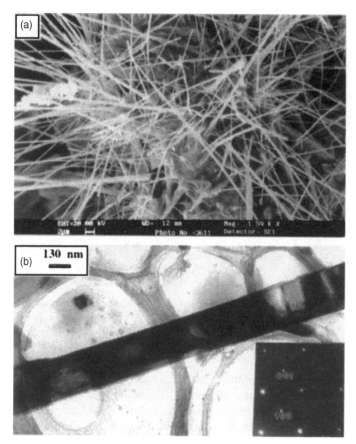

Figure 3.64 (a) *SEM image of MoO₃ nanorods and* (b) *TEM image of a nanorod. The inset in* (b) *shows the SAED pattern along the [010] direction*
(Reproduced from ref. 328)

manipulated in the nanowires, indicating that nanowires as small as 10 nm in diameter retain ferroelectricity. The coercive field for polarization reversal is $\sim 7\,\text{kV cm}^{-1}$, and the retention time for the induced polarization exceeds 5 days. These nanowires should be useful to investigate nanoscale ferroelectricity, and in nanoscale nonvolatile memory applications.

High aspect ratio single crystal $BaWO_4$ nanowires, with diameters as small as 3.5 nm and lengths greater than 50 μm, have been synthesized in cationic reverse micelles, formed by an equimolar mixture of undecylic acid and decylamine, using $BaCl_2$ and Na_2WO_4 as precursors.[343]

Nanowire arrays of $La_{1-x}Ca_xMnO_3$ have been prepared by a sol–gel process in which the respective nitrate starting materials were taken within a nano-channel alumina template and heated to 800 °C for about 2 h.[344] Single crystalline $La_{0.5}(Ba, Sr)_{0.5}MnO_3$ nanowires with a cubic perovskite structure were synthe-sized hydrothermally at reaction temperature of 270 °C.[345,346] Typical lengths of

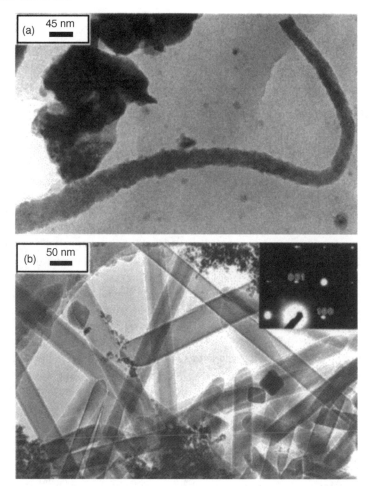

Figure 3.65 *TEM images of* (a) *intermediate nanostructure of IrO$_2$ obtained on heat-treatment of oxide-coated carbon nanotubes and* (b) *nanorods of IrO$_2$ obtained after the complete removal of carbon template. SAED pattern of a nanorod is shown as the inset*
(Reproduced from ref. 328)

the nanowires are in the range of several to several tens of micrometres, and the widths are 30–150 nm. La$_{0.5}$Ba$_{0.5}$MnO$_3$ nanowires show the typical behaviour of an ordinary ferromagnet and those of La$_{0.5}$Sr$_{0.5}$MnO$_3$ exhibit a spin freezing behaviour.[345,346] La$_{0.5}$Ba$_{0.5}$MnO$_3$ nanowires also show an enhanced MR at low temperatures, and the disappearance of the MR peak is ascribed to the effect of grain boundaries and porosity. La$_{0.7}$Ca$_{0.3}$MnO$_3$ nanowire arrays with enhanced ferromagnetism have been synthesized.[347,348] Nanowires of several other oxide materials such as CoFe$_2$O$_4$,[349] YBa$_2$Cu$_3$O$_7$,[350] ZnGa$_2$O$_4$,[351] PZT,[352] LnPO$_4$[353] and WO$_2$Cl$_2$[354] have been reported.

5 Metal Nitride Nanowires

BN

Tubular BN filaments have been grown by the reaction of N_2 or NH_3 with ZrB_2 or HfB_2.[355] To obtain the filaments, metal diborides of Zr, Hf, Ti, V, Na, Ta heated to ~1100 °C in a flow of Ar are treated with NH_3, along with a mixture of B_2H_6 with H_2. A chemical method has been developed for synthesizing BN nanowires through the reaction of a mixture gas of N_2 and NH_3 over catalytic α-FeB nanoparticles at 1100 °C. The boron content in the product comes from the catalyst itself, suggesting a VLS growth mechanism.[356] Nanobamboo structures of BN are obtained on heating a mixture of B and Fe_2O_3 in flowing NH_3. The reaction temperature and the B/Fe_2O_3 ratio play important roles in determining the diameter and morphology of the product. The growth of the nanobamboo structures is ascribed to a catalytic growth within the framework of the VLS catalytic growth mechanism.[357] Nanobamboos, nanocables and nanobells of BN have been synthesized by a CVD method using $B_4N_3O_2H$, or commercial BN powders enriched with oxygen under flowing $N_2 + H_2O$ or $N_2 + NH_3$ at 1700 °C.[358]

BN nanowires have been prepared by employing several methods, starting with a mixture of active carbon and boric acid. The mixture is heated in a NH_3 atmosphere at 1300 °C.[359,360] Figure 3.66(a), gives a TEM image of single crystal nanowires with a diameter of around 75 nm. The reaction was also carried out in the presence of catalytic Fe particles, which yielded nanobamboo structures. The procedure had two objectives, one to achieve the reduction of boric acid by carbon and the other to provide catalytic particles for the growth of the BN nanostructures. A nanobamboo structure with tiny hair-like features attached to the outer surface is shown in Figure 3.66(b). Conversely, heating boric acid with only catalytic Fe particles dispersed over silica resulted mainly in plates and whiskers of BN (Figure 3.66c). TEM images of the nanowires with a diameter of 60 nm are given in Figure 3.66(d).

Cubic BN nanorods have been prepared by the reaction between BCl_3 and Li_3N at low pressures (1–2 MPa) in an autoclave at 450 °C for 18–24 h.[361] Pyrolysis of $CH_3CN.BCl$ at ~900–1000 °C over Co powder generates graphitic $B_xC_yN_z$ nanofibres with various morphologies.[362] HREM and EELS studies suggest that the stoichiometry of the filaments is $[BC_2N_z]_n$ ($z = 0.3$–0.6). The growth of the nanostructures is explained in terms of a slip mechanism proposed previously for the bamboo-shaped structures found in arc-discharge carbon experiments. The borazine 2-$[(CH_3)_2N]$-4,6-$(CH_3NH_2)B_3N_3H_3$ and polymers derived from it are attractive precursors for high-performance BN fibres, the crystallization of the fibres being carried out to completion at 1600 °C.[363] Continuous BN fibres have been fabricated by melt spinning and pyrolysis of the borazine.[364] Longitudinal mechanical properties depend on the mechanical stress and temperature applied during the conversion process.

High purity BN nanofibres with diameters ranging from 30 to 100 nm have been prepared by pyrolysing a mixture of BN, B_2O_3 and B in a molar ratio 2:1:1 in a N_2 atmosphere at ~2000 K.[365] Figure 3.67(a) shows the TEM image of the nanofibres–some of them exhibit periodic changes in the diameter, causing a

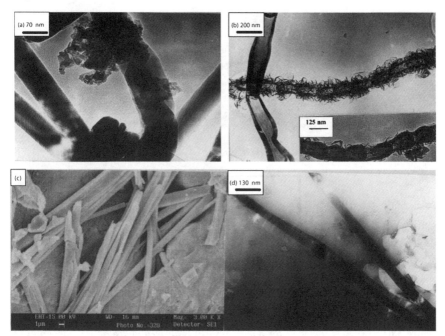

Figure 3.66 (a) *TEM image of nanowires obtained by the reaction of H₃BO₃ and activated carbon with NH₃, (b) bamboo structure of BN with hair-like species attached to the outer surface, obtained by the reaction of H₃BO₃ and activated carbon with NH₃ in the presence of catalytic Fe particles. Inset in (b) shows the magnified image of the hair like bamboo structure. (c) SEM image of plates and whiskers and (d) TEM image of nanowires: (c) and (d) obtained by the reaction of H₃BO₃ and catalytic Fe particles (dispersed over silica) with NH₃* (Reproduced from ref. 359)

stacked cone-morphology (black arrow in the figure). The BN fibres could adsorb 2.9 wt% hydrogen at room temperature and ~10 MPa.[365] Figure 3.67(b) shows the experimental adsorption isotherm for H_2 at different pressures for a nanofibre sample. The hydrogen uptake amount increases with the rising pressure. The uptake of 2.9 wt% is close to that of bamboo-like nanotubes (2.6 wt%), but 1.1 wt% higher than that of carbon MWNTs (1.8 wt%) with close tips under the same pressure. The open-ended edge layers on the nanofibre exterior surface may contribute to the hydrogen uptake capability under high pressures. These measurements suggest that investigations of BN nanofibres as potential hydrogen storage materials may be worthwhile.

Using a Ni catalyst, BCN nanofibres were obtained from a mixture of N_2, H_2, CH_4 and B_2H_6 by means of HFCVD. The Ni catalyst particles play a important role in the formation of the BCN nanofibres with different compositions.[366] Highly oriented BCN nanofibres are also produced by bias-assisted HFCVD from the same mixture of gases.[367,368] Figure 3.68(a) shows a SEM image of BCN nanofibres on the Ni substrate. The nanofibres are perpendicular to the substrate surface and are of similar height. The diameters of the fibres are in the 50–400 nm range,

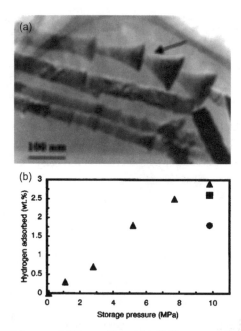

Figure 3.67   (a) *TEM image of BN nanofibres.* (b) *Hydrogen uptake in BN nanofibres (▲) under different pressure at room temperature. The highest uptake capacities of BN bamboo-like nanotubes (■) and MWNTs (●) under ~10 MPa are shown for comparison*
(Reproduced from ref. 365)

and the average density of nanofibres is estimated to be $10^8\,cm^{-2}$. BN and BCN nanofibrous structures have also been prepared by pyrolysing the 1:1 addition compound of BH_3 with $N(CH_3)_3$ and pyridine over cobalt powder in flowing Ar at 1000 °C.[369]

The field-emission behaviour of BCN nanofibres has been examined. A high emission current density of ~20–80 mA cm^{-2} at a low electric field of 5–6 V μm^{-1} has been obtained.[367] Figure 3.68(b) shows the electron emission current as a function of the applied field, and its Fowler–Nordheim (FN) plot (inset). The field-emission characteristics suggest possible applications in vacuum cold-cathode flat-panel displays and microelectronics.

Raman spectra of BCN nanostructures show two sharp bands characteristic of the graphitic structure. The sharp bands and the weak D band in the Raman spectrum indicate a high degree of graphitization. PL measurements of the BCN nanostructures reveal them to be semiconductors with a bandgap of around 1.0 eV.[368] The π-electronic structures and absorption spectra of BN ribbons have been calculated using the tight-binding model. The ribbons are predicted to exhibit prominent absorption bands owing to the divergent density of states. The spectra are affected by the geometry. Important differences occur between zigzag and armchair ribbons, which include the frequency range of absorption bands and the selection rules. Optical measurements have to be carried out to verify the predicted

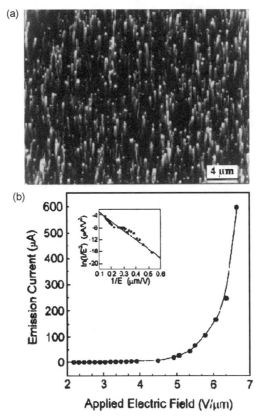

Figure 3.68 (a) *SEM image of oriented BCN nanofibres perpendicular to the substrate.* (b) *Field-emission current vs. applied electric field for a BCN nanofibre sample and the FN plot (inset)* (Reproduced from ref. 367)

spectra and electronic structures.[370] BN whiskers of 500 nm diameter prepared by a thermal reaction process exhibit a bandgap in the UV–Visible absorption spectrum and multiple fine absorption peaks due to electron–phonon coupling.[371] Strong UV and visible luminescence has also been observed from these whiskers.

AlN

AlN has many attractive properties, including high thermal conductivity, low coefficient of thermal expansion, high electrical resistivity, good mechanical strength and excellent chemical stability. AlN nanowires can be prepared by a relatively low-cost, high efficiency method wherein a mixture of Al and Al_2O_3 powders is heated at 1100 °C in flowing NH_3 along with carbon nanotubes.[372,373] Single-crystalline AlN nanowires so obtained have diameters in the 18–35 nm range, depending on the diameter of the CNTs. Single-crystal AlN nanowires have been fabricated by a two-step sublimation process wherein AlN powder was treated by a mechanical deformation process, consisting of ball milling for 72 h, followed by heating in a NH_3 atmosphere at 1500 °C.[374] The diameter of the nanowires varies in the range 10–60 nm, with lengths going up to 360 μm. Hexagonal AlN nanowires have also been synthesized by heating a mixture of Al, SiO_2 and Fe_2O_3/Al_2O_3 at 1100 °C in a flow of NH_3 gas for 1 h. The presence of Si effectively reduces the melting point of Fe to form the activated liquid catalyst needed for the VLS mechanism to operate.[375]

Carbothermal procedures involving the reaction of a mixture of Al powder and carbon in a NH_3 atmosphere yields AlN nanowires. Notably, however, AlN nanowires are also obtained by the direct reaction of Al with NH_3 or N_2.[376]

GaN

GaN is an important wide bandgap semiconductor with various applications in electronic and optoelectronic devices. Single-crystal GaN nanowires with the wurtzite structure have been prepared by several methods, the simplest being the thermal evaporation of GaN powders at 1200 °C in an Ar atmosphere without the use of a template or a patterned catalyst.[377] The nanowires grow by the vapour–solid mechanism. The as-synthesized nanowires have a diameter of 30 nm and lengths of several microns. Dielectric measurements at different frequencies reveal that the dielectric constants of the nanowires are much larger than those of powders at low frequencies due to space-charge polarization and rotation direction polarization. HFCVD, wherein a solid source of Ga_2O_3 and C mixture was heated in a NH_3 atmosphere, has been employed to synthesize bulk quantities of GaN nanowires of 5–12 nm diameter and several micrometres in length.[378] PL spectra of the nanowires show a broad band around 420 nm. Hydride vapour phase epitaxy has been carried out employing a sapphire substrate, involving $GaCl_3$ mixed with flowing NH_3 at 478 °C, to obtain nanorods.[379,380] Cathodoluminescence measurements on individual GaN nanorods show a blue-shift with decreasing diameter of the nanorods due to quantum confinement.

GaN nanowires of 10–50 nm diameter are formed over Si wafers and quartz, by the use of molten Ga and other catalysts, in the temperature range of 850–1000 °C, by the VLS mechanism.[381,382] PL measurements show a strong band-edge emission and a weak yellow luminescence at 2.3 eV. Field-emission studies on the nanowires show significant emission currents at low electric field with current densities of 20 μA cm^{-2} at a field of 14 V μm^{-1}.[381]

The synthesis of straight and smooth GaN nanowires 10–40 nm in diameter and 500 μm long, starting from Ga and NH_3 at 920–940 °C, has been achieved, by the use of catalytic nickel oxide nanoparticles dispersed over $LaAlO_3$ substrates.[383] Laser-assisted catalytic growth of a GaN/Fe target has been employed successfully to obtain nanowires of 10 nm diameter and 1 μm length, wherein the *in situ* generation of the fine liquid nanoclusters of the catalyst causes the growth of the nanowires.[384] The growth direction of the GaN nanowires is ⟨100⟩.

GaN nanorods 4–50 nm in diameter and up to 25 μm long were obtained through a carbon nanotube confined reaction wherein Ga_2O vapour was reacted with NH_3 gas.[385] GaN nanowires are also obtained by heating Ga(acac)$_3$ in the presence of carbon nanotubes or activated carbon in NH_3 vapour at 910 °C. The reaction can be carried out by heating $Ga(acac)_3$ in NH_3 vapour over catalytic Fe/Ni particles dispersed on silica.[360,386,387] In the above reactions, $Ga(acac)_3$ generates fine Ga_2O particles *in situ*, which react with the NH_3 vapour. The diameter of the nanowires could be reduced by employing (SWNTs or a lower proportion of the catalyst. Figure 3.69(a) shows a SEM image of the nanowires obtained by the reaction of GaO_x and NH_3 in the presence of catalytic Fe particles dispersed over silica. The yield of the nanowires was high, the diameter and length being 30–50 nm and 1–2 μm, respectively. Figure 3.69(b) shows a TEM image of the single-crystalline nanowires (30–60 nm diameter) obtained by using MWNTs. Figure 3.69(c) and 3.69(d) show the TEM and HREM images of a GaN nanowire obtained by use of Fe catalyst dispersed over silica. The HREM image reveals the characteristic 0.259 nm spacing between (002) planes. The SAED pattern (inset) shows Bragg spots that correspond to the (002) reflections of the wurtzite structure. The growth direction of the nanowire, represented by the arrow in the figure, is perpendicular to the (002) planes.

High quality, ultrafine GaN nanowires have been synthesized by the VLS growth mechanism by using HFCVD, wherein liquid Ga placed over a p-type Si wafer

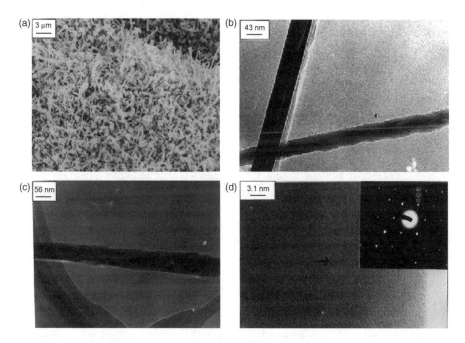

Figure 3.69 (a) *SEM image of GaN nanowires obtained by using catalytic Fe nano-particles dispersed on silica.* (b) *TEM image of single-crystalline GaN nanowires obtained by using MWNTs as templates.* (c) *TEM and* (d) *HREM images of GaN nanowire obtained by using catalytic Fe nanoparticles dispersed on silica. Inset in* (d) *shows the ED pattern corresponding to the [002] reflections of the wurtzite structure The arrow in* (d) *is the growth direction* (Reproduced from ref. 386)

covered with a Ni film was heated at 700 °C in NH_3.[388] CVD has also been employed with a Si/SiO_2 substrate coated with 30 Å Au *via* e-beam evaporation. GaN nanowires 20–100 nm in diameter were grown on this substrate by using a Ga source heated to 900 °C in flowing NH_3.[389] Such direct and precise control of the diameter of the nanowires is desirable to exploit the electronic and optoelectronic properties that are closely related to the diameter. The method has been modified by employing e-beam lithography to create catalyst islands in specific patterned regions to enable the growth of the single GaN nanowire between two catalyst islands. Position-controlled growth is important for making nanowire integrated systems since one does not need to locate the nanowires individually. Many devices can be made in parallel on one chip by exploiting the advantage of batch processing. Raman spectra of GaN nanowires reveal significant broadening of the Raman modes, indicative of phonon confinement effects associated with the nanoscale dimension.[381,390]

Novel nanostructures of GaN such as nanobelts are obtained over large areas of Si substrates by the reaction of Ga/Ga_2O_3 and ammonia, using Ni/Fe and boron oxide as catalysts. The width of the nanobelts ranges from 100 nm to 1 μm and the thickness is about $\frac{1}{10}$th of the width.[391,392] Solvothermal reaction between $GaCl_3$ and NaN_3 gives an azide precursor that decomposes to amorphous/nanocrystalline GaN at or below 260 °C in superheated toluene and THF solvents near their critical points. The nanoscale structures include nanoparticles, nanorods and faceted crystallites.[393] Nanowires and nanobelts of GaN have been synthesized over Si and $LaAlO_3$ substrates, starting from ball-milled Ga_2O_3 and NH_3.[394] Figure 3.70(a) shows an SEM image of the nanobelts, with typical widths in the range 100–500 nm, obtained by this method. The lengths vary in the 10–50 μm range, and the thickness in the 10–30 nm range. The width is, generally, not uniform along the entire length. The inset in Figure 3.70(a) is the image of a belt under high magnification, revealing the morphology more clearly. Figure 3.70(b) shows a nano-ring along with the other nanostructures, formed by the twist of the nanobelts, with typical diameters of 1–3 μm. Figure 3.70(c) shows a HREM lattice image revealing the [001] lattice plane corresponding to 0.32 nm. There are many defects (marked by arrows) arising from the growth processes.

A pyrolysis route has been employed for the synthesis of nanorods of GaN, wherein gallium dimethylamide and ferrocene were pyrolysed in NH_3 using a two-stage furnace.[395] The controlled growth of oriented GaN nanopillars and randomly distributed nanowires has been accomplished by MOCVD using $(N_3)_2Ga[(CH_2)_3NMe_2]$ in the presence of pure N_2 gas at high flow rates of 100 sccm at 950 °C. The room-temperature PL spectrum of a suspension of the GaN nanopillars showed a strong broad emission peak around 430 nm.[396]

Nanocomposites of GaN nanorods coated with graphitic carbon layers (usually less than five) have been synthesized by the conventional thermal CVD method.[397] The GaN nanorods were first synthesized by reacting Ga_2O vapour with ammonia gas at 950 °C, which on CVD under flowing methane at 900 °C for 15 min resulted in the graphitic layers. GaN nanorods have also been coated with insulating BN layers by heating Ga_2O_3, Ga, amorphous B powder and an iron oxide catalyst in N_2 followed by NH_3.[398] Large-scale GaN nanowires of ~50 nm diameter have been

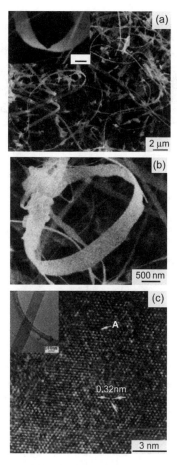

Figure 3.70 (a) *SEM images of the GaN nanobelts. Inset: image of a belt with a higher magnification. Scale bar corresponds to 200 nm. (b) A nanoring. (c) HREM image of a selected GaN nanobelt. Inset: TEM image of the nanobelt* (Reproduced from ref. 394)

prepared starting from Ga and Ga_2O_3 using catalytic In nanoparticles, within the nanochannels of AAMs.[399]

Optical characterization of wurtzite GaN nanowires has been carried out. The energy gap of GaN exhibits a blue-shift from the bulk value of the gap. The effect of temperature on the energy gap is weaker in the GaN nanowires than in the bulk, and the infrared response can be used to evaluate the free-carrier density in GaN nanowires.[400] Photoconduction studies have also been carried out on GaN nanowires.[401]

Ultraviolet–blue laser action has been reported in monocrystalline GaN nanowires, using both near-field and far-field optical microscopy to characterize the waveguide mode structure and the spectral properties of the radiation at room temperature.[402] The far-field image in Figure 3.71(a) shows optically pumped

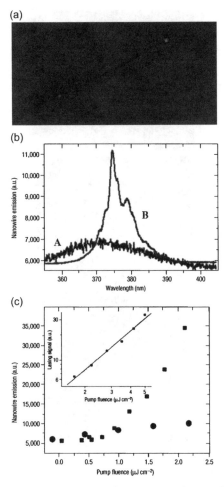

Figure 3.71 *GaN nanowire laser: (a) Far-field image of a single GaN nanolaser. The nanowires were excited by about 3 µJ cm^{-2}. Laser emission is seen at the ends of the nanowire. (b) Photoluminescence spectrum (curve A) using 1 mW continuous-wave excitation and lasing (curve B) using about 1 µJ cm^{-2} pulsed excitation. (c) Power dependence of the lasing near threshold (filled squares) and of photoluminescent emission from a non-lasing region (filled circles). Inset: logarithmic plot of lasing power dependence* (Reproduced from ref. 402)

(310 nm, 4.0 eV) laser emission from a single, isolated GaN nanowire. The localization of bright emission at the ends of the nanowires suggests strong waveguiding behaviour and that the cavity modes are Fabry–Perot (axial) rather than whispering gallery modes. Below the lasing threshold, the image has minimal contrast, and the PL spectrum is broad and featureless (Figure 3.71b, curve A). Near the threshold (in this case ~700 nJ cm^{-2}), several sharp (< 1.0 nm) features appear in the spectrum, indicating the onset of stimulated emission. Additional laser modes appear as the excitation intensity is increased (Figure 3.71b, curve B). At higher

pump fluences, the maximum laser emission was detected at lower energy (>380 nm), indicating a shifting of the gain curve due to bandgap renormalization. This is probably the result of the formation of an electron–hole plasma, shown to be the dominant lasing mechanism for GaN at high temperatures because of its weakly bound excitons (~25 meV) and Coulombic screening at high excitation intensities. The dependence of the laser emission on pump fluence is shown in Figure 3.71(c). Below the threshold, the PL dependence is linear, but a superlinear increase in emission intensity with pump fluence occurs around $700\,nJ\,cm^{-2}$. This is characteristic of stimulated emission, and a log–log plot of the power dependence above the threshold shows an approximate quadratic dependence on the pump fluence (inset). The power dependence of PL from nonlasing GaN material is linear even at high excitation fluence (Figure 3.71c, filled circles).

Doped GaN nanowires are of interest because of their optical and magnetic properties. Ferromagnetic Mn-doped GaN nanowires have been prepared by reacting a mixture of acetylacetonates with NH_3 at 950 °C in the presence of multi-walled and single-walled carbon nanotubes, the nanowires prepared with the latter being considerably smaller in diameter (25 nm).[403] The coercivity in the GaMnN nanowires was higher than those reported for thin films. PL spectra of the Mn doped samples (diameters up to 50 nm) show a band centred around 420 nm, the red-shift arising from hole-doping (Figure 3.72a). The characteristic Mn^{2+} emission, however, around 610 nm is due to the ${}^4T_1 \rightarrow {}^6A_1$ internal Mn^{2+} ($3d^5$) transition. The FWHM of the Mn^{2+} emission band decreases with increasing Mn

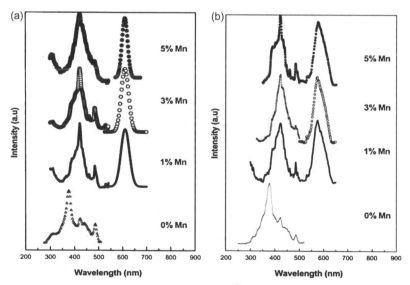

Figure 3.72 (a) *Photoluminescence spectra of the GaMnN nanowires prepared with carbon MWNTs.* (b) *Photoluminescence spectra of the GaMnN nanowires prepared with carbon SWNTs. The spectrum of the undoped nanowires is also shown in each case. Mn^{2+} emission for the 25 nm nanowires exhibits a blue-shift compared to the bands of the larger diameter nanowires, as seen in* (b) (Reproduced from ref. 403a)

concentration. The PL spectrum of the 25 nm nanowires of GaN (Figure 3.72b) shows the main feature at 410 nm. The characteristic Mn^{2+} emission is blue-shifted in the spectra of the thinner nanowires. The intensity of the Mn emission also increases with the Mn content. Mg-doped (p-type) as well as Si-doped (n-type) GaN nanowires have been prepared by a method similar to that of Mn-doped GaN nanowires.[403] The Mg- and Si-doped GaN nanowires show PL bands at 442 and 390 nm, respectively. Optical emission properties of Mg-doped GaN nanowires coated with a phosphor such as YAG:Ce have been investigated (Figure 3.73). The PL spectrum of the Mg-doped nanowires reveals a band at 442 nm [curve (a) shown by filled circles in Figure 3.73], attributed to the donor–acceptor (D-A) pair type recombination. The blue emission excites the yellow phosphor coating, which emits yellow fluorescence [curve (b), full line in Figure 3.73]. The mixture of blue emission from the Mg-doped nanowire and yellow emission from the phosphor results in white emission [curve (c), broken line in Figure 3.73].

Field-effect transistors (FETs) based on individual GaN nanowires have been fabricated.[404] Gate-dependent electrical transport measurements show that the GaNNWs are n-type and that the conductance of NW-FETs can be modulated by more than three orders of magnitude. Electron mobilities are comparable to or larger than thin film materials with similar carrier concentration and thus demonstrate the high quality of these nanowire building blocks and their potential for nanoscale electronics. GaN NW FETs (Figure 3.74a) have been prepared by dispersing a suspension of the GaN NWs in ethanol onto the surface of an oxidized silicon substrate, where the underlying conducting silicon was used as a global back gate. Source and drain electrodes were defined by electron beam lithography followed by electron beam evaporation of Ti/Au (50/70 nm), and electrical transport

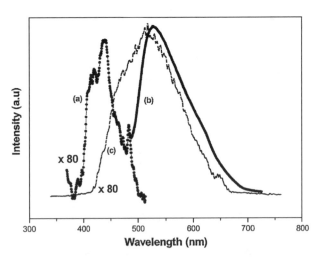

Figure 3.73   *White-light emitting Mg-doped nanowire coated with the YAG:Ce phosphor: curve (a) in filled circles is the blue emission from the Mg-doped GaN nanowire, curve (b), a full curve, is the yellow emission from the YAG:Ce phosphor and curve (c), shown as a broken curve, is the white light emission (Reproduced from ref. 403b)*

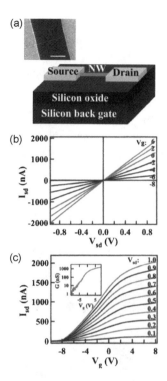

Figure 3.74 (a) *Schematic of a NW FET and (inset) FE-SEM image of a GaN NW FET; scale bar 2 μm (b) Gate-dependent I–V$_{sd}$ data recorded on a 17.6 nm diameter GaN NW. The gate voltages for each I–V$_{sd}$ curve are indicated.* (c) I–V$_g$ data recorded for V$_{sd}$ = 0.1–1 V. Inset: conductance, G, vs. *gate voltage* (Reproduced from ref. 404)

measurements made at room temperature. Figure 3.74(b) shows a set of typical current *vs.* source-drain voltage (*I–V*$_{sd}$) data obtained from a single GaN NW-FET at different gate voltages (*V*$_g$). The two-terminal *I–V*$_{sd}$ are all linear, thus indicating that the metal electrodes make ohmic contacts to the GaN NWs. The n-type behaviour in nominally undoped GaN is due to the nitrogen vacancies and/or oxygen impurities. The transfer characteristics of the GaNNW FETs have also been examined. The *I vs. V*$_g$ for a GaN NW device recorded at different source-drain voltages (Figure 3.74c) are characteristic of an n-channel metal-oxide-semiconductor FET. Single-crystalline GaN nanowire heterostructures have been grown by metal-organic chemical vapour deposition and the PL data exhibit blue light emission, showing promise in nanophotonics.[405]

InN

InN fibres have been obtained well below the decomposition temperature of InN, by the use of azido-indium precursors. Polycrystalline InN fibres of diameters of ~20 nm and lengths in the range of 100–1000 nm have thus been obtained, the growth process being attributed to the solution–liquid–solid (SLS) mechanism.[406] Large-scale fabrication of single-crystalline InN nanowires (10–100 nm diameter) with the hexagonal wurtzite structure was achieved through the reaction of mixtures of In metal and In_2O_3 powders with flowing ammonia at 700 °C *via* a VS process.[407] InN nanowires have been obtained on gold-patterned silicon substrates in a controlled manner by the thermal evaporation of pure In in a NH_3/N_2 mixture.[408] Figure 3.75(a) shows typical SEM images of nanowires grown in this manner, to demonstrate how the InN nanowires grow only on the Au-coated areas. The higher magnification SEM image in Figure 3.75(b) shows the diameter of the nanowires to be in the 40–80 nm range, which is related to the size of the aggregated Au clusters caused by the heating of the substrate. Typical wire lengths go up to 5 μm. Figure 3.75(c) shows a HREM image of an individual nanowire of 40 nm diameter. The inset shows the SAED pattern of the nanowire, which can be indexed to the reflections of the hexagonal structure along the $\langle 100 \rangle$ direction. Accordingly, the HREM image shows the lattice planes with an

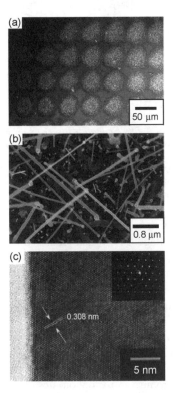

Figure 3.75   *SEM images of InN nanowires on a Si substrate:* (a) *Image taken at a lower magnification. Bright circles are InN nanowires grown on patterned-gold film. The other areas remain blank. The array of Au circles was formed by sputtering through a shadow mask.* (b) *Top view of the InN nanowires on one gold circle taken at higher magnification.* (c) *HREM image of a typical InN nanowire. Inset: the SAED pattern*
(Reproduced from ref. 408)

interlayer spacing of 0.308 nm. The $\langle 110 \rangle$ direction is parallel to the long axis of the wire, indicating that the nanowire grows along this direction. TEM observations reveal the presence of Au nanoparticles at the tips of the nanowires, which is consistent with the VLS mechanism. InN nanowires have been prepared by the reaction of indium acetate with hexamethyldisilazane as well as by decomposition of atomized indium acetylacetonate vapour over Au islands coated on Si(100) substrates.[409a]

Single-crystalline GaN, AlN and InN nanowires have been prepared on gold islands deposited on Si(100) substrates by nebulized spray pyrolysis (NSP).[409a] A dilute solution ($\sim$1 mM) containing the metal acetylacetonate in methanol is atomized and reacted with NH_3. InN nanowires grow in the narrow temperature range 585–600 °C, whereas GaN nanowires grow between 750 and 900 °C. The diameter of the nanowires is solely dependent on the particle size of the catalyst (Au), whereas the length of the nanowires can be dictated by controlling the time of the deposition. Figure 3.76 shows the SEM (Figure 3.76a and 3.76b) and TEM (Figure 3.76c and 3.76d) image of the nanowires obtained on a patterned Au-coated Si substrate. The low magnification SEM image reveals that the nanowires grow only on the places where Au is present (the white circles with average diameter 0.6 mm). High magnification SEM images on those places show the presence of a large amount of nanowires ($\sim$98% morphological yield) that are hexagonal InN, as confirmed by XRD. The nanowires are 10–50 nm in diameter and extend up to few

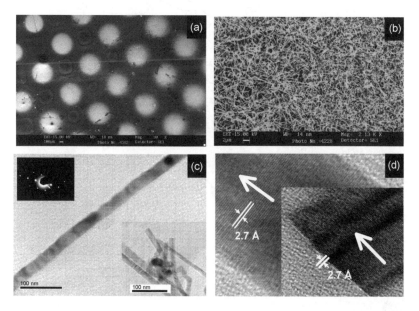

Figure 3.76 (a,b) *SEM image of the InN nanowires obtained by NSP; (c,d) TEM and HREM images, respectively of the nanowires. Inset in* (c) *shows the SAED pattern and the inset in* (d) *shows the HREM image at the tip of the nanowire* (Reproduced from ref. 409a)

microns long. The nanowires are single crystalline, as can be seen from the HREM image (Figure 3.76d), which shows fringes with a spacing of 0.27 nm, corresponding to the (101) planes of the wurtzite structure, as well as from the spots corresponding to the (100), (002), and (101) planes in the SAED pattern (inset in Figure 3.76c). The nanowires grow in the ⟨101⟩ direction (white arrow in Figure 3.76d shows the growth direction of the nanowire). GaN nanowires also have similar morphology to InN nanowires. GaN, AlN and InN nanowires have also been obtained by the thermal decomposition of the urea complexes of the metal trichlorides over Au islands deposited on silicon.[409b]

The Raman spectrum of InN nanowires shows bands at 445, 489, 579 cm^{-1}, attributed to the A_1 (transverse optical), E_2, and A_1 (longitudinal optical) phonon modes of the wurtzite structure (Figure 3.77a).[408] The Raman spectrum of an InN thin film is also shown for comparison. Figure 3.77(b) shows a typical PL spectrum measured from a large quantity of InN nanowires at room temperature, with a strong broad peak at 1.85 eV. The large half-width is primarily due to thermal excitations and the broad size distribution of the nanowires.[408] The characteristic absorption band of InN nanostructures has been established recently to be around 0.7 eV rather than the 1.85 eV reported earlier.[409]

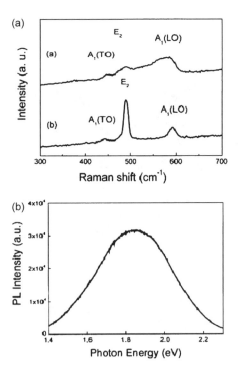

Figure 3.77 (a) *Raman spectrum of (top) InN nanowires and (bottom) InN thin films.* (b) *Room-temperature PL spectrum of InN nanowires using an Ar laser at 514 nm as the excitation source* (Reproduced from ref. 408)

Si$_3$N$_4$ and Si$_2$N$_2$O

Silicon nitride and β-sialon whiskers have been synthesized by the carbothermal reduction of silica under flowing N$_2$ + 3% H$_2$ gas mixture at around 1300 °C for 10 h.[410] Silicon nitride nanorods have been prepared by using CNTs as templates.[411] The products of the reaction of CNTs with a mixture of Si and SiO$_2$ powder under nitrogen are β-Si$_3$N$_4$, α-Si$_3$N$_4$ and Si$_2$N$_2$O nanorods of 4–40 nm diameter and lengths of several microns. Coaxial nanowires, 45 nm in diameter and 15 μm long, have been obtained by the reaction of silicon oxide nanoparticles with active carbon at 1450 °C in flowing nitrogen.[412] The coaxial nanowires consist of a silicon nitride core, an amorphous outershell of silicon and silicon dioxide. Si$_3$N$_4$ nanowires 10–70 nm diameter were synthesized by heating Si powders or Si/SiO$_2$ mixtures, with or without metal catalyst, at 1200 °C in N$_2$ or NH$_3$ or Ar flow.[413] The Si$_3$N$_4$ nanowires thus obtained are single crystals covered by an amorphous SiO$_2$ layer.

Nanowires of α-Si$_3$N$_4$ as well as of Si$_2$N$_2$O have been obtained by the reaction of NH$_3$ with SiO$_2$ gel and carbon nanotubes or activated carbon in the presence of a Fe catalyst.[360,414] When the reaction of arc-generated MWNTs with silica gel was carried out at 1360 °C for longer periods ($\geq$7 h), the product contained mainly Si$_3$N$_4$ nanowires with a minor component of Si$_2$N$_2$O. The composition of the final product depended on the duration over which the reaction was carried out. In the presence of catalytic Fe particles, silica gel reacts with the nanotubes within 4 h, giving a good yield of nanowires of pure Si$_3$N$_4$ with no Si$_2$N$_2$O. While the 0.1% Fe catalyst gives a mixture of the α-phase with the hexagonal β-phase, the 0.5% Fe catalyst yields almost monophasic α-Si$_3$N$_4$ nanowires. Figure 3.78(a) and 3.78(b) show SEM images of the Si$_3$N$_4$ nanowires obtained with the use of Fe catalysts. The reaction of silica and NH$_3$ with MWNTs prepared by the decomposition of ferrocene at 1360 °C yields a mixture of α- and β-Si$_3$N$_4$. Figure 3.78(c) shows the SEM image of the product obtained by this reaction. The nanowires have a large diameter (5–7 μ) and are of the order of hundreds of microns long.

6 Metal Carbide Nanowires

BC

Boron carbide nanowires (BCNWs) have been obtained, along with nanoparticles, by plasma-enhanced CVD of *closo*-1,2-dicarbadodecanborane.[415] The deposition temperature was between 1100 and 1200 °C and the carrier gas was argon. Aligned, monodisperse boron carbide nanofibres were generated using a porous alumina template and the single molecular precursor 6,6′-(CH$_2$)$_6$(B$_{10}$H$_{13}$)$_2$.[416] Shown in Figure 3.79 is a SEM image of the nanofibres with diameters of $\sim$250 nm and lengths of $\sim$45 μm. The fibres, heated at 1000 °C, were amorphous and became crystalline at 1025 °C. Thermal evaporation of C/B/B$_2$O$_3$ powders in an Ar atmosphere with or without a catalyst produces high-purity BCNWs by the VS mechanism.[417] The nanowires synthesized without catalyst had diameters in

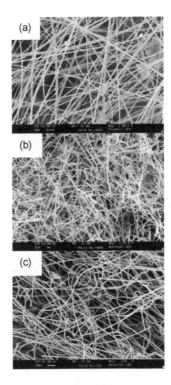

Figure 3.78 *SEM images of the Si₃N₄ nanowires prepared by the reaction of multi-walled carbon nanotubes with silica at 1360 °C for 4 h in the presence of 0.1% Fe and 0.5% Fe catalysts, respectively, are shown in (a) and (b). (c) SEM image of the Si₃N₄ nanowires prepared by the reaction of aligned multi-walled carbon nanotubes with silica at 1360 °C for 4 h*
(Reproduced from ref. 414)

Figure 3.79 *SEM image of aligned boron carbide nanofibres obtained upon pyrolysis of a filled template at 1025 °C, followed by the dissolution of the alumina matrix*
(Reproduced from ref. 416)

the range of 50–200 nm. Those grown using an iron catalyst form by the VLS mechanism, having diameters ranging from 10–30 nm. As early as 1995, Lieber and co-workers[418] demonstrated the synthesis of carbide nanorods by a carbon nanotube confined reaction. They synthesized nanorods of TiC, NbC, Fe_3C, SiC and BC_x with diameters between 2 and 30 nm and lengths of up to 20 μm. B-doped CNTs have been prepared through a partial substitution reaction, where some carbon atoms of CNTs are substituted by B atoms.[419] CNTs react with B powder at 1150 °C to yield boron carbide nanorods with the formula B_4C.[420] The transition from linear to helical growth of amorphous BCNWs in the presence of Fe catalyst has been investigated by HREM.[421]

SiC

SiC nanowires (SiCNWs) have been synthesized at 900 °C by the laser ablation of a SiC target, by means of the VLS growth mechanism, and characterized by SEM, HREM and Raman scattering.[422] A SEM image of the SiCNWs obtained by this route is shown in Figure 3.80. β-SiC nanorods, with and without amorphous silica layers, have been obtained by the carbothermal reduction of sol–gel derived silica xerogels containing carbon nanoparticles at 1800 and 1650 °C, respectively, in an Ar atmosphere.[423] Bulk quantities of β-SiC nanowires have been produced from a mixture of activated carbon and sol–gel derived silica embedded with Fe nanoparticles.[424] The nanowires consist of a 10–30 nm diameter single-crystalline core wrapped with an amorphous silicon oxide layer, and are up to several tens of microns long.

Several methods have been employed by Gundiah *et al.*[414] to prepare SiC nanowires, including heating silica gel or fumed silica with activated carbon in a reducing atmosphere, the carbon particles being generated *in situ* in one of the methods. The simplest method to obtain β-SiC nanowires involves heating silica gel with activated carbon at 1360 °C in H_2 or NH_3. In Figure 3.81(a) we show the

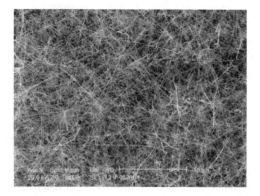

Figure 3.80 *SEM image of SiC nanowires obtained by the laser ablation of a SiC target* (Reproduced from ref. 422)

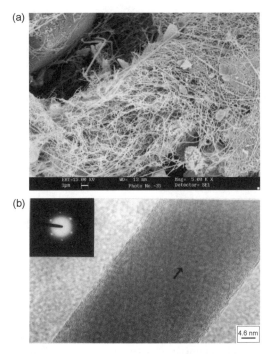

Figure 3.81 *β-SiC nanowires obtained by heating a gel containing activated carbon and silica at 1360 °C for 7 h: (a) SEM image and (b) HREM image of a single nanowire. Inset: the SAED pattern. The arrow denotes the normal to the (111) plane and the direction of growth of the nanowire*
(Reproduced from ref. 414)

SEM image of SiCNWs obtained by heating the gel containing activated carbon and silica at 1360 °C in NH_3 for 7 h. The HREM image of a single nanowire in Figure 3.81(b) reveals the (111) planes. The inset SAED pattern shows Bragg spots corresponding to the (111) planes, with some streaking due to the presence of stacking faults.

Silicon–carbon nanotubes and nanowires of various shapes and structures have been obtained through the reaction of Si (prepared by the disproportionation of SiO) with MWNTs at different temperatures.[425] β-SiC nanowires obtained by the reaction of Si with CCl_4 and Na at 700 °C have been characterized.[426] Heating a mixture of SiC powder and Al catalyst at 1700 °C in Ar produces SiC nanowires by the VLS growth mechanism.[427] These had diameters in the range 10–50 nm and lengths in the 1–2 μm range. Nanojunctions of SiC were obtained by a similar procedure.[428]

A CVD technique has been used to synthesize helical SiC nanowires covered with a silicon oxide sheath.[429] The SiC core had a diameter between 10 and 40 nm (with a helical periodicity of 40 to 80 nm) and was covered by a uniform layer of 30–60 nm thick amorphous silica. The VLS process has been used to synthesize BN-coated β-SiC nanowires.[430] Oriented SiCNWs are obtained on reacting

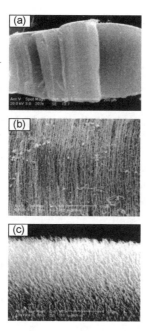

Figure 3.82 *SEM images of oriented SiC nanowires:* (a) *Low-magnification images of an oriented SiC array.* (b) *High-magnification image of the oriented SiC array. The particles present on the array surface are silicon-containing particles formed and deposited during the reaction.* (c) *Bottom surface of the SiC nanowire array, showing high density of well-separated, oriented nanowire tips*
(Reproduced from ref. 431)

aligned CNTs with SiO by a carbon nanotube confined reaction.[431] SEM images of the nanowires (10–40 nm diameter, 2 mm length) are shown in Figure 3.82. Field-emission current densities of the nanowires were $\sim 10\,\mu A\ cm^{-2}$ at applied fields as low as $2.5–3.5\ V\,\mu m^{-1}$. Aligned SiCNWs have also been synthesized from Si substrates *via* a catalytic reaction with a methane–H_2 mixture at $1100\,^{\circ}C$.[432] Thermal evaporation of SiC powders over a Fe catalyst yields needle-shaped SiCNWs.[433]

Self-assembled multicoaxial SiC nanowires within wires have been synthesized by CVD.[434] The recrystallization of the copper substrate by thermal treatment before the vapour deposition causes thermal grooving along grain boundaries, which provides a natural template for the self-assembled growth of nanowires. The core of the wires so produced is made up of numerous faceted SiC nanowires, typically 20–50 nm in diameter, encapsulated by larger hollow wires that are typically 0.2–1.0 nm in diameter. Optical and electrical transport properties of SiC NWs have been examined; the nanowires are n-type semiconductors.[435] The nanowires may be useful for applications as sensors, detectors and actuators. Field-emission properties of SiC nanowires have been reported.[433]

7 Metal Chalcogenide nanowires

CdS

Thermal evaporation of CdS nanoparticles results in the nucleation and growth of CdS nanowires by the VS route.[436] CdS nanowires are also obtained by the thermal evaporation of CdS powders in the presence of a Au catalyst.[437] The nanowires have diameters of 60–80 nm, with lengths going up to several tens of micrometres. CdS nanowires thus prepared show strong red emission with a maximum centred around 750 nm, attributed to the surface states.[437] Uniform arrays of uniform CdS nanowires have been fabricated in AAMs, by electrochemical deposition.[438] The length of the nanowires by this method is in the 1 μm range whereas the diameter is as low as 9 nm. Resonant Raman spectra of CdS nanowires show strong features that depend on the particle size. Single-crystalline CdS nanowires are also obtained by the electrochemical deposition of $CdCl_2$ and thioacetamide or similar precursors within the pores of AAM templates.[439,440] The Raman spectra of CdS nanowires prepared by electrochemical deposition of $CdCl_2$ and thioacetamide show Raman bands at 304, 606 and 908 cm^{-1}, corresponding to the first-, second-, and third-order transverse optical (TO) phonon modes of CdS, respectively. Other AAM template based methods for the synthesis of CdS nanowires include a sol–gel templated synthesis,[441] use of a Au/Si substrate[442] and sulfphurization of Cd electrodeposited inside the pores.[443]

Mesoporous CdS nanorods have been prepared by an ion-exchange process starting from CdS nanocrystallites as precursors.[444] The resonant Raman spectrum of mesoporous CdS nanorods shows bands at 296 and 594 cm^{-1} due to the fundamental and the overtone of the longitudinal optical phonon mode. The specific surface areas of mesoporous CdS nanorods, calculated by Barrett–Joyner–Halenda, method were fairly high (*e.g.* 57.5 m^2 g^{-1}).

A non-aqueous synthesis, using ethylenediamine as a solvent at 120–190 °C and starting from elemental precursors, yields CdS, CdSe and CdTe nanorods.[445] Nanowires of CdS with high aspect ratios have been obtained by the use of Triton X-100 [t-octyl-C_6H_4-$(OCH_2CH_2)_xOH$] ($x = 9$, 10) as the surfactant.[19,20] Figure 3.83(a) shows SEM images of the hexagonal CdS nanowires obtained by using a low concentration of Triton X-100 as the surfactant. The TEM images of the CdS nanowires shown in Figure 3.83(b)–(d) reveal the diameter of the nanowires to be in the 80–160 nm range with lengths of up to 4 μm. Most of the nanowires are polycrystalline, as revealed in the ED pattern in the inset of Figure 3.83(c), corresponding to a [100] spacing of 3.66 Å. CdS nanowires prepared by the surfactant method show a blue-shift of the absorption maximum compared with the bulk, which shows a band around 600 nm.[20] CdS, CdSe, ZnSe and PbSe nanorods have been prepared by using a monodentate ligand such as *n*-butylamine as the solvent.[446] CdS nanowires have been prepared *via* an *in situ* micelle–template interface reaction route by adjusting the concentration of the surfactant.[447] CdS nanowires, prepared thus, show an absorption peak at 452 nm, which is quantum-confined in comparison to the bulk. Size-tunable CdS nanorods have been synthesized *via* the reaction at low temperature (25–65 °C) of air-insensitive inorganic

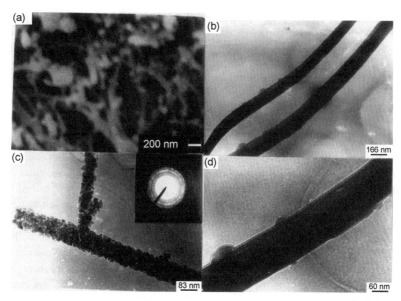

Figure 3.83 (a) *SEM image of CdS nanowires (80–160 nm diameter) obtained by using Triton X-100 as the surfactant; (b)–(d) TEM images of the CdS nanowires. Inset in (c) shows the ED pattern of the nanowires* (Reproduced from ref. 20)

precursors cadmium acetate and sodium sulphide in on aqueous phase with $(EO)_x(PO)_y(EO)_x$ triblock copolymers as surfactant templates.[448]

A single source cadmium thiosemicarbazide complex has been used to prepare CdS nanorods.[449] The nanorods show a band edge at 490 nm, which is slightly blue-shifted in comparison to the bulk, while the PL spectrum shows an emission maximum at 515 nm.[449] Long CdS nanowires with typical widths of 70–80 nm and lengths ranging from several micrometers to tens of micrometres can be obtained by *in situ* growth on a substrate of a carboxylic acid-functionalized self-assembled monolayer.[450] CdS nanorods are formed under solvothermal conditions, starting from thiourea and cadmium nitrate in ethylenediamine.[451] A solvothermal route has been developed for the preparation of multi-armed CdS nanorods by the reaction of $CdCl_2$ and thiourea in the presence of dodecylthiol and ethylenediamine at $160\,^\circ C$.[452] Multi-armed CdS nanorods show a strong emission peak at 718 nm, due to a self-activated emission of CdS. This band shifts to the red with an increase in the length of the arms of the nanorods and to the blue with decreasing arm length.[432] Nanohelices of CdS have been obtained by the mineralization of supramolecular organic ribbons.[453]

Ordered arrays of $Cd_{1-x}Mn_xS$ magnetic nanowires of high crystallinity have been obtained by incorporating semimagnetic semiconductor nanowires into SiO_2 MCM-41 hosts.[454] The PL spectrum of $Cd_{0.985}Mn_{0.015}S$ in MCM-41 shows an IR band at 1.38 eV and a red band at 2.02 eV, which could possibly arise from the $^4T_1 \rightarrow {}^6A_1$ internal Mn^{2+} ($3d^5$) transitions for the ions in different crystallographic environments.[454] The Raman spectrum of the CdS nanowires in AAMs reveals a small blue-shift in the excitation energy with decreasing particle size, demonstrating quantum confinement effects. The lack of polarization dependence is considered to be due to microcrystallites with low aspect ratios in the nanowires.[455]

CdSe and CdTe

CdSe nanorods and dendritic fractals have been synthesized through a solution-phase hydrothermal synthesis method, starting from cadmium nitrate and sodium selenite.[456] CdSe nanorods and fractals show absorption edges at 1.68 and 1.78 eV in comparison to the bulk band gap of 1.74 eV.[456] The Raman spectrum of CdSe fractals shows a strong peak at 209 cm^{-1}, which is due to the characteristic vibration of Cd-Se-Cd.[456] CdSe and CdTe nanowires with the zinc blende structure have been prepared through a poly(vinyl alcohol)-assisted solvothermal method at ~170 °C using ethylenediamine.[457] Monodisperse CdSe nanorods are obtained at a relatively low temperature of 160 °C, starting from cadmium naphthenate and Se.[458] The PL spectrum of the CdSe nanorods shows emission maxima between 500 and 600 nm.

CdSe nanorods get aligned in a nematic liquid-crystalline phase, and the superlattice structure of the nanorods formed upon deposition on a substrate is determined by the liquid crystalline phase.[459] Since the liquid-crystalline phase is affected by external fields or pretreated surfaces, it may be possible to achieve a high degree of control over the deposited films of the nanorods.

Nanowire arrays of group II–VI semiconductors such as CdSe and CdTe have been prepared by the direct current electrodeposition in porous AAMs.[460–462] Polycrystalline nanowires of CdSe with high aspect ratios have been prepared by employing Triton X-100 as the surfactant.[19,20] These CdSe nanowires show a blue-shift of the absorption maximum compared to the bulk. 2D arrays of CdSe nano-pillars with large height-to-width ratio have been fabricated by using e-beam lithography as well as electrochemical deposition techniques.[463]

Colloidal CdSe quantum rods were synthesized by the injection of precursor molecules into a hot surfactant.[464] The radiative lifetime, polarization, and the global Stokes shift of colloidal CdSe quantum rods with aspect ratios from 1.9 to 3.8 measured at room temperature indicate a transformation of the electronic structure from a 0D quantum-dot system into a 1D quantum-wire system.[464] CdS/CdSe core/sheath nanowires with diameters of 7.7/0.75 nm were prepared by treating CdS nanowires with Se in tributylphosphine at 100 °C for 24 h.[465] The absorption spectrum of core/sheath CdS/CdSe nanowires in comparison to the CdS nanowires shows little difference, with a peak at 472 nm being common to both.[465] This is consistent in that the major component of the core/shell particle dominates the absorption spectra. The PL spectra of CdS-only and CdS/CdSe core/sheath nanowires differ, with the sheath completely quenching the fluorescence of CdS nanowires, as found in the CdS/CdSe core/shell system. When excess Cd^{2+} cations are added to a dispersion of the CdS/CdSe nanowires in water, the fluorescence is restored to virtually the previous level.[465]

Metal–CdSe–metal (metal = Au, Ni) nanowires were grown by the electro-chemical replication of porous Al_2O_3 and polycarbonate track etch membranes with pore diameters of 350 and 700 nm, respectively.[466] Nanoparticles of CdTe spontaneously reorganize into crystalline nanowires upon controlled removal of the protective shell of organic stabilizer.[467] Accumulation of Cd^{2+} at the surface may destroy sites where radiationless recombination of charge carriers can occur. A red-shift of the emission band arises during the spontaneous reorganization of CdTe nanoparticles into the nanowire.[467] Shape-control as well as shape-evolution of colloidal semiconductor nanocrystals such as CdSe are possible, depending on the nature of the precursor and the monomer concentration.[468]

Alivisatos and co-workers synthesized arrow-, teardrop- and tetrapod- shaped CdSe nanocrystals by thermal decomposition of the organometallic precursors.[469] This has been followed by the synthesis of tetrapod structures of chalcogenides by other workers. Branched tetrapods with CdSe central tetrapods and terminal CdTe branches possess unusual charge-separating poperties.[470] Both CdSe nanorods and CdTe tetrapods emit well-defined bandgap luminescence under similar conditions. Electrons seem to localize in the CdSe tetrapods.

PbS and PbSe

Rod-like PbS nanocrystals have been prepared *via* a biphasic solvothermal inter-face reaction route at 140–160 °C.[471] The as-prepared nanorods have diameters of 30–160 nm and lengths of up to several micrometres. PbS nanowires 6 nm diameter and several micrometres in length have been synthesized inside the channels of mesoporous silica SBA-15.[472] The PL spectrum of PbS-wire-SBA-15 shows an emission maximum around 665 nm, corresponding to an energy gap of 1.87 eV. Compared to the bulk (0.41 eV), nanosized PbS exhibits a large blue-shift in the PL spectrum, indicating the quantum size effect. A soft-template reaction is found to be effective in preparing PbS nanowires with an average size of 8×350 nm by the reaction of $Pb(NO_3)_2$ and Se powder at 60 °C under ambient pressure.[473] PbSe nanowires of high aspect ratios are obtained by the room temperature reaction of Se, $PbCl_2$ and KBH_4 in ethylenediamine solvent.[474]

Bismuth Chalcogenides

Bi_2S_3 nanorods, 10 nm in diameter and length up to 300 nm long have been obtained from a formaldehyde solution of bismuth nitrate and thiourea, on microwave irradiation.[475] A solvothermal decomposition process has also been successful through a reaction between $BiCl_3$ and thiourea in polar solvents at 140 °C for 6–12 h.[476] A hydrothermal reaction between $Bi(NO_3)_3.5H_2O$ and $Na_2S.9H_2O$ in the temperature range 100–170 °C also yields Bi_2S_3 nanorods.[477] A sonochemical method has been employed to prepare Bi_2S_3 nanorods, starting from an aqueous solution of bismuth nitrate and sodium thiosulfate, in the presence of complexing agents such as ethylenediaminetetraacetic acid, triethanolamine and sodium tartrate.[478] Uniform urchin-like patterns of Bi_2S_3 nanorods are obtained in ethylene glycol at 197 °C after 30 min.[479] The reaction involves decomposition of a

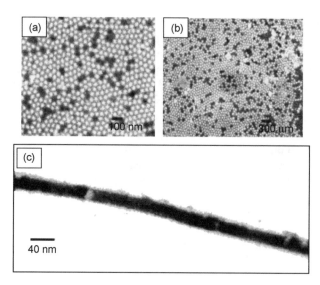

Figure 3.84   (a) *and* (b) *SEM images of the Bi₂Te₃ nanowire array composite obtained by electrodeposition. Bright regions in the image are the filled pores.* (c) *Bright-field TEM images of a single Bi₂Te₃ nanowire* (Reproduced from ref. 481)

Bi-thiourea complex, starting from $Bi(NO_3)_3.5H_2O$ and thiourea as precursors, resulting in nanorods of Bi_2S_3. The diameters of individual nanorods are in the 50 nm range. Biomolecule-assisted synthesis of ordered structures of Bi_2S_3 nanorods has been reported.[480]

Dense continuous Bi_2Te_3 nanowires have been fabricated by the direct electro-deposition into the pores of an AAO template.[481–483] In Figure 3.84(a) and 3.84(b) we show the SEM images of the bottom surface of the nanowire arrays prepared by the above method. The images reveal the filling of the pores to be ~20%, whereas the nanowires have nucleated in >80% of the pores. A bright-field TEM image of a single Bi_2Te_3 nanowire is shown in Figure 3.84(c).

CuS and CuSe

Nanowires of CuS and CuSe with high aspect ratios have been prepared by employing Na-AOT [sodium bis(2-ethylhexyl)sulphosuccinate] or Triton X-100 as the surfactant.[20,484] Figure 3.85(a) and 3.85(b) show the SEM and TEM images of CuS nanowires obtained by using AOT as the surfactant. The average diameter of the wire-like structures varies between 5 and 20 nm, with the length extending up to 0.2 μm or longer. Figure 3.85(c) and 3.85(d) shows the SEM and TEM images of the CuS nanowires obtained when using Triton X-100 as the surfactant. The average diameter is in the 5–15 nm range and the length varies between 150 and 900 nm. The inset in Figure 3.85(d) shows the ED pattern of a CuS nanowire, corresponding to a lattice spacing of 2.8 Å between the (103) planes. The CuS nanowires are single-crystalline. The use of Triton X-100 appears to yield better

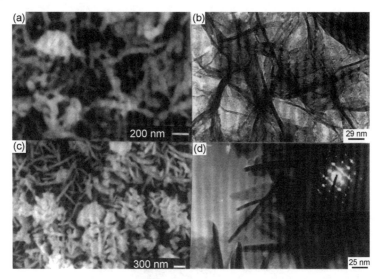

Figure 3.85 (a) *and* (b) *SEM and TEM images of CuS nanowires obtained by using AOT as the surfactant;* (c) *and* (d) *SEM and TEM images of CuS nanowires obtained by using Triton X-100 as the surfactant. Inset in* (d) *shows the SAED pattern* (Reproduced from ref. 20)

CuS nanowires. Figure 3.86(a) and 3.86(b) shows SEM and TEM images of the hexagonal CuSe nanowires obtained with Triton X-100. The inset in Figure 3.86(b) shows the ED pattern of the CuSe nanowires, with a lattice spacing of 1.90 Å corresponding to the (110) planes of CuSe. The average diameter of the nanowires is around 70 nm, with the lengths varying between 1 and 2 μm. The HREM image of a CuSe nanowire in Figure 3.86(c) reveals an interlayer spacing of 0.33 nm between (101) planes. Electronic absorption spectra of the 5–20 nm CuS nanowires exhibit a broad band between 300 and 600 nm, with a distinct band around 400 nm and a long-wavelength band peaking around 1000 nm. The band in the near-infrared is characteristic of CuS, arising from an electron-acceptor state lying within the bandgap. CuSe nanowires, with an average diameter of ~70 nm show a broad feature, centred around 540 nm, along with an intense band above 1000 nm.[484]

A surfactant-assisted method for the preparation of core-shell Cu_xS nanowires has been reported.[485] The growth of the nanowires was carried out by exposing a Na-AOT-covered copper surface to H_2S for 12 h. The growth of highly oriented and uniform-sized Cu_2S nanowire arrays occurs on copper substrates in a H_2S/O_2 atmosphere by an oxide-assisted mechanism.[486] Field-emission measurements on the Cu_2S nanowire arrays show nonlinearity in the Fowler–Nordheim plot, indicating that the nanowires can have potential as cold cathodes.[486]

ZnS and ZnSe

Nanowires of ZnS with high aspect ratios are obtained by using a surfactant such as AOT or Triton X-100, similar to the method used in the preparation of CuS and

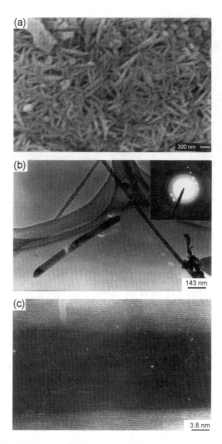

Figure 3.86   (a) *and* (b) *SEM and TEM images of CuSe nanowires obtained by using Triton*
X-100 as the surfactant. (c) *HREM image of a CuSe nanorod, showing lattice*
resolution. Inset in (b) *is the SAED pattern*
(Reproduced from ref. 20)

CuSe nanowires.[20,484] Figure 3.87 shows the TEM images of the ZnS nanowires
with a hexagonal structure, obtained using Triton X-100. The diameter is in the
range 10–50 nm, and the length in the range 8–10 μm. The inset in Figure 3.87(c)
shows the ED pattern of ZnS nanowires, corresponding to a lattice spacing of
2.854 Å between (101) planes. ZnS nanowires with an average diameter in the
10–50 nm range exhibit a absorption band between 325 and 365 nm, with a peak
around 340 nm. The PL spectrum of the ZnS nanowires shows a broad band
that peak's around 430 nm and extends up to 550 nm.[484] Decomposition of zinc
xanthate provides a convenient means of producing uniform, thin ZnS nanorods
and nanowires.[487]

Winding ZnS nanowires have been prepared by reacting Zn^{2+} and S^{2-} in a
reverse micellar template.[488] Nanowire formation occurs through the process of the
directional aggregation and oriented growth of ZnS nanoparticles. ZnS nanorods have been fabricated by annealing precursor ZnS nanoparticles, which were

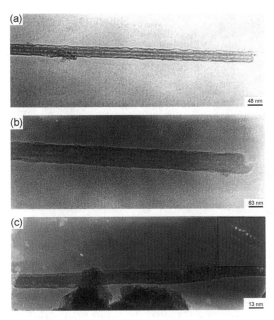

Figure 3.87 *TEM images of ZnS nanowires obtained by using Triton X-100 as the surfactant. Inset in (c) is the SAED pattern, showing the single-crystalline nature of the nanowires*
(Reproduced from ref. 20)

prepared by one-step, solid-state reaction of $ZnCl_2$ and Na_2S through grinding by hand at ambient temperature, in an NaCl flux.[489] The nanorods thus prepared have diameters of 40–80 nm, and lengths of up to several micrometres. Raman spectra of the nanorods show a Raman band at $349\,cm^{-1}$ due to the LO phonon mode. Apart from this, there is a weak Raman band at $276\,cm^{-1}$, corresponding to the TO phonon mode.[489] Thermal evaporation of ZnS particles, in the presence of a Au catalyst, onto a silicon substrate results in the growth of single-crystalline ZnS nanowires.[490] The PL spectra of ZnS nanowires prepared by thermal evaporation of ZnS show blue and green light emissions that are attributed to surface states and the presence of Au ions.[490] A general CVD procedure for the synthesis of 1D metal sulphides, including ZnS and CdS, has been described.[491] A virus-based scaffold for the synthesis of nanowires of ZnS, CdS and other materials has been reported, with the means of modifying substrate specificity through biological methods.[492]

Single-crystal ZnS nanosheets with micrometre-scale lateral dimensions and flake-like ZnO dendrites have been synthesized from a lamellar molecular precursor, $ZnS.(NH_2CH_2CH_2NH_2)_{0.5}$, by a template method.[493] By using as-synthesized ZnO nanobelts as templates, nanostructured rectangular porous ZnS nanocables have been synthesized, by the chemical reaction of ZnO with H_2S.[494] The PL spectrum of the nanocables shows a small blue-shift, possibly indicating a small quantum-confinement effect. Lasing in ZnS nanowires[495] as well as optical properties and Raman spectra of cross-sectional ZnS nanowires[496,497] have been reported.

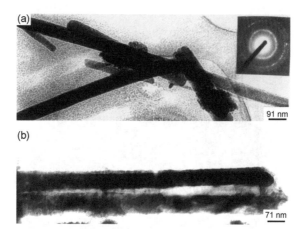

Figure 3.88   (a) *and* (b) *TEM images of ZnSe nanowires (50–150* nm *diameter) obtained using Triton X-100 as the surfactant. Inset in* (a): *ED pattern of the nanowires* (Reproduced from ref. 20)

ZnSe nanowires have been prepared by a surfactant-assisted method, starting from ZnO and Se powder using Triton X-100 as the surfactant.[20,484,498] This method gives hexagonal ZnSe nanowires with a diameter in the 10–50 nm range, the lengths of up to 8–10 μm. TEM images of the nanowires are shown in Figure 3.88(a) and 3.88(b). The inset in Figure 3.88(a) shows the SAED pattern with a spacing of 1.97 Å, corresponding to the (110) planes. Size-dependent periodically twinned ZnSe nanowires have been fabricated by the VLS process.[499] ZnSe tetrapods have been obtained by Sn-catalysed thermal evaporation.[500] Laser-ablation has also been used to produce the nanowires.[501] ZnSe nanowires prepared by the surfactant route show a broad absorption band around 480 nm. The PL spectrum of the nanowires shows a blue-shift from the bulk, with a peak around 425 nm and a shoulder at 435 nm, respectively.[484] The Raman spectrum of ZnSe nanowires shows bands at 257 and 213 cm^{-1}, due to the longitudinal and transverse optic phonons, blue-shifted with respect to the bulk ZnSe modes because of compressive strain. The surface phonon mode is seen at 237 cm^{-1}.[498]

NbS$_2$ and NbSe$_2$

Nanorods of NbS$_2$ have been prepared by the thermal decomposition of NbS$_3$ in H$_2$ at ~1000 °C.[502] The nanorods extend up to several microns (Figure 3.89a). Nanorods of NbS$_2$ have also been synthesized by the use of carbon nanotubes as templates by the treatment of the NbCl$_4$-coated carbon nanotubes with H$_2$S.[503] NbS$_2$ nanowires have been generated by heating Nb and S powders in sealed tubes in the presence of iodine.[504]

NbSe$_2$ nanorods have been prepared by the decomposition of the triselenide at 970 K, under a gas flow of Ar.[505] A representative HREM image of the NbSe$_2$ nanorod (Figure 3.89b) reveals that the nanorod has a diameter of ~20 nm,

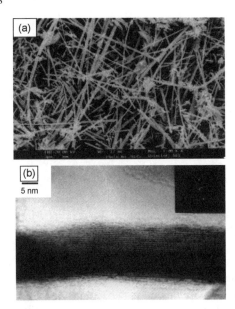

Figure 3.89   (a) *SEM image of NbS₂ nanorods*, (b) *HREM image of a NbSe₂ nanorod. Inset in (b): ED pattern of the nanorod*
(Reproduced from ref. 502, 505 respectively)

consisting of 25 layers with an interlayer spacing of ~6.2 Å. The inset in the same figure shows an ED pattern with Bragg spots corresponding to the $d(002)$ reflection. No stacking faults or dislocations are seen in the nanorods of $NbSe_2$. The PL spectrum of the $NbSe_2$ nanostructures show a band at 820 nm, a blue-shift of about ~5 nm from the bulk. The nanostructures of $NbSe_2$ are metallic from 300 K down to low temperatures and become superconducting at 8.3 K.[505] The observed T_c is close to the bulk value, probably because of the relatively large diameters of the nanostructures. The temperature variation of resistance in the metallic regime is, however, considerably steeper in the bulk sample.

Other Chalcogenides

A low-temperature solution route has been employed for the synthesis of MoS_2 fibres, wherein the reaction is carried out by stirring a solution of $(NH_4)_2Mo_3S_{13}$ with ethylene diamine, followed by annealing of the product obtained at 400 °C in N_2.[506] The electronic absorption spectrum of MoS_2 fibres displays a strong, broad absorption with a maximum centred at 386 nm and two weaker absorbances at 610 and 652 nm, the features being comparable to the bulk sample.[506] β-La_2S_3 nanorods, with diameters of 30–50 nm and lengths of up to a micrometre, have been synthesized by a solvothermal route.[507] The method involves the reaction of the precursors $LaCl_3$ and thiourea in an autoclave sealed and maintained at 260–280 °C for 10 h. The PL spectrum of β-La_2S_3 nanorods shows a strong, wide emission band, peaking at about 420 nm, on excitation at 250 nm.[507]

Fe$_7$S$_8$ nanowires have been prepared recently by a solvothermal procedure by the reaction of FeCl$_2$.4H$_2$O with thioacetamide in the presence of ethylenediamine (en).[508a] Nanowires of the Fe$_7$S$_8$.en adduct first formed are decomposed at a low temperature to obtain the pure sulphide nanowires. GeS and GeSe nanowires have been prepared by using the organoammonium derivatives as precursors.[508b]

A solvothermal route involving the reaction of Cu, In and S powders in ethylenediamine solvent at 280 °C has been developed for the synthesis of CuInS$_2$ nanorods.[509] A low-temperature hydrothermal preparation has been carried out, starting from CuCl, SbCl$_3$ and NH$_2$CSNH$_2$, to produce famatinite (Cu$_3$SbS$_4$) nanofibres and tetradrite (Cu$_{12}$Sb$_4$S$_{13}$) nanoflakes.[510] Nanometre-wide, micrometre-long fibres of LiMo$_3$Se$_3$ have been obtained by the reaction of the elements in sealed, evacuated quartz ampoules at 1000 °C, followed by Li intercalation in the Mo$_3$Se$_3$ chains.[511] A soft chemical route for making MMo$_3$Se$_3$ nanowires with different M cations has been is reported.[16a] This involves a cation-exchange process at low temperatures, facilitated by the addition of 12-crown-4, which extracts the Li cations from the starting material and incorporates the other cations in solution as their respective alkali halides. A redox templating procedure has been developed to synthesize continuous metal nanowires of Au, Ag, Pt, Pd, of 10–100 nm diameter, using LiMo$_3$Se$_3$ nanowires as sacrificing templates with a reducing capability.[512]

8 GaAs, InP and Other Semiconductor Nanowires

GaAs

Laser ablation has been widely used to synthesize semiconducting nanowires. In this method, ablation of a suitable target by a laser source produces nanometre-diameter catalytic clusters that define the size and direct the growth of the crystalline nanowires by the VLS mechanism. Equilibrium phase diagrams are used to predict catalysts and growth conditions. Using this method, Duan and Lieber[513] synthesized nanowires of group III–V semiconductors such as GaAs, GaAsP, InAs, InP, InAsP, II–VI semiconductors such as ZnS, ZnSe, CdS, CdSe and IV–IV alloys of SiGe. GaAs nanowires (GaAsNWs) were fabricated by laser ablation of GaAs powders mixed with Ga$_2$O$_3$ by the oxide-assisted process without the use of any metal catalyst.[514] Figure 3.90(a) shows a typical SEM image of GaAsNWs, which have diameters of ~5 nm and lengths of up to 10 μm. The HREM image of a single nanowire, along with the electron diffraction pattern, shown in Figure 3.90(b) reveals that the nanowire is single-crystalline and grows along the $\langle\bar{1}11\rangle$ direction. A thin amorphous oxide layer covers the nanowire. By a similar approach, nanowires of GaP and GaN were also synthesized. The GaAs nanowires show a significant shift and broadening in the Raman and PL spectra, arising from stress, impurities, and defects.[515]

Low-temperature solution–liquid–solid growth has been used to synthesize GaAsNWs in the diameter range 6–16 nm with lengths of several microns, using In as the catalyst.[516] The nanowires had a narrow diameter distribution that depended on the size of the metal particles. Ordered GaAsNWs have been grown by MBE on

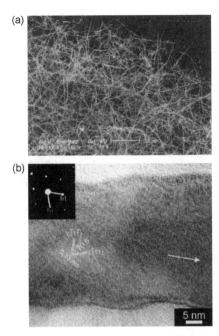

Figure 3.90 (a) *A SEM image of GaAs nanowires synthesized by the oxide-assisted method.* (b) *A HREM of a GaAs nanowire. The growth axis, denoted by the white arrow, is close to the [$\bar{1}1\bar{1}$] direction. The inset is the corresponding ED pattern recorded along the [110] zone axis perpendicular to the nanowire growth direction*
(Reproduced from ref. 514)

GaAs (111)B substrates by Au-catalysed VLS growth in porous alumina templates.[517] The diameter distribution obtained was narrow. PL measurements on individual GaAsNWs synthesized by the laser-assisted catalytic growth show large blue-shifts compared to the bulk, due to quantum confinement.[518]

Using porous membranes, uniform arrays of GaAs, CdS, CdSe, CdS$_x$Se$_{1-x}$, and Cd$_x$Zn$_{1-x}$S nanowires as well as of metallic (Ni, Fe) nanowires have been synthesized.[519] The *I–V* characteristics of two terminal devices made from the nano-arrays exhibit periodic conductance oscillations and Coulomb staircases. These observations are likely to be associated with the ultra-small tunnel junctions formed naturally in the arrays. Single electron tunnelling in the presence of interwire coupling in these arrays leads to spontaneous electrostatic polarization of the wires.

InP and GaP

Single-crystalline InP nanowires with n- and p-type doping are obtained by laser-assisted catalytic growth. They have diameters of the order of 10 nm and lengths of up to tens of microns.[520] Gate-voltage-dependent transport measurements demonstrate that both the n- and p-type nanowires can be synthesized. The doped nanowires function as nanoscale FETs, and can be assembled into crossed-wire p-n junctions that can exhibit rectifying behaviour. Electric-field-directed assembly enables integrated device arrays from nanowire building blocks. Aqueous-solution growth of InP, GaP and other nanowires employing inorganic–surfactant mesostructures has been described.[521]

Lieber and co-workers[31a] have investigated the hierarchical assembly of 1D nanostructures into well-defined functional networks. InPNWs nanowires have been assembled into parallel arrays with control of the average separation. By combining fluidic alignment with surface-patterning, they have controlled the periodicity. Complex crossed nanowire arrays can be prepared with the layer-by-layer assembly, with different flow directions for sequential steps, and their transport properties studied. The diameters and lengths of InPNWs grown by laser-assisted catalytic growth can be controlled using monodisperse nanocluster catalysts.[522] InPNWs with nearly monodisperse diameters of 10, 20 and 30 nm were grown from nanocluster catalysts of the same diameters. A schematic of the growth is shown in Figure 3.91(a). The length of the nanowires can be controlled by varying the ablation time as shown in Figure 3.91(b). Figure 3.91(c) and 3.91(d) show TEM images of a nanocluster catalyst at a nanowire end and a crystalline wire core produced using 20 and 10 nm diameter Au nanoclusters, respectively. A bottom-up method to produce ordered and uniformed arrays of InP nanowires, employing nanoimprint nanolithography, has been demonstrated.[523]

PL imaging and spectroscopy have been employed to investigate isolated, individual InPNWs synthesized by laser-assisted catalytic growth.[524] The emission maxima shift to higher energies with decreasing wire diameter for diameters less than 20 nm due to radial quantum confinement. Polarization-sensitive measurements on isolated InPNWs reveal a striking anisotropy in the PL intensity recorded parallel and perpendicular to the long axis of a nanowire.[525] The order of magnitude polarization anisotropy is explained in terms of the large dielectric contrast between these free-standing nanowires and the surrounding environment as opposed to quantum confinement effects.

Semiconductor nanowire superlattices of group III–V and group IV materials have been generated within the nanowires by repeated modulation of the vapour-phase semiconductor reactants during growth of the wires.[526] Compositionally modulated superlattices consisting of 2 to 21 layers of GaAs and GaP have been prepared. n-Si/p-Si and n-InP/p-InP modulation doped nanowires have been synthesized. These have been characterized by single-nanowire photoluminescence, electrical transport and electroluminescence measurements.

Laser-assisted catalytic growth has been used to synthesize single crystalline gallium phosphide nanowires (GaPNWs).[527] Ga and P were ablated with a laser and the nanowires formed on Au nanoclusters supported on a SiO_2 substrates. The nanowires grow along the [111] direction. GaPNWs of 22 nm in diameter and hundreds of microns in length were synthesized by the laser ablation of a mixture of GaP and Ga_2O_3 using oxide-assisted growth.[528] GaPNWs are also formed by the sublimation of ball-milled GaP powders onto Ni-coated alumina substrates.[529] Figure 3.92(a) shows a SEM image of the nanowires and Figure 3.92(b)–(f) show the TEM images.

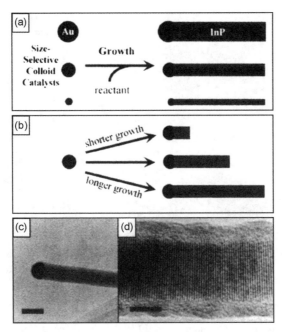

Figure 3.91 (a) *Schematic depicting the use of monodisperse colloid catalysts for the synthesis of diameter-selective InP nanowires.* (b) *Schematic of the effect of the variation of growth time on nanowire length.* (c) *TEM image showing the nanocluster catalyst at the end of an InP nanowire grown from a 20 nm Au cluster (scale bar is 50 nm).* (d) *HREM showing the crystalline core of the InP nanowire grown from a 10 nm colloid (scale bar is 5 nm). The (111) lattice planes resolved perpendicular to the growth axis have an average spacing of 0.59 ± 0.05 nm, which is in good agreement with the bulk value f or zinc blende InP of 0.5869*
(Reproduced from ref. 522)

9 Miscellaneous Nanowires

Nanorods and nanowires of AgCl, AgBr and AgI have been prepared in water/oil microemulsions.[530] The molar ratio of reactants plays a role in determining the morphology of the AgI nanomaterials. Nanowires of $CaCO_3$, $BaCO_3$ and $CaSO_4$ have been prepared in nonionic surfactant reverse micelles (Triton X-100–cyclohexane–water).[531] The $CaCO_3$ nanowires are 5–30 nm in diameter and have lengths of more than 10 μm. The growth process of $CaCO_3$ nanowires was monitored by TEM. Figure 3.93 shows TEM images of the $CaCO_3$ samples recorded at different times.

Reaction of trioctylphosphine oxide and trioctylphosphine with Fe-containing precursors gives FeP nanowires with large aspect ratios.[532]

Epitaxial $CoSi_2$ nanowires are obtained on Si(100) and Si-O substrates by using conventional optical lithography and silicon processing steps.[533] $NiSi_2$ nanowires have been produced by the decomposition of silane on Ni surfaces.[534]

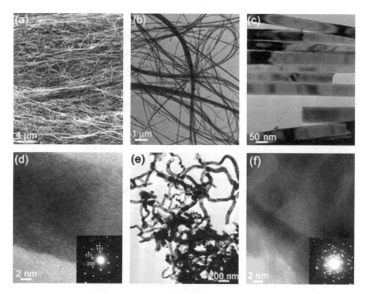

Figure 3.92   (a) *SEM image of GaP nanowires grown on Ni-catalysed alumina substrates* via *the sublimation of GaP powders.* (b) *TEM image showing the general morphology of the nanowires.* (c) *Higher magnification image, showing all the nanowires to be straight and cylindrical.* (d) *HREM of a single nanowire with the SAED (inset) pattern showing the single-crystalline zinc blende structure.* (e) *TEM image showing curled GaP nanowires obtained.* (f) *HREM image showing a polycrystalline nanowire with the SAED pattern (inset)* (Reproduced from ref. 529)

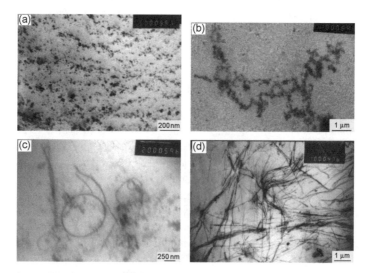

Figure 3.93   *TEM images of CaCO₃ samples prepared in Triton X-100 reverse micelles after aging for* (a) *10 min,* (b) *1 h,* (c) *4 h and* (d) *48 h* (Reproduced from ref. 531)

Highly ordered Prussian blue ($Fe_4^{3+}[Fe^{2+}(CN)_6]_3.xH_2O$) nanowire arrays, ~50 nm in diameter and up to 4 μm long, have been fabricated by an electro-deposition in AAO films.[535]

10 Useful Properties and Potential Applications

In earlier sections we referred to some of the useful properties of inorganic nanowires that are of potential technological value. Of the various applications, those related to the electronics have been of interest to several workers. Thus, semiconducting nanowire building blocks have been used to assemble functional device elements and element arrays.[536] Nanowire junction arrays were configured as OR, AND and NOR logic-gate structures with substantial gain and have been used to implement basic computation.[537] Functional resonant tunnelling diodes have been obtained *via* bottom-up assembly of designed segments of different III–V semiconductor nanowires.[538] These studies were carried out in 2002 or earlier.

Recent studies related to applications are the following. Field-effect transistors based on nanowires of Si,[539,540] NiSi[541] and ZnO.[542a,b] Other transistor-related studies are: coaxially gated in-wire thin-film transistors of Au/CdS/Au@(SiO$_2$)$_{10}$ nanowires,[543] and In$_2$O$_3$ nanowire-based vertical field-effect transistor.[544] Resonance tunnelling diode structures of CdTe nanowires have been made by conductive AFM.[545] A strategy for parallel and scalable integration of nanowire devices over large areas without the need to register individual nanowire-electrode interconnects has been developed.[546] A general approach for addressing integrated nanosystems in which signals are restored has been discussed.[547] Achievements and challenges of nanocomputers have been reviewed.[548]

Pt/ZnO nanowire Schottky diodes have been formed.[549] Nanowire transistors with ferroelectric gate dielectrics, exhibiting enhanced performance and memory effects have been described.[550]

Nanoscale light-emitting diodes with colours spanning from the UV to near-IR region have been fabricated by employing a solution-based approach wherein the emissive semiconductor nanowires were assembled with non-emissive Si nanowires in a crossed nanowire architecture.[551] Properties and functions of oxide nanoribbons that act as subwavelength optical waveguides have been described and their possible use as nanoscale photonic elements assessed.[552]

Lasing action of ZnO nanowires was referred to earlier. Nanoribbons of ZnS also exhibit good lasing characteristics.[553] Self-aligned, gated arrays of silicon nanowires (or carbon nanotubes) are potentially useful as electron sources for field emission displays.[554] Dye-sensitized solar cells based on TiO$_2$ nanowires have been fabricated.[555]

Biological applications of high-aspect ratio nanoparticles have been reviewed.[556] Electrical detection of DNA and DNA sequence variations based on nanowire nanosensors has been reported.[557] Receptor-mediated self-assembly of multi-component magnetic nanowires may have applications in some areas, including biosensors.[558] Nanomotors based on nanorods have received much attention.[559]

References

1. S. Iijima, *Nature*, 1991, **354**, 56.
2. P. Yang, Y. Wu and R. Fan, *Int. J. Nanosci.*, 2002, **1**, 1.
3. Y. Xia, P. Yang, Y. Sun, Y. Wu, B. Mayers, B. Gates, Y. Yin, F. Kim and H. Yan, *Adv. Mater.*, 2003, **15**, 353.
4. C. N. R. Rao, F. L. Deepak, G. Gundiah and A. Govindaraj, *Prog. Solid State Chem.*, 2003, **31**, 5.
5. Z. L. Wang (ed.), *Nanowires and Nanobelts*, Kluwer Academic Publishers, Boston, 2003, Vols 1 and 2.
6. (a) R. S. Wagner, *Whisker Technology* ed. A. P. Levitt, Wiley-Interscience, New York, 1970 (b) R. S. Wagner, and W. C. Ellis *Appl. Phys. Lett.*, 1964, **4**, 89.
7. P. S. Shah, T. Hanrath, K. P. Johnston and S. A. Korgel, *J. Phys. Chem. B*, 2004, **108**, 9574.
8. Y. Wu and P. Yang. *J. Am. Chem. Soc.*, 2001, **123**, 3165.
9. S. T. Lee, N. Wang, Y. F. Zhang and Y. H. Tang, *MRS Bull.*, 1999, 36.
10. N. Wang, Y. H. Tang, Y. F. Zhang, C. S. Lee and S. T. Lee, *Phys. Rev. B*, 1998, **58**, R16024.
11. J. J. Stejny, R.W. Trinder and J. Dlugosz, *J. Mater. Sci.*, 1981, **16**, 3161.
12. J. J. Stejny, R.W. Dlugosz and A. Keller, *J. Mater. Sci.*, 1979, **14**, 1291.
13. H. R. Kruyt and A. E. V. Arkel, *Kolloid-Z*, 1928, **32**, 29.
14. B. Gates, B. Mayers, B. Cattle and Y. Xia, *Adv. Funct. Mater.*, 2002, **12**, 219.
15. B. Mayers and Y. Xia, *J. Mater. Chem.*, 2002, **12**, 1875.
16. (a) B. Messer, J. H. Song, M. Huang, Y. Wu, F. Kim and P. Yang, *Adv. Mater.* 2000, **12**, 1526; (b) J. Song, B. Messer, Y. Wu, H. Kind and P. Yang, *J. Am. Chem. Soc.* 2001, **123**, 9714.
17. (a) C. R. Martin, *Science*, 1994, **266**, 1961; (b) D. Almawlawi, C. Z. Liu and M. Moskovits, *J. Mater. Res.*, 1994, **9**, 1014; (c) M. Zheng, L. Zhang, X. Zhang, J. Zhang and G. Li, *Chem. Phys. Lett.*, 2001, **334**, 298.
18. (a) Y. Xia, P. Yang, Y. Sun, Y. Wu, B. Mayers, B. Gates, Y. Yin, F. Kim and H. Yan, *Adv. Mater.*, 2003, **15**, 353; (b) H. Ringsdorf, B. Schlarb and J. Verzmer, *Angew. Chem. Int. Ed.*, 1988, **27**, 113.
19. C. N. R. Rao, A. Govindaraj, F. L. Deepak, N. A. Gunari and M. Nath, *Appl. Phys. Lett.*, 2001, **78**, 1853.
20. A. Govindaraj, F. L. Deepak, N. A. Gunari and C. N. R. Rao, *Isr. J. Chem.* 2001, **41**, 23.
21. Y. Yin, Y. Lu, Y. Sun and Y. Xia, *Nano Lett.*, 2002, **2**, 427.
22. B. Gates, Y. Wu, Y. Yin, P. Yang and Y. Xia, *J. Am. Chem. Soc.*, 2001, **123**, 11500.
23. A. Govindaraj, B. C. Satishkumar, M. Nath and C. N. R. Rao, *Chem. Mater.* 2000, **12**, 202.
24. T. J. Trentler, K. M. Hickman, S. C. Geol, A. M. Viano, P. C. Gibbons and W. E. Buhro, *Science*, 1995, **270**, 1791.
25. (a) T. J. Trentler, S. C. Goel, K. M. Hickman, A. M. Viano, M. Y. Chiang, A. M. Beatty, P. C. Gibbons and W. E. Buhro, *J. Am. Chem. Soc.*, 1997, **119**, 2172; (b) P. D. Markowitz, M. P. Zach, P. C. Gibbons, R. M. Penner and

W. E. Buhro, *J. Am. Chem. Soc.*, 2001, **123**, 4502; (c) O. R. Lourie, C. R. Jones, B. M. Bartlett, P. C. Gibbons, R. S. Ruoff and W. E. Buhro, *Chem. Mater.*, 2000, **12**, 1808.

26 J. D. Holmes, K. P. Johnston, R. C. Doty and B. A. Korgel, *Science*, 2000, **287**, 1471.

27. (a) L. Manna, E. C. Scher and A. P. Alivisatos, *J. Am. Chem. Soc.*, 2000, **122**, 12700; (b) C.-Y. Chao, C.-C. Chen and Z.-H. Lang, *Chem. Mater.*, 2000, **12**, 1516.

28. M. H. Huang, Y. Wu, H. Feick, N. Tran, E. Weber and P. Yang, *Adv. Mater.*, 2001, **13**, 113.

29. M. H. Huang, S. Mao, H. Feick, H. Yun, Y. Wu, H. Kind, E. Weber, R. Russo and P. Yang, *Science*, 2001, **292**, 1897.

30. B. Messer, J. H. Song and P. Yang, *J. Am. Chem. Soc.*, 2000, **122**, 10232.

31. (a) Y. Huang, X. Duan, Q. Q. Wei and C. M. Lieber, *Science*, 2001, **291**, 630; (b) Y. Wu, H. Yan, M. Huang, B. Messer, J. H. Song and P. Yang, *Chem. Eur. J.*, 2002, **8**, 1260.

32. F. Kim, S. Kwan, J. Arkana and P. Yang, *J. Am. Chem. Soc.*, 2001, **123**, 4360.

33. D. P. Yu, Z. G. Bai, Y. Ding, Q. L. Hang, H. Z. Zhang, J. J. Wang, Y. H. Zou, W. Qian, G. C. Xiong, H. T. Zhou and S. Q. Feng, *Appl. Phys. Lett.*, 1998, **72**, 3458.

34. H. Z. Zhang, D. P. Yu, Y. Ding, Z. G. Bai, Q. L. Hang and S. Q. Feng, *Appl. Phys. Lett.*, 1998, **73**, 3396.

35. S. Fan, J. Cao, H. Dang, Q. Gu and J. Zhao, *J. Mater Sci. Eng. C*, 2001, **15**, 295.

36. Y. J. Chen, J. B. Li and J. H. Dai, *Chem. Phys. Lett.*, 2001, **344**, 450.

37. G. Gundiah, F. L. Deepak, A. Govindaraj and C.N.R. Rao, *Chem. Phys. Lett.*, 2003, **381**, 579.

38. M. K. Sunkara, S. Sharma, R. Miranda, G. Lian and E. C. Dickey, *Appl. Phys. Lett.*, 2001, **79**, 1546.

39. (a) Y. Wu, Y. Cui, L. Huynh, C. J. Barrelet, D. C. Bell and C. M. Lieber, *Nano Lett.*, 2004, **4**, 433; (b) D. Wang, F. Qian, C. Yang, Z. Zhong and C. M. Lieber, *Nano Lett.*, 2004, **4**, 871.

40. T. Ono, H. Saitoh and M. Esashi, *Appl. Phys. Lett.*, 1997, **70**, 1852.

41. Y. F. Zhang, Y. H. Tang, N. Wang, D. P. Yu, C. S. Lee, I. Bello and S. T. Lee, *Appl. Phys. Lett.*, 1998, **72**, 1835.

42. G. W. Zhou, Z. Zhang, Z. G. Bai, S. Q. Feng and D. P. Yu, *Appl. Phys. Lett.*, 1998, **73**, 677.

43 Y. F. Zhang, Y. H. Tang, H. Y. Peng, N. Wang, C. S. Lee, I. Bello and S. T. Lee, *Appl. Phys. Lett.*, 1999, **75**, 1842.

44. A. M. Morales and C. M. Lieber, *Science*, 1998, **279**, 208.

45. N. Wang, Y. F. Zhang, Y. H. Tang, C. S. Lee and S. T. Lee, *Appl. Phys. Lett.*, 1998, **73**, 3902.

46. N. Wang, Y. H. Tang, Y. F. Zhang, C. S. Lee, I. Bello and S. T. Lee, *Chem. Phys. Lett.*, 1999, **299**, 237.

47. Y.-H. Yang, S.-J. Wu, H.-S. Chiu, P.-I. Lin and Y.-T. Chen, *J. Phys. Chem. B*, 2004, **108**, 846.

48. J. Westwater, D. P. Gosain, S. Tomiya, S. Usui and H. Ruda, *J. Vac. Sci. Technol. B* 1997, **15**, 554.

49. Y. Cui, L. J. Lauhon, M. S. Gudisen, J. Wang and C. M. Lieber, *Appl. Phys. Lett.*, 2001, **78**, 2214.

50. Z. Q. Liu, W. Y. Zhou, L. F. Sun, D. S. Tang, X. P. Zou, Y. B. Li, C. Y. Wang, G. Wang and S. S. Xie, *Chem. Phys. Lett.*, 2001, **341**, 523.

51. T. I. Kamins, X. Li and R. S. Williams, *Appl. Phys. Lett.*, 2003, **82**, 263.

52. Y. F. Zhang, Y. H. Tang, N. Wang, C. S. Lee, I. Bello and S. T. Lee, *J. Cryst. Growth*, 1999, **197**, 136.

53. W. S. Shi, H. Y. Peng, Y. F. Zhang, N. Wang, N. G. Shang, Z. W. Pan, C. S. Lee and S. T. Lee, *Adv. Mater.*, 2000, **12**, 1343.

54. Z. Zhang, X. H. Fan, L. Xu, C. S. Lee and S. T. Lee, *Chem. Phys. Lett.*, 2001, **337**, 18.

55. H. Y. Peng, Z. W. Pan, L. Xu, X. H. Fan, N. Wang, C. S. Lee and S. T. Lee, *Adv. Mater.*, 2001, **13**, 317.

56. C. P. Li, X. H. Sun, N. B. Wong, C. S. Lee, S. T. Lee and B. K. Teo, *Chem. Phys. Lett.*, 2002, **365**, 22.

57. N. R. B. Coleman, M. A. Morris, T. R. Spalding and J. D. Holmes, *J. Am. Chem. Soc.*, 2001, **123**, 187.

58. K. Q. Peng, Y. J. Yan, S. P. Gao and J. Zhu, *Adv. Mater.*, 2002, **14**, 1164.

59. C. P. Li, N. Wang, S. P. Wong, C. S. Lee and S. T. Lee, *Adv. Mater.*, 2002, **14**, 218.

60. Y. Chen, D. A. A. Ohlberg and R. S. Williams, *J. Appl. Phys.*, 2002, **91**, 3213.

61. D. Lee and S. Kim, *Appl. Phys. Lett.*, 2003, **82**, 2619.

62. C. Preinesberger, S. K. Becker, S. Vandré, T. Kalka and M. Dähne, *J. Appl. Phys.* 2002, **91**, 1695.

63. Y. Shan, A. K. Kalkan, C.-Y. Peng, and S. J. Fonash, *Nano Lett.*, 2004, **4**, 2085.

64. B. Li, D. Yu and S. L. Zhang, *Phys. Rev. B*, 1999, **59**, 1645.

65. Y. Cui, X. Duan, J. Hu and C. M. Lieber, *J. Phys. Chem. B*, 2000, **104**, 5213.

66. G. W. Zhou, H. Li, H. P. Sun, D. P. Yu, Y. Q. Wang, X. J. Huang, L. Q. Chen and Z. Zhang, *Appl. Phys. Lett.*, 1999, **75**, 2447.

67. D. D. D. Ma, C. S. Lee and S. T. Lee, *Appl. Phys. Lett.*, 2001, **79**, 2468.

68. Y. Cui and C. M. Lieber, *Science*, 2001, **291**, 851.

69. J. Hu, M. Ouyang, P. Yang and C. M. Lieber, *Nature*, 1999, **399**, 48.

70. S. F. Hu, W. Z. Wong, S. S. Liu, Y. C. Wu, C. L. Sung, T. Y. Huang and T. J. Yang, *Adv. Mater.*, 2002, **14**, 736.

71. S. W. Chung, J. Y. Yu and J. R. Heath, *Appl. Phys. Lett.*, 2000, **76**, 2068.

72. J. Y. Yu, S. W. Chung and J. R. Heath, *J. Phys. Chem. B*, 2000, **104**, 11864.

73. D. D. D. Ma, C. S. Lee, Y. Lifshitz and S. T. Lee, *Appl. Phys. Lett.*, 2002, **81**, 3233.

74. X. T. Zhou, J. Q. Hu, C. P. Li, D. D. D. Ma, C. S. Lee and S. T. Lee, *Chem. Phys. Lett.*, 2003, **369**, 220.

75. Z. Li, Y. Chen, X. Li, T. I. Kamins, K. Nauka and R. S. Williams, *Nano Lett.*, 2004, **4**, 245.

76. Y. Wu and P. Yang, *Adv. Mater.*, 2001, **13**, 520.

77. Y. Wu and P. Yang, *Chem. Mater.* 2000, **12**, 605.

78. D. Wang and H. Dai, *Angew. Chem. Int. Ed.*, 2002, **114**, 4977.

79. S. Mathur, H. Shen, V. Sivakov and U. Werner, *Chem. Mater.*, 2004, **16**, 2449.

80. T. Harnath and B. A. Korgel, *J. Am. Chem. Soc.*, 2002, **124**, 1424.

81. N. B. R. Coleman, K. M. Ryan, T. R. Spalding, J. D. Holmes and M. A. Morris, *Chem. Phys. Lett.*, 2001, **343**, 1.

82. J. R. Heath and F. K. Legoues, *Chem. Phys. Lett.*, 1993, **208**, 263.

83. Q. Wan, T. H. Wang, T. Feng, X. H. Liu and C. C. Lin, *Appl. Phys. Lett.* 2002, **81**, 3281.

84. Q. Wan, G. Li, T. H. Wang and C. L. Lin, *Solid State Commun.*, 2003, **123**, 503.

85. Y. F. Zhang, Y. H. Tang, N. Wang, C. S. Lee, I. Bello and S. T. Lee, *Phys. Rev. B*, 2000, **61**, 4518.

86. M. P. Halsall, H. Omi and T. Ogino, *Appl. Phys. Lett.*, 2002, **81**, 2448.

87. G. Gu, M. Burghard, G. T. Kim, G. S. Düsberg, P. W. Chiu, V. Krstic, S. Roth and W. Q. Han, *J. Appl. Phys.*, 2001, **90**, 5747.

88. L. Cao, Z. Zhang, G. Li, J. Zhang and W. Wang, *Adv. Mater.*, 2001, **13**, 1701.

89. Y. Q. Wang, X. F. Duan, L. M. Cao and W. K. Wang, *Chem. Phys. Lett.* 2002, **359**, 273.

90. L. M. Cao, K. Hahn, C. Scheu, M. Rühle, Y. Q. Wang, Z. Zhang, C. X. Gao, Y. C. Li, Y. Zhang, M. He, L. L. Sun and W. K. Wang, *Appl. Phys. Lett.*, 2002, **80**, 4226.

91. L. Cao, K. Hahn, Y. Wang, C. Scheu, Z. Zhang, C. Gao, Y. Li, X. Zhang, L. Sun, W. Wang and M. Rühle, *Adv. Mater.*, 2002, **14**, 1294.

92. X. M. Meng, J. Q. Hu, Y. Jiang, C. S. Lee and S. T. Lee, *Chem. Phys. Lett.* 2003, **370**, 825.

93. Z. Wang, Y. Shimizu, T. Sasaki, K. Kawaguchi, K. Kimura and N. Koshizaki, *Chem. Phys. Lett.*, 2003, **368**, 663.

94. Y. Zhang, H. Ago, M. Yumura, S. Ohshima, K. Uchida, T. Komatsu and S. Iijima, *Chem. Phys. Lett.*, 2004, **385**, 177.

95. S. H. Yun, A. Dibos and J. Z. Wu and D.-K. Kim, *Appl. Phys. Lett.*, 2004, **84**, 2892.

96. C. J. Otten, O. R. Lourie, M. F. Yu, J. M. Cowley, M. J. Dyer, R. S. Ruoff and W. E. Buhro, *J. Am. Chem. Soc.*, 2002, **124**, 4564.

97. Y. Wu, B. Masser and P. Yang, *Adv. Mater.*, 2001, **13**, 1487.

98. J. L. Li, X. J. Liang, J. F. Jia, X. Liu, J. Z. Wang, E. G. Wang and Q. K. Xue, *Appl. Phys. Lett.*, 2001, **79**, 2826.

99. K. Soulantica, A. Maisonnat, F. Senocq, M. C. Fromen, M. J. Casanove and B. Chaudret, *Angew. Chem. Int. Ed.*, 2001, **40**, 2984.

100. W. K. Hsu, M. Terrones, H. Terrones, N. Grobert, A. I. Kirkland, J. P. Hare, K. Prassides, P. D. Townsend, H. W. Kroto and D. R. M. Walton, *Chem. Phys. Lett.*, 1998, **284**, 177.

101. Y. T. Peng, G. W. Meng, L. D. Zhang, Y. Qin, X. Y. Gao, A. W. Zhao and Q. Feng, *Adv. Funct. Mater.*, 2002, **12**, 719.

102. Y. Wang, X. Jiang, T. Herricks and Y. Xia, *J. Phys. Chem. B*, 2004, **108**, 8631.

103. S. Yuan, L. Ren and F. Li, *Phys. Rev. B*, 2004, **69**, 092509.

104. X. Y. Zhang and J. Y. Dai, *Nanotechnology*, 2004, **15**, 1166.

105. Y. Zhang, G. Li, Y. Wu, B. Zhang, W. Song and L. Zhang, *Adv. Mater.*, 2002, **14**, 1227.

106. Y. T. Cheng, A. M. Weiner, C. A. Wong and M. P. Balogh, *Appl. Phys. Lett.*, 2002, **81**, 3248.

107. A. J. Yin, J. Li, W. Jian, A. J. Bennett and J. M. Xu, *Appl. Phys. Lett.*, 2001, **79**, 1039.

108. J. Wang, X. Wang, Q. Peng and Y. Li, *Inorg. Chem.*, 2004, **43**, 7552.

109. G. M. Veith, R. E. Sykora and N. J. Dudney, *Chem. Mater.*, 2004, **16**, 3348.

110. H. Yu, P. C. Gibbons and W. E. Buhro, *J. Mater. Chem.*, 2004, **14**, 595.

111. C. H. Kiang, J. S. Choi, T. T. Tran and A. D. Bacher, *J. Phys. Chem. B*, 1999, **103**, 7449.

112. P. Chiu and I. Shih, *Nanotechnology*, 2004, **15**, 1489.

113. Z. Y. Jiang, Z. X. Xie, S. Y. Xie, X. H. Zhang, R. B. Huang and L. S. Zheng, *Chem. Phys. Lett.*, 2003, **368**, 425.

114. L. Ren, H. Zhang, P. Tan, Y. Chen, Z. Zhang, Y. Chang, J. Xu, F. Yang and D. Yu, *J. Phys. Chem. B*, 2004, **108**, 4627.

115. X. Gao, T. Gao and L. Zhang, *J. Mater. Chem.*, 2003, **13**, 6.

116. A. Abdelonas, W. L. Gong, W. Lutze, J. A. Shelrutt, R. Franco and I. Moura, *Chem. Mater.*, 2000, **15**, 1510.

117. U. K. Gautam, M. Nath and C. N. R. Rao, *J. Mater. Chem.*, 2003, **13**, 2845.

118 U. K. Gautam and C. N. R. Rao, *J. Mater. Chem.*, 2004, **14**, 2530.

119. H. Yu, P. C. Gibbons and W. E. Buhro, *J. Mater. Chem.*, 2004, **14**, 595.

120. Y.-J. Zhu, W.-W. Wang, R.-J. Qi and X.-L. Hu, *Angew. Chem. Int. Ed.*, 2004, **43**, 1410.

121. Z. Liu, Z. Hu, Q. Xie, B. Yang, J. Wu and Y. Qian, *J. Mater. Chem.*, 2003, **13**, 159.

122. B. Mayers, B. Gates, Y. Yin and Y. Xia, *Adv. Mater.*, 2001, **13**, 1380.

123. B. D. Busbee, S. O. Obare and C. J. Murphy, *Adv. Mater.*, 2003, **15**, 414.

124. A. Swami, A. Kumar, P. R. Selvakannan, S. Mandal, R. Pasricha and M. Sastry, *Chem. Mater.*, 2003, **15**, 17.

125. C. N. R. Rao, G. U. Kulkarni, P. J. Thomas, V. V. Agarwal and P. Saravanan, *J. Phys. Chem. B*, 2003, **107**, 7391.

126. P. Göring, E. Pippel, H. Hofmeister, R. B. Wehrspohn, M. Steinhart and U. Gölsele, *Nano Lett.* 2004, **4**, 1121.

127. X. Y. Zhang, L. D. Zhang, Y. Lei, L. X. Zhao and Y. Q. Mao, *J. Mater. Chem.* 2001, **11**, 1732.

128. N. I. Kovtyukhova, B. R. Martin, J. K. N. Mbindyo, P. A. Smith, B. Razavi, T. S. Mater and T. E. Mallouk, *J. Phys. Chem. B*, 2001, **105**, 8762.

129. P. A. Smith, C. D. Nordquist, T. N. Jackson, T. S. Mayer, B. R. Martin, J. Mbindyo and T. E. Mallouk, *Appl. Phys. Lett.*, 2002, **77**, 1399.

130. J. Mbindyo, T. E. Mallouk, J. B. Mattzela, I. Kratochvilova, B. Razavi, T. N. Jackson and T. S. Mayer, *J. Am. Chem. Soc.*, 2002, **124**, 4020.

131. M. Wirtz and C. R. Martin, *Adv. Mater.*, 2003, **15**, 455.

132. J. K. N. Mbindyo, B. D. Reiss, B. R. Martin, C. D. Keating, M. J. Natan and T. E. Mallouk, *Adv. Mater.*, 2001, **13**, 249.

133. O. Harnack, W. E. Ford, A. Yasuda and J. M. Wessels, *Nano Lett.*, 2002, **2**, 913.

134. R. Djalali, Y. F. Chen and H. Matsui, *J. Am. Chem. Soc.*, 2002, **124**, 13660.

135. T. Hassenkam, K. Norgaard, L. Iversen, C. J. Kiely, M. Brust and T. Bjornholm, *Adv. Mater.*, 2002, **14**, 1126.

136. S. Fullam, D. Cottell, H. Rensmo and D. Fitzmaurice, *Adv. Mater.*, 2000, **2**, 1430.

137. Y. Kondo and K. Takayanagi, *Science*, 2000, **289**, 606.

138. M. Kaempfe, H. Graener, A. Kiesow and A. Heilmann. *Appl. Phys. Lett.*, 2001, **79**, 1876.

139. S. Link, C. Burda, B. Nikoobakht and M. A. El-Sayed, *J. Phys. Chem. B*, 2000, **104**, 6152.

140. K. Sarah, S. Angelo, C. C. Waraksa and T. E. Mallouk, *Adv. Mater.*, 2003, **15**, 400.

141. J. N. O'Shea, M. A. Phillips, M. D. R. Taylor, P. Moriarty, M. Brust and V. R. Dhanak, *Appl. Phys. Lett.*, 2002, **81**, 5039.

142. T. O. Hutchison, Y. P. Liu, C. Kiely, C. J. Kiely and M. Brust, *Adv. Mater.*, 2001, **13**, 1800.

143. T. Hassenkam, K. Moth-Poulsen, N. Stuhr-Hansen, K. Nørgaard, M. S. Kabir and T. Bjørnholm, *Nano Lett.*, 2004, **4**, 19.

144. K. G. Thomas, S. Barazzouk, B. I. Ipe, S. T. S. Joseph and P. V. Kamat, *J. Phys. Chem. B*, 2004, **108**, 13066.

145. G. Ramanath, J. D'Arcy-Gall, T. Maddanimath, A. V. Ellis, P. G. Ganesan, R. Goswami, A. Kumar, and K. Vijayamohanan, *Langmuir* 2004, **20**, 5583.

146. H. Mehrez, A. Wlasenko, B. Lorade, J. Taylor, P. Grütter and H. Guo, *Phys. Rev. B*, 2002, **65**, 195419–1.

147. A. Beitsch and B. Michel, *Appl. Phys. Lett.*, 2002, **80**, 3346.

148. S. Link, D. J. Hathcock, B. Nikoobakht and M. A. El-Sayed, *Adv. Mater.*, 2003, **15**, 393.

149. N. I. Kovtyukhova and T. E. Mallouk, *Chem. Eur. J.*, 2002, **8**, 4355.

150. N. Li, J. Zheng, M. Gleiche, H. Fuchs, L. Chi, O. Vidoni, T. Reuter and G. Schmid, *Nano Lett.*, 2002, **2**, 1097.

151. K. Gall, J. Diao and M. L. Dunn, *Nano Lett.*, 2004, **4**, 2431.

152. J. Diao, K. Gall and M. L. Dunn, *Nano Lett.*, 2004, **4**, 1863.

153. N. R. Jana, L. Gearheart and C. J. Murphy, *Chem. Commun.*, 2001, 617.

154. X. Jiang, Y. Xie, J. Lu, L. Zhu, W. He and Y. Qian, *J. Mater. Chem.*, 2001, **11**, 1775.

155. Y. Sun and Y. Xia, *Adv. Mater.*, 2002, **14**, 833.

156. Y. Xiong, Y. Xie, C. Wu, J. Yang, Z. Li and F. Xu, *Adv. Mater.*, 2003, **15**, 405.

157. Y. Sun, B. Gates, B. Mayers and Y. Xia, *Nano Lett.*, 2002, **2**, 165.

158. S. Liu, J. Yue and A. Gedanken, *Adv. Mater.*, 2001, **13**, 656.

159. Y. Zhou, S. H. Yu, C. Y. Wang, X. G. Li, Y. R. Zhu and Z. Y. Chen, *Adv. Mater.*, 1999, **11**, 850.

160. B. H. Hong, S. C. Bae, C. W. Lee, S. Jeong and K. S. Kim, *Science*, 2001, **294**, 348.

161. D. Zhang, L. Qi, J. Ma and H. Cheng, *Chem. Mater.*, 2001, **13**, 2753.
162. L. Huang, H. Wang, Z. Wang, A. Mitra, K. N. Bozhilov and Y. Yan, *Adv. Mater.*, 2002, **14**, 61.
163. J. Sloan, D. M. Wright, H. G. Woo, S. Bailey, G. Brown, A. P. E. York, K. S. Coleman, J. L. Hutchison and M. L. H. Green, *Chem. Commun.*, 1999, 699.
164. M. J. Edmondson, W. Zhou, S. A. Sieber, I. P. Jones, I. Gameson, P. A. Anderson and P. P. Edwards, *Adv. Mater.*, 2001, **13**, 1608.
165. G. Sauer, G. Brehm, S. Schneider, K. Nielsch, R. B. Wehrspohn, J. Choi, H. Hofmeister and U. Gösele, *J. Appl. Phys.*, 2002, **91**, 3243.
166. J. Choi, G. Sauer, K. Nielsch, R. B. Wehrspohn and U. Gösele, *Chem. Mater.*, 2003, **15**, 776.
167. J. P. Kottmann, O. J. F. Martin, D. R. Smith and S. Schultz, *Phys. Rev. B*, 2001, **64**, 235402-1.
168. V. Rodrigues, J. Bettini, A. R. Rocha, L. G. C. Rego and D. Ugarte, *Phys. Rev. B*, 2002, **65**, 153402-1.
169. J. Zhao, C. Buia, J. Han and J. P. Lu, *Nanotechnology*, 2003, **14**, 501.
170. D. Zhang, L. Qi, J. Yang, J. Ma, H. Cheng and L. Huang, *Chem. Mater.*, 2004, **16**, 872.
171. Z. Wang, J. Liu, X. Chen, J. Wan and Y. Qian, *Chem. Eur. J.*, 2005, **11**, 160.
172. Y. Wu, T. Livneh, Y. X. Zhang, G. Cheng, J. Wang, J. Tang, M. Moskovits and G. D. Stucky, *Nano Lett.*, 2004, **4**, 2337.
173. R.-L. Zong, J. Zhou, Q. Li, B. Du, B. Li, M. Fu, X.-W. Qi, L.-T. Li and S. Buddhudu, *J. Phys. Chem. B*, 2004, **108**, 16713.
174. J. Luo, Z. Huang, Y. Zhao, L. Zhang and J. Zhu, *Adv. Mater.*, 2004, **16**, 1512.
175. A. Tao, F. Kim, C. Hess, J. Goldberger, R. He, Y. Sun, Y. Xia and P. Yang, *Nano. Lett.*, 2003, **3**, 1229.
176. L. Vayssieres, L. Rabenberg and A. Manthiram, *Nano Lett.*, 2002, **2**, 1393.
177. K. S. Napolsky, A. A. Eliseev, A. V. Knotko, A. V. Lukashin, A. A. Vertegel and Y. D. Tretyakov, *Mat. Sci. Eng. C*, 2003, **23**, 151.
178. D. Spišák and J. Hafner, *Phys. Rev. B*, 2002, **65**, 235405-1.
179. B. C. Satishkumar, A. Govindaraj, P. V. Vanitha, A. K. Raychaudhuri and C. N. R. Rao, *Chem. Phys. Lett.*, 2002, **362**, 301.
180. T. G. Sorop, C. Untiedt, F. Luis, M. Kröll, M. Rasa and L. J. De Jongh, *Phys. Rev. B*, 2003, **67**, 014402-1.
181. (a) L. Mohaddes-Ardabili, H. Zheng, S. B. Ogale, B. Hannoyer, W. Tian, J. Wang, S. E. Lofland, S. R. Shinde, T. Zhao, Y. Jia, L. Salamanca-Riba, D. G. Schlom, M. Wuttig and R. Ramesh, *Nat. Mater.*, 2004, **3**, 533; (b) J. B. Wang, X. Z. Zhou, Q. F. Liu, D. S. Xue, F. S. Li, B. Li, H. P. Kunkel and G. Williams, *Nanotechnology.*, 2004, **15**, 485.
182. T. T. Albrecht, J. Schotter, G. A. Kästle, N. Emley, T. Shibauchi, L. K. Elbaum, K. Guarini, C. T. Black, M. T. Tuominen and T. P. Russell, *Science*, 2000, **290**, 2126.
183. H. Cao, Z. Xu, H. Sang, D. Sheng and C. Tie, *Adv. Mater.*, 2001, **13**, 121.
184. S. Ge, C. Li, X. Ma, W. Li, L. Xi and C. X. Li, *J. Appl. Phys.*, 2001, **90**, 509.

185. J. Bao, C. Tie, Z. Xu, Q. Ma, J. Hong, H. Sang and D. Sheng, *Adv. Mater.*, 2002, **14**, 44.

186. S. M. York and F. M. Leibsle, *Phys. Rev. B*, 2001, **64**, 033411–1.

187. L. Vila, L. Piraus, J. M. George and G. Faini, *Appl. Phys. Lett.*, 2002, **80**, 3805.

188. B. Hausmanns, T. P. Krome, G. Dunpich, E. F. Wassermann, D. Hinzke, U. Nowak and K. D. Usadel, *J. Magn. Magn. Mater.*, 2002, **240**, 297.

189. L. Vila, P. Vincent, L. D.-De Pra, G. Pirio, E. Minoux, L. Gangloff, S. Demoustier-Champagne, N. Sarazin, E. Ferain, R. Legras, L. Piraux and P. Legagneux, *Nano Lett.*, 2004, **4**, 521.

190. L. Menon, S. Bandopadhyay, Y. Liu, H. Zeng and D. J. Sellmyer, *J. Nanosci. Nanotechnol.*, 2001, **1**, 149.

191. Y. W. Wang, L. D. Zhang, G. W. Meng, X. S. Peng, Y. X. Jin and J. Zhang, *J. Phys. Chem. B*, 2002, **106**, 2502.

192. B. K. Pradhan, T. Kyotani and A. Tomita, *Chem. Commun.*, 1999, 1317.

193. K. Nielsch, R. B. Wehrspoon, R. Barthel, J. Kirschner, U. Gösele, S. F. Fischer and H. Kronmüller, *Appl. Phys. Lett.*, 2001, **79**, 1360.

194. R. Ferre, K. Ounadjela, J. M. George, L. Piraux and S. Dubois, *Phys. Rev. B*, 1997, **56**, 14066.

195. H. Zeng, R. S. Skomski, L. Menon, Y. Liu, S. Bandopadhyay and D. J. Sellmyer, *Phys. Rev. B*, 2002, **65**, 134426–1.

196. N. Cordente, M. Respaud, F. Senoaq, M. J. Casanove, C. Amiens and B. Chaudret, *Nano Lett.*, 2001, **1**, 565.

197. L. Sun, P. C. Searson and C. L. Chien, *Appl. Phys. Lett.*, 2001, **79**, 4429.

198. F. Elhoussine, S. M. Tempfi, A. Encinas and L. Piraux, *Appl. Phys. Lett.*, 2002, **81**, 1681.

199. J. J. Mock, S. J. Oldenburg, D. R. Smith, D. A. Schultz and S. Schultz, *Nano Lett.*, 2002, **2**, 465.

200. Z. K. Wang, M. H. Kuok, S. C. Ng, D. J. Lockwood, M. G. Cotton, K. Nielsch, R. B. Wehrspoon and U. Gösele, *Phys. Rev. Lett.*, 2002, **89**, 027201–1.

201. A. N. Abdi and J. P. Bucher, *Appl. Phys. Lett.*, 2003, **82**, 430.

202. Y. G. Guo, L. J. Wan, C. F. Zhu, D. L. Yang, D. M. Chen and C. L. Bai, *Chem. Mater.*, 2003, **15**, 664.

203. R. Adelung, F. Ernst, A. Scott, M. T. Azar, L. Kipp, M. Skibowski, S. Hollensteiner, E. Spiecker, W. Jäger, S. Gunst, A. Klein, W. Jägermann, V. Zaporojtchenko and F. Faupel, *Adv. Mater.*, 2002, **14**, 1056.

204. M. Y. Yen, C. W. Chiu, C. H. Hsia, F. R. Chen, J. J. Kai, C. Y. Lee and H. T. Chio, *Adv. Mater.*, 2003, **15**, 235.

205. C. F. Monson and A. T. Woolley, *Nano Lett.*, 2003, **3**, 359.

206. Z. Zhang, S. Dai, D. A. Blom and J. Shen, *Chem. Mater.*, 2002, **14**, 965.

207. H. Choi and S.-H. Park, *J. Am. Chem. Soc.*, 2004, **126**, 6248.

208. T. Gao, G. Meng, J. Zhang and S. I. Zhang, *Appl. Phys. A*, 2002, **74**, 403.

209. Y. Wang, J. Yang, C. Ye, X. Fang and L. Zhang, *Nanotechnology*, 2004, **15**, 1437.

210. Y. Wang, L. Zhang, G. Meng, C. Liang, G. Wang and S. Sun, *Chem. Commun.*, 2001, 2632.

211. S. R. C. Vivekchand, G. Gundiah, A. Govindaraj and C.N.R. Rao, *Adv. Mater.*, 2004, **16**, 1842.

212. J. T. L. Thong, C. H. Oon, M. Yeadon and W. D. Zhang, *Appl. Phys. Lett.* 2002, **81**, 4823.

213. Y. Li, X. Li, Z. X. Deng, B. Zhou, S. Fan, J. Wang and X. Sun, *Angew. Chem. Int. Ed.*, 2002, **41**, 333.

214. J. Wang and Y. Li, *Adv. Mater.*, 2003, **15**, 445.

215. M. P. Zach, K. H. Ng and R. M. Penner, *Science*, 2000, **290**, 2120.

216. E. C. Walter, M. P. Zach, F. Favier, B. J. Murray, K. Inazu, J. C. Hemminger and R. M. Penner, *ChemPhysChem.*, 2003, **4**, 131.

217. H. Wang, J. Wang, M. Tian, L. Bell, E. Hutchinson, M. M. Rosario and Y. Liu, *Appl. Phys. Lett.*, 2004, **84**, 5171.

218. D. Wang, W. L. Zhou, B. F. McMaughy, J. E. Hampsey, R. H. S. Schmehl, C. O'Connor, J. Brinker and Y. Lu, *Adv. Mater.*, 2003, **15**, 130.

219. J. Jorritsma, M. A. M. Gijs, J. M. Kerkhof and J. G. H. Stienen, *Nanotechnology*, 1996, **7**, 263.

220. L. H. Chan, K. H. Hong, S. H. Lai, X. W. Liu and H. C. Shih, *Thin Solid Films*, 2003, **423**, 27.

221. K. B. Lee, S. M. Lee and J. Cheon, *Adv. Mater.*, 2001, **13**, 517.

222. T. Kyotani, L. F. Tsai and A. Tomita, *Chem. Commun.*, 1997, 701.

223. Y. Sakamoto, A. Fukuoka, T. Higuchi, N. Shimomura, S. Inagaki and M. Ichikawa, *J. Phys. Chem. B*, 2004, **108**, 853.

224. J. Chen, T. Herricks, M. Geissler and Y. Xia, *J. Am. Chem. Soc.*, 2004, **126**, 10854.

225. D. H. Qin, L. Cao, Q. Y. Sun, Y. Huang and H. L. Li, *Chem. Phys. Lett.*, 2002, **358**, 484.

226. S. L. Tang, W. Chen, M. Lu, S. G. Yang, F. M. Zhang and Y. W. Du, *Chem. Phys. Lett.*, 2004, **384**, 1.

227. H. Zhu, S. Yang, G. Ni, D. Yu and Y. Du, *Scrip. Mater.*, 2001, **44**, 2291.

228. Y. W. Wang, L. D. Zhang, G. W. Meng, X. S. Peng, Y. X. Jin and J. Zhang, *J. Phys. Chem. B*, 2002, **106**, 2502.

229. K. Liu, K. Nagodawithana, P. C. Searson and C. L. Chien, *Phys. Rev. B*, 1995, **51**, 7381.

230. A. Fukuoka, Y. Sakamoto, S. Guan, S. Inagaki, N. Sugimoto, Y. Fukushima, K. Hirahara, S. Iijima and M. Ichikawa, *J. Am. Chem. Soc.*, 2001, **123**, 3373.

231. Y. Sun, Z. Tao, J. Chen, T. Herricks and Y. Xia, *J. Am. Chem. Soc.*, 2004, **126**, 5940.

232. Y. Yin, G. Zhang and Y. Xia, *Adv. Funct. Mater.*, 2002, **12**, 293.

233. (a) J. Zhang and L. Zhang, *Chem. Phys. Lett.*, 2002, **363**, 293; (b) K. P. Kalyanikutty, F. L. Deepak, A. Govindaraj and C. N. R. Rao, *Mater. Res. Bull.*, 2005, **40**, 831.

234. P. Yang and C. M. Lieber, *Science*, 1996, **273**, 1836.

235. V. Valcárcel, A. Souto and F. Guitián, *Adv. Mater.*, 1998, **10**, 138.

236. Z. Yuan, H. Huang and S. Fan, *Adv. Mater.*, 2002, **14**, 303.

237. Z. L. Xiao, C. Y. Han, U. Welp, H. H. Wang, W. K. Kwok, G. A. Willing, J. M. Hiller, R. E. Cook, D. J. Miller and G. W. Crabtree, *Nano. Lett.*, 2002, **2**, 1293.

238. G. Gundiah, F. L. Deepak, A. Govindaraj and C. N. R. Rao *Top. Cat.* 2003, **24**, 137.

239. X. S. Peng, L. D. Zhang, G. W. Meng, X. F. Wang, Y. W. Wang, C. Z. Wang and G. S. Wu, *J. Phys. Chem. B.*, 2002, **106**, 11163.

240. H. Z. Zhang, Y. C. Kong, Y. Z. Wang, X. Du, Z. G. Bai, J. J. Wang, D. P. Yu, Y. Ding, Q. L. Hang and S. Q. Feng, *Solid. State Commun.*, 1999, **109**, 677.

241. C. H. Liang, G. W. Meng, G. Z. Wang, Y. W. Wang, L. D. Zhang and S. Y. Zhang, *Appl. Phys. Lett.*, 2001, **78**, 3202.

242. B. C. Kim, K. T. Sun, K. S. Park, K. J. Im, T. Noh, M. Y. Sung, S. Kim, S. Nahm, Y. N. Choi and S. S. Park, *Appl. Phys. Lett.*, 2002, **80**, 479.

243. S. Sharma and M. K. Sunkara, *J. Am. Chem. Soc.*, 2002, **124**, 12288.

244. G. Gundiah, A. Govindaraj and C. N. R. Rao, *Chem. Phys. Lett.*, 2002, **351**, 189.

245. K.-W. Chang and J.-J. Wu, *Adv. Mater.*, 2004, **16**, 545.

246. C. Liang, G. Meng, Y. Lei, F. Phillipp and L. Zhang, *Adv. Mater.*, 2001, **13**, 1330.

247. M. J. Zheng, L. D. Zhang, G. H. Li, X. Y. Zhang and X. F. Wang, *Appl. Phys. Lett.*, 2001, **79**, 839.

248. (a) K. C. Kam, F. L. Deepak, A. K. Cheetham and C. N. R. Rao, *Chem. Phys. Lett.*, 2004, **397**, 329; (b) K. P. Kalyanikutty, G. Gundiah, C. Edem, A. Govindaraj and C. N. R. Rao, *Chem. Phys. Lett.*, 2005, **408**, 389.

249. C. Li, D. Zhang, S. Han, X. Liu, T. Tang and C. Zhou, *Adv. Mater.*, 2003, **15**, 143.

250. C. Xiangfeng, W. Caihong, J. Dongli and Z. Chenmou, *Chem. Phys. Lett.*, 2004, **399**, 461.

251. D. Zhang, Z. Liu, C. Li, T. Tang, X. Liu, S. Han, B. Lei and C. Zhou, *Nano Lett.*, 2004, **4**, 1919.

252. Z. R. Dai, Z. W. Pan and Z. L. Wang, *Solid State Commun.*, 2001, **118**, 351.

253. Z. R. Dai, J. L. Gole, J. D. Stout and Z. L. Wang, *J. Phys. Chem. B*, 2002, **106**, 1274.

254. Z. R. Dai, Z. W. Pan and Z. L. Wang, *J. Am. Chem. Soc.*, 2002, **124**, 8673.

255. Z. L. Wang and Z. W. Pan, *Adv. Mater.*, 2002, **14**, 1029.

256. Y. Chen, X. Cui, K. Zhang, D. Pan, S. Zhang, B. Wang and J. G. Hou, *Chem. Phys. Lett.*, 2003, **369**, 16.

257. J. Q. Hu, X. L. Ma, N. G. Shang, Z. Y. Xie, N. B. Wong, C. S. Lee and S. T. Lee, *J. Phys. Chem. B*, 2002, **106**, 3823.

258. S. H. Sun, G. W. Meng, Y. W. Wang, T. Gao, M. G. Zhang, Y. T. Tian, X. S. Peng and L. D. Zhang, *Appl. Phys. A*, 2003, **76**, 287.

259. W. Wang, C. Xu, X. Wang, Y. Liu, Y. Zhan, C. Zheng, F. Song and G. Wang, *J. Mater. Chem.*, 2002, **12**, 1922.

260. Y. Liu, C. Zheng, W. Wang, C. Yin and G. Wang, *Adv. Mater.*, 2001, **13**, 1883.

261. M. Zheng, G. Li, X. Zhang, S. Huang, Y. Lei and L. Zhang, *Chem. Mater.*, 2001, **13**, 3859.

262. Y. Zhang, A. Kolmakov, S. Chretien, H. Metiu and M. Moskovits, *Nano Lett.*, 2004, **4**, 403.

263. Z. Q. Liu, S. S. Xie, L. F. Sun, D. S. Tang, W. Y. Zhou, C. Y. Wang, W. Liu, Y. B. Li, X. P. Zou and G. Wang, *J. Mater. Res.*, 2001, **16**, 683.

264. Z. W. Pan, Z. R. Dai, C. Ma and Z. L. Wang, *J. Am. Chem. Soc.*, 2002, **124**, 1817.

265. B. Zheng, Y. Wu, P. Yang and J. Liu, *Adv. Mater.*, 2002, **14**, 122.

266. J.-J. Wu, T.-C. Wong and C.-C. Yu, *Adv. Mater.*, 2002, **14**, 1643.

267. Z. L. Wang, R. P. Gao, J. L. Gole and J. D. Stout, *Adv. Mater.*, 2000, **12**, 1938.

268. J. S. Wu, S. Dhara, C. T. Wu, K. H. Chen, Y. F. Chen and L. C. Chen, *Adv. Mater.*, 2002, **14**, 1847.

269. D. P. Yu, Q. L. Hang, Y. Ding, H. Z. Zhang, Z. G. Bai, J. J. Wang, Y. H. Zou, W. Qian, G. C. Xiong and S. Q. Feng, *Appl. Phys. Lett.*, 1998, **73**, 3076.

270. L. Liu, M. Singh, V. T. John, G. L. McPherson, J. He, V. Agarwal and A. Bose, *J. Am. Chem. Soc.*, 2004, **126**, 2276.

271. F. L. Deepak, G. Gundiah, Md. M. Seikh, A. Govindaraj and C. N. R. Rao, *J. Mater. Res.*, 2004, **19**, 2216.

272. Z. G. Bai, D. P. Yu, H. Z. Zhang, Y. Ding, Y. P. Wang, X. Z. Gai, Q. L. Hang, G. C. Xiong and S. Q. Feng, *Chem. Phys. Lett.*, 1999, **303**, 311.

273. (a) X. C. Wu, W. H. Song, B. Zhao, Y. P. Sun and J. J. Du, *Chem. Phys. Lett.*, 2001, **349**, 210; (b) K. P. Kalyanikutty, G. Gundiah, A. Govindaraj and C. N. R. Rao, *J. Nanosci. Nanotechnol.*, 2005, **5**, 425.

274. Y. Zhang, J. Zhu, Q. Zhang, Y. Yan, N. Wang and X. Zheng, *Chem. Phys. Lett.*, 2000, **317**, 504.

275. J.-Q. Hu, Q. Li, X.-M. Meng, C.-S. Lee and S.-T. Lee, *Adv. Mater.*, 2002, **14**, 1396.

276. B. B. Lakshmi, C. J. Patrissi and C. R. Martin, *Chem. Mater.*, 1997, **9**, 2544.

277. X. Y. Zhang, L. D. Zhang, W. Chen, G. W. Meng, M. J. Zheng and L. X. Zhao, *Chem. Mater.*, 2001, **13**, 2511.

278. Y. Lei, L. D. Zhang, G. W. Meng, G. H. Li, X. Y. Zhang, C. H. Liang, W. Chen and S. X. Wang, *Appl. Phys. Lett.*, 2001, **78**, 1125.

279. Y. Lei and L. D. Zhang, *J. Mater. Res.*, 2001, **16**, 1138.

280. Z. Miao, D. Xu, J. Ouyang, G. Guo, X. Zhao and Y. Tang, *Nano Lett.*, 2002, **2**, 717.

281. D. K. Yi, S. J. Yoo and D.-Y. Kim, *Nano Lett.*, 2002, **2**, 1101.

282. S. J. Limmer and G. Cao, *Adv. Mater.*, 2003, **15**, 427.

283. Y. X. Zhang, G. H. Li, Y. X. Jin, Y. Zhang, J. Zhang and L. D. Zhang, *Chem. Phys. Lett.*, 2002, **365**, 300.

284. X. Gao, H. Zhu, G. Pan, S. Ye, Y. Lan, F. Wu and D. Song, *J. Phys. Chem. B*, 2004, **108**, 2868.

285. Z.-Y. Yuan, J.-F. Colomer and B.-L. Su, *Chem. Phys. Lett.*, 2002, **363**, 362.

286. Y. Zhu, H. Li, Y. Koltypin, Y. R. Hacohen and A. Gedanken, *Chem. Commun.*, 2001, 2616.

287. G. H. Du, Q. Chen, P. D. Han, Y. Yu and L.-M. Peng, *Phys. Rev. B*, 2003, **67**, 035323–1.
288. A. R. Armstrong, G. Armstrong, J. Canales and P. G. Bruce, *Angew. Chem. Int. Ed.*, 2004, **43**, 2286.
289. X. Wang and Y. Li, *Chem. Commun.*, 2002, 764.
290. X. Wang and Y. Li, *J. Am. Chem. Soc.*, 2002, **124**, 2880.
291. X. Wang and Y. Li, *Chem. Eur. J.*, 2003, **9**, 300.
292. Q. Li, J. B. Olson and R. M. Penner, *Chem. Mater.*, 2004, **16**, 3402.
293. M. Imperor-Clerc, D. Bazin, M.-D. Appay, P. Beaunier and A. Davidson, *Chem. Mater.*, 2004, **16**, 1813.
294. W. Wang, C. Xu, G. Wang, Y. Liu and C. Zheng, *Adv. Mater.*, 2002, **14**, 837.
295. X. Jiang, T. Herricks and Y. Xia, *Nano Lett.*, 2002, **2**, 1333.
296. C. Lu, L. Qi, J. Yang, D. Zhang, N. Wu and J. Ma, *J. Phys, Chem., B*, 2004, **108**, 17825.
297. Z.-Z. Chen, E.-W. Shi, Y.-Q. Zheng, W.-J. Li, B. Xiao and J.-Y. Zhuang, *J. Cryst. Growth.*, 2003, **249**, 294.
298. W. Wang, Y. Zhan and G. Wang, *Chem. Commun.*, 2001, 727.
299. W. Wang, G. Wang, X. Wang, Y. Zhan, Y. Liu and C. Zheng, *Adv. Mater.*, 2002, **14**, 67.
300. H. Yan, R. He, J. Pham and P. Yang, *Adv. Mater.*, 2003, **15**, 402.
301. J. J. Wu and S. C. Liu, *Adv. Mater.*, 2002, **14**, 215.
302. J. J. Wu, S. C. Liu, C. T. Wu, K. H. Chen and L. C. Chen, *Appl. Phys. Lett.*, 2002, **81**, 1312.
303. S. C. Liu and J. J. Wu, *J. Mater. Chem.*, 2002, **12**, 3125.
304. S. C. Lyu, Y. Zhang, H. Ruh, H. J. Lee, H. W. Shim, E. K. Suh and C. S. Lee, *Chem. Phys. Lett.*, 2002, **363**, 134.
305. W. I. Park, D. H. Kim, S. W. Jung and G. C. Yi, *Appl. Phys. Lett.*, 2002, **80**, 4232.
306. Y. Li, G. W. Meng, L. D. Zhang and F. Phillipp, *Appl. Phys. Lett.*, 2000, **76**, 2011.
307. M. J. Zheng, L. D. Zhang, G. H. Li and W. Z. Shen, *Chem Phys. Lett.*, 2002, **363**, 123.
308. E. C. Greyson, Y. Babayan and T. W. Odom, *Adv. Mater.*, 2004, **16**, 1348.
309. S. Y. Bae, H. W. Seo, H. C. Choi, J. Park and J. Park, *J. Phys. Chem. B*, 2004, **108**, 12318.
310. R. Yang, Y. Ding and Z. L. Wang, *Nano Lett.*, 2004, **4**, 1309.
311. L. Guo, J. X. Cheng, X. Y. Li, Y. J. Yan, S. H. Yang, C. L. Yang, J. N. Wang and W. K. Ge, *Mat. Sci. Eng. C*, 2001, **16**, 123.
312. (a) D. S. Boyle, K. Govender and P. O'Brien, *Chem. Commun.*, 2002, 80; (b) L. Vayssieres, *Adv. Mater.*, 2003, **15**, 464.
313. J. Zhang, L. Sun, H. Pan, C. Liao and C. Yan, *New J. Chem.*, 2002, **26**, 33.
314. Z. W. Pan, Z. R. Dai and Z. L. Wang, *Science*, 2001, **291**, 1947.
315. J. Y. Lao, J. Y. Huang, D. Z. Wang and Z. F. Ren, *J. Mater. Chem.*, 2004, **14**, 770.
316. W. I. Park, Y. H. Jun, S. W. Jung and G. C. Yi, *Appl. Phys. Lett.*, 2003, **82**, 964.
317. I. Shalish, H. Temkin and V. Narayanamurthi, *Phys. Rev. B*, 2004, **69**, 245401.

318. C. J. Lee, T. J. Lee, S. C. Lyu, Y. Zhang, H. Ruh and H. J. Lee, *Appl. Phys. Lett.*, 2002, **81**, 3648.

319. S. H. Jo, D. Banerjee and Z. F. Ren, *Appl. Phys. Lett.*, 2004, **85**, 1407.

320. J. C. Johnson, K. P. Knutsen, H. Yan, M. Law, Y. Zhang, P. Yang and R. J. Saykally, *Nano Lett.*, 2004, **4**, 197.

321. H. Kind, H. Yan, B. Messer, M. Law and P. Yang, *Adv. Mater.*, 2002, **14**, 158.

322. Q. H. Li, Q. Wan, Y. X. Liang and T. H. Wang, *Appl. Phys. Lett.*, 2004, **84**, 4556.

323. J. C. Johnson, H. Yan, R. D. Schaller, P. B. Petersen, P. Yang and R. J. Saykally, *Nano Lett.*, 2002, **2**, 279.

324. J. C. Johnson, H. Yan, R. D. Schaller, L. H. Haber, R. J. Saykally and P. Yang, *J. Phys. Chem. B*, 2001, **105**, 11387.

325. Q. Wan, C. L. Lin, X. B. Yu and T. H. Wang, *Appl. Phys. Lett.*, 2004, **84**, 124.

326. Q. Wan, Q.H. Li, Y. J. Chen, T. H. Wang, X. L. He, J. P. Li and C. L. Lin, *Appl. Phys. Lett.*, 2004, **84**, 3654.

327. J. Muster, G. T. Kim, V. Krstić, J. G. Park, Y. W. Park, S. Roth and M. Burghard, *Adv. Mater.*, 2000, **12**, 420.

328. B. C. Satishkumar, A. Govindaraj, M. Nath and C. N. R. Rao, *J. Mater. Chem.*, 2000, **10**, 2115.

329. J. Liu, Q. Li, T. Wang, D. Yu and Y. Li, *Angew. Chem. Int. Ed.*, 2004, **43**, 5048.

330. X. Chen, X. Wang, Z. Wang, J. Wan, J. Liu and Y. Qian, *Nanotechnology*, 2004, **15**, 1685.

331. G. Gu, B. Zheng, W. Q. Han, S. Roth and J. Liu, *Nano Lett.*, 2002, **2**, 849.

332. X. -L. Li, J.-F. Liu and Y.-D. Li, *Inorg. Chem.*, 2003, **42**, 921.

333. H. Qi, C. Wang and J. Liu, *Adv. Mater.*, 2003, **15**, 411.

334. X. W. Lou and H. C. Zeng, *J. Am. Chem. Soc.*, 2003, **125**, 2697.

335. J. Zhou, N. S. Xu, S. Z. Deng, J. Chen and J. C. She, *Chem. Phys. Lett.*, 2003, **382**, 443.

336. Y. Xiong, Z. Li, X. Li, B. Hu and Y. Xie, *Inorg. Chem.*, 2004, **43**, 6540.

337. Y. Y. Fu, R. M. Wang, J. Xu, J. Chen, Y. Yan, A. V. Narlikar and H. Zhang, *Chem. Phys. Lett.*, 2003, **379**, 373.

338. L. Xu, W. Zhang, Y. Ding, Y. Peng, S, Zhang, W. Yu and Y. Qian, *J. Phys. Chem. B*, 2004, **108**, 10859.

339. J. Wang, Q. Chen, C. Zeng and B. Hou, *Adv. Mater.*, 2004, **16**, 137.

340. D. Zhang, Z. Liu, S. Han, C. Li, B. Lei, M. P. Stewart, J. M. Tour and C. Zhou, *Nano Lett.*, 2004, **4**, 2151.

341. J. J. Urban, J. E. Spanier, L. Ouyang, W. S. Yun and H. Park, *Adv. Mater.*, 2003, **15**, 423.

342. W. S. Yun, J. J. Urban, Q. Gu and H. Park, *Nano Lett.*, 2002, **2**, 447.

343. H. Shi, L. Qi, J. Ma and H. Cheng, *Chem. Commun.*,, 2002, 1704.

344. X. Ma, H. Zhang, J. Xu, J. Niu, Q. Yang, J. Sha and D. Yang, *Chem. Phys. Lett.*, 2002, **363**, 579.

345. D. Zhu, H. Zhu and Y. Zhang, *Appl. Phys. Lett.*, 2002, **80**, 1634.

346. D. Zhu, H. Zhu and Y. Zhang, *J. Cryst. Growth*, 2003, **249**, 172.

347. T. Zhang, C. G. Jin, T. Qian, X. L. Lu, J. M. Bai and X. G. Li, *J. Mater. Chem.*, 2004, **14**, 2787.

348 K. S. Shankar, S. Kar, A. K. Raychaudhuri and G. N. Subbanna, *Appl. Phys. Lett.*, 2004, **84**, 993.

349. G. Ji, S. Tang, B. Xu, B. Gu and Y. Du, *Chem. Phys. Lett.*, 2003, **379**, 484.

350. J. A. Bonetti, D. S. Caplan, D. J. V. Harlingen and M. B. Weissman, *Phys. Rev. Lett.*, 2004, **93**, 087002–1.

351. S. Y. Bae, H. W. Seo, C. W. Na and J. Park, *Chem. Commun.*, 2004, 1834.

352. X. Y. Zhang, X. Zhao, C. W. Lai, J. Wang, X. G. Tang and J. Y. Dai, *Appl. Phys. Lett.*, 2004, **85**, 4190.

353. W.-B. Bu, Z.-L. Hua, L.-X. Zhang, H.-R. Chen, W.-M. Huang and J.-L. Shi, *J. Mater. Res.*, 2004, **19**, 2807.

354. A. R. Armstrong, J. Canales and P. G. Bruce, *Angew. Chem. Int. Ed.*, 2004, **43**, 4899.

355. P. Gleize, S. Herreyre, P. Gadelle, M. Mermoux, M. C. Cheynet and L. Abello, *J. Mater Sci. Lett.*, 1994, **13**, 1413.

356. K. F. Huo, Z. Hu, F. Chen, J. J. Fu, Y. Chen, B. H. Liu, J. Ding, Z. L. Dong and T. White, *Appl. Phys. Lett.*, 2002, **80**, 3611.

357. C. C. Tang, M. Lamy de la Chappelle, P. Li, Y. M. Liu, H. Y. Dang and S. S. Fan, *Chem. Phys. Lett.*, 2001, **342**, 492.

358. R. Ma, Y. Bando and T. Sato, *Adv. Mater.*, 2002, **14**, 366.

359. F. L. Deepak, C. P. Vinod, K. Mukhopadhyay, A. Govindaraj and C. N. R. Rao, *Chem. Phys. Lett.*, 2002, **353**, 345.

360. F. L. Deepak, G. Gundiah, A. Govindaraj and C. N. R. Rao, *Bull. Pol. Acad. Sci.*, 2002, **50**, 165.

361. X. P. Hao, D. L. Cui, X. G. Xu, M. Y. Yu, Y. J. Bai, Z. G. Liu and M. H. Jiang, *Mater Res. Bull.*, 2002, **37**, 2085.

362. M. Terrones, A. M. Benito, C. Manteca-Diego, W. K. Hsu, O. I. Osman, J. P. Hare, D. G. Reid, H. Terrones, A. K. Cheetham, K. Prassides, H. W. Kroto and D. R. M. Walton, *Chem. Phys. Lett.*, 1996, **257**, 576.

363. B. Toury, S. Bernard, D. Cornu, F. Chassagneux, J.-M. Letoffe and P. Miele, *J. Mater. Chem.*, 2003, **13**, 274.

364. H. Vincent, F. Chassagneux, C. Vincent, B. Bonnetot, M. P. Berthet, A. Vuillermoz and J. Bouix, *J. Mat. Sci. Eng A*, 2003, **340**, 181.

365. R. Ma, Y. Bando, T. Sato, D. Goldberg, H. Zhu, C. Xu and D. Wu, *Appl. Phys. Lett.*, 2002, **81**, 5225.

366. X. D. Bai, J. Yu, S. Liu and E. G. Wang, *Chem. Phys. Lett.*, 2000, **325**, 485.

367. X. D. Bai, J. D. Guo, J. Yu, E. G. Wang, J. Yuan and W. Zhou, *Appl. Phys. Lett.*, 2000, **76**, 2624.

368. J. Yu, J. Ahn, S. F. Yoon, Q. Zhang, Rusli, B. Gan, K. Chew, M. B. Yu, X. D. Bai and E. G. Wang, *Appl. Phys. Lett.*, 2000, **77**, 1949.

369. R. Sen, B. C. Satishkumar, A. Govindaraj, K. R. Harikumar, G. Raina, J.-P. Zhang, A. K. Cheetham and C. N. R. Rao, *Chem. Phys. Lett.*, 1998, **287**, 671.

370. R. B. Chen, C. P. Chang, F. L. Shyu and M. F. Lin, *Solid State. Commun*, 2002, **123**, 365.

371. Y.-C. Zhu, Y. Bando, D.-F. Xue, T. Sekiguchi, D. Golberg, F.-F. Xu and Q.-L. Liu, *J. Phys. Chem. B*, 2004, **108**, 6193.

372. J. Liu, X. Zhang, Y. Zhang, R. He and J. Zhu, *J. Mater. Res.*, 2001, **16**, 3133.

373. Y. Zhang, J. Liu, R. He, Q. Zhang, X. Zhang and J. Zhu, *Chem. Mater.*, 2001, **13**, 3899.

374. H. Chen, X. K. Lu, S. Q. Zhou, X. H. Hao and Z. X. Wang, *Mod. Phys. Lett.*, 2001, **15**, 1455.

375. C. C. Tang, S. S. Fan, M. Lamy de la Chapelle and P. Li, *Chem. Phys. Lett.*, 2001, **333**, 12.

376. Q. Wu, Z. Hu, X. Wang, Y. Hu, Y. Tian and Y. Chen, *Diamond Relat. Mater.*, 2004, **13**, 38.

377. S. M. Zhou, Y. S. Feng and L. D. Zhang, *Chem. Phys. Lett.*, 2003, **369**, 610.

378. H. Y. Peng, X. T. Zhou, N. Wang, Y. F. Zheng, L. S. Liao, W. S. Shi, C. S. Lee and S. T. Lee, *Chem. Phys. Lett.*, 2000, **327**, 263.

379 H.-M. Kim, D. S. Kim, D. Y. Kim, T. W. Kang, Y.-H. Cho and K. S. Chung, *Appl. Phys. Lett.*, 2002, **81**, 2193.

380. H.-M. Kim, D. S. Kim, Y. S. Park, D. Y. Kim, T. W. Kang and K. S. Chung, *Adv. Mater.*, 2002, **14**, 991.

381. C.-C. Chen, C.-C. Yeh, C.-H. Chen, M.-Y. Yu, H.-L. Liu, J.-J. Wu, K.-H. Chen, L.-C. Chen, J.-Y. Peng and Y.-F. Chen, *J. Am. Chem. Soc.*, 2001, **123**, 2791.

382. C.-C. Chen and C.-C. Yeh, *Adv. Mater.*, 2000, **12**, 738.

383. X. Chen, J. Li, Y. Cao, Y. Lan, H. Li, M. He, C. Wang, Z. Zhang and Z. Qiao, *Adv. Mater.*, 2000, **12**, 1432.

384. X. Duan and C. M. Lieber, *J. Am. Chem. Soc.*, 2000, **122**, 188.

385 W. Han, S. Fan, Q. Li and Y. Hu, *Science*, 1997, **277**, 1287.

386. F. L. Deepak, A. Govindaraj and C. N. R. Rao, *J. Nanosci. Nanotechnol.*, 2001, **1**, 303.

387. A. Govindaraj, F. L. Deepak, T. Poovarasan and C. N. R. Rao, *Ind. J. Phys.*, 2002, **76**, 1.

388. X. Chen, J. Xu, R. M. Wang and D. Yu, *Adv. Mater.*, 2003, **15**, 419.

389. S. Han, W. Jin, T. Tang, C. Li, D. Zhang, X. Liu, J. Han and C. Zhou, *J. Mater. Res.*, 2003, **18**, 245.

390. S. C. Lyu, O. H. Cha, E.-K. Suh, H. Ruh, H. J. Lee and C. J. Lee, *Chem. Phys. Lett.*, 2003, **367**, 136.

391. S. Y. Bae, H. W. Seo, J. Park, H. Yang and S. A. Song, *Chem. Phys. Lett.*, 2002, **365**, 525.

392. S. Y. Bae, H. W. Seo, J. Park, H. Yang, J. C. Park and S. Y. Lee, *Appl. Phys. Lett.*, 2002, **81**, 126.

393. L. Grocholl, J. Wang and E. G. Gillan, *Chem. Mater.*, 2001, **13**, 4290.

394. J. K. Jian, X. L. Chen, M. He, W. J. Wang, X. N. Zhang and F. Shen, *Chem. Phys. Lett.*, 2003, **368**, 416.

395. W.-Q. Han and A. Zettl, *Appl. Phys. Lett.*, 2002, **80**, 303.

396. A. Wohlfart, A. Devi, E. Maile and R. A. Fischer, *Chem. Commun.*, 2002, 998.

397. W. Han and A. Zettl, *Adv. Mater.*, 2002, **14**, 1560.

398. W. Han and A. Zettl, *Appl. Phys. Lett.*, 2002, **81**, 5051.

399. J. Zhang, L. D. Zhang, X. F. Wang, C. H. Liang, X. S. Peng and Y. W. Wang, *Appl. Phys. Lett.*, 2001, **115**, 5714.

400. M. W. Lee, H. Z. Twu, C.-C. Chen and C.-H. Chen, *Appl. Phys. Lett.*, 2001, **79**, 3693.

401. S. Han, W. Jin, D. Zhang, T. Tang, C. Li, X. Liu, Z. Liu, B. Lei and C. Zhou, *Chem. Phys. Lett.*, 2004, **389**, 176.

402. J. C. Johnson, H.-J. Choi, K. P. Knutsen, R. D. Schaller, P. Yang and R. J. Saykally, *Nat. Mater.*, 2002, **1**, 106.

403. (a) F. L. Deepak, P. V. Vanitha, A. Govindaraj and C. N. R. Rao, *Chem. Phys. Lett.*, 2003, **374**, 314; (b) F. L. Deepak, A. Govindaraj and C. N. R. Rao, unpublished results.

404. Y. Huang, X. Duan, Y. Cui and C. M. Lieber, *Nano Lett.*, 2002, **2**, 101.

405. F. Qian, Y. Li, S. Gradečak, D. Wang, C. J. Barrelet and C. M. Lieber, *Nano Lett.*, 2004, **4**, 1975.

406. S. D. Dingman, N. P. Rath, P. D. Markowitz, P. C. Gibbons and W. E. Buhro, *Angew. Chem. Int. Ed.*, 2000, **39**, 1470.

407. J. Zhang, L. Zhang, X. Peng and X. Wang, *J. Mater. Chem.*, 2002, **12**, 802.

408. C. H. Liang, L. C. Chen, J. S. Hwang, K. H. Chen, Y. T. Hung and Y. F. Chen, *Appl. Phys. Lett.*, 2002, **81**, 22.

409. (a) K. Sardar, F. L. Deepak, A. Govindaraj, M. M. Seikh, and C. N. R. Rao, *Small*, 2005, **1**, 91; (b) K. Sardar, M. Dan, B. Schwenzer and C. N. R. Rao, *J. Mater. Chem.*, 2005, **15**, 2175.

410. M. J. Wang and H. Wada, *J. Mater. Sci.*, 1990, **25**, 1690.

411. W. Han, S. Fan, Q. Li, B. Gu, X. Zhang and D. Yu, *Appl. Phys. Lett.*, 1997, **71**, 2271.

412. X. C. Wu, W. H. Song, B. Zhao, W. D. Huang, M. H. Pu, Y. P. Sun and J. J. Du, *Solid State Commun.*, 2000, **115**, 683.

413. Y. Zhang, N. Wang, R. He, J. Liu, X. Zhang and J. Zhu, *J. Cryst. Growth*, 2001, **233**, 803.

414. G. Gundiah, G. V. Madhav, A. Govindaraj, M. Md. Sheikh and C. N. R. Rao, *J. Mater. Chem.*, 2002, **12**, 1606.

415. D. Zhang, D. N. Mcilroy, Y. Geng and M. G. Norton, *J. Mater Sci. Lett.*, 1999, **18**, 349.

416. M. J. Pender and L. G. Sneddon, *Chem. Mater.*, 2000, **12**, 280.

417. R. Ma and Y. Bando, *Chem. Mater.*, 2002, **14**, 4403.

418. H. Dai, E. W. Wong, Y. Z. Liu, S. Fan and C. M. Lieber, *Nature*, 1995, **375**, 769.

419. (a) W. Han, Y. Bando, K. Kurashima and T. Sato, *Chem. Phys. Lett.*, 1999, **299**, 368; (b) W. Han, P. K. Redlich, F. Ernst and M. Rühle, *Chem. Mater.*, 1999, **11**, 3620.

420. J. Wei, B. Jiang, Y. Li, C. Xu, D. Wu and B. Wei, *J. Mater. Chem.*, 2002, **10**, 3121.

421. D. N. Mcllroy, D. Zhang, Y. Kranov and G. M. Norton, *Appl. Phys. Lett.*, 2001, **79**, 1540.

422. W. Shi, Y. Zheng, H. Peng, N. Wang, C. S. Lee and S. T. Lee, *J. Am. Chem. Soc.*, 2000, **83**, 3228.

423. G. W. Meng, L. D. Zhang, C. M. Mo, S. Y. Zhang, Y. Qin, S. P. Feng and H. J. Li, *J. Mater. Res.*, 1998, **13**, 2533.

424. C. H. Liang, G. W. Meng, L. D. Zhang, Y. C. Wu and Z. Cui, *Chem. Phys. Lett.*, 2000, **329**, 323.

425. X. H. Sun, C. P. Li, W. K. Wong, N. B. Wong, C. S. Lee, S. T. Lee and B. K. Teo, *J. Am. Chem. Soc.*, 2002, **124**, 14464.

426. J. Q. Hu, Q. Y. Lu, K. B. Tang, B. Deng, R. R. Jiang, Y. T. Qian, W. C. Yu, G. E. Zhou, X. M. Lu and J. X. Wu, *J. Phys. Chem. B*, 2000, **104**, 5251.

427. S. Z. Deng, Z. S. Wu, J. Zhou, N. S. Xu, J. Chen and J. Chen, *Chem. Phys. Lett.*, 2002, **356**, 511.

428. S. Z. Deng, Z. S. Wu, J. Zhou, N. S. Xu, J. Chen and J. Chen, *Chem. Phys. Lett.*, 2002, **364**, 608.

429. H. F. Zhang, C. M. Wang and L. S. Wang, *Nano Lett.*, 2002, **2**, 941.

430. C. C. Tang, Y. Bando, T. Sato, K. Kurashima, X. X. Ding, Z. W. Gan and S. R. Qi, *Appl. Phys. Lett.*, 2002, **80**, 4641.

431. Z. Pan, H. L. Lai, F. C. K. Au, X. Duan, W. Zhou, W. Shi, N. Wang, C. S. Lee, N. B. Wong, S. T. Lee and S. Xie, *Adv. Mater.*, 2000, **12**, 1186.

432. H. Y. Kim, J. Park and H. Yang, *Chem. Commun.*, 2003, 256.

433. Z. S. Wu, S. Z. Deng, N. S. Xu, J. Chen, J. Zhou and J. Chen, *Appl. Phys. Lett.*, 2002, **80**, 3829.

434. G. W. Ho, A. S. W. Wong, A. T. S. Wee and M. E. Welland, *Nano Lett.*, 2004, **4**, 2023.

435. H.-K. Seong, H.-J. Choi, S.-K. Lee, J.-I. Lee and D.-J. Choi, *Appl. Phys. Lett.*, 2004, **85**, 1256.

436. C. Ye, G. Meng, Y. Wang, Z. Jiang and L. Zhang, *J. Phys. Chem. B*, 2002, **106**, 10338.

437. Y. Wang, G. Meng, L. Zhang, C. Liang and J. Zhang, *Chem. Mater.*, 2002, **14**, 1773.

438. D. Routkevitch, T. Bigioni, M. Moskovits and J. M. Xu, *J. Phys. Chem. B*, 1996, **100**, 14037.

439. D. Xu, Y. Xu, D. Chen, G. Guo, L. Gui and Y. Tang, *Adv. Mater.*, 2000, **12**, 520.

440. Y. Li, D. Xu, Q. Zhang, D. Chen, F. Huang, Y. Xu, G. Guo and Z. Gu, *Chem. Mater.*, 1999, **11**, 3433.

441. H. Cao, Y. Xu, J. Hong, H. Liu, G. Yin, B. Li, C. Tie and Z. Xu, *Adv. Mater.*, 2001, **13**, 1393.

442. Y. Yang, H. Chen, Y. Mei, J. Chen, X. Wu and X. Bao, *Solid. State Commun.*, 2002, **123**, 279.

443. J. Fan, T. Gao, G. Meng, Y. Wang, X. Liu and L. Zhang, *Mater. Lett.*, 2002, **57**, 656.

444. L. Yang, J. Yang, Z.-H. Wang, J.-H. Zeng, L. Yang and Y.-T. Qian, *J. Mater. Res.*, 2003, **18**, 396.

445. Y.-D. Li, H.-W Liao, Y. Ding, Y.-T. Qian, L. Yang and G.-E. Zhou, *Chem. Mater.*, 1998, **10**, 2301.

446. J. Yang, C. Xue, S.-H. Yu, J.-H. Zeng and Y.-T. Qian, *Angew. Chem. Int. Ed.*, 2002, **41**, 4697.

447. Y. Xiong, Y. Xie, J. Yang, R. Zhang, C. Wu and G. Du, *J. Mater. Chem.*, 2002, **12**, 3712.

448. C.-S. Yang, D. D. Awschalom and G. D. Stucky, *Chem. Mater.*, 2002, **14**, 1277.

449. P. S. Nair, T. Radhakrishnan, N. Revaprasadu, G. A. Kolawole and P. O'Brien, *Chem. Commun*, 2002, 564.

450. W. Wang, J. Zhai and F. Bai, *Chem. Phys. Lett.*, 2002, **366**, 165.

451. J. Yang, J.-H. Zeng, S.-H. Yu, L. Yang, G. E. Zhou and Y.-T. Qian, *Chem. Mater.*, 2000, **12**, 3259.

452. F. Gao, Q. Lu, S. Xie and D. Zhao, *Adv. Mater.*, 2002, **14**, 1537.

453. E. D. Sone, E. R. Zubarev and S. I. Stupp, *Angew. Chem. Intl. Ed.*, 2002, **41**, 1705.

454. L. Chen, P. J. Klar, W. Heimbrodt, F. Brieler and M. Fröba, *Appl. Phys. Lett.*, 2000, **76**, 3531.

455. D. Routkevitch, T. L. Haslett, L. Ryan, T. Bigioni, C. Douketis and M. Moskovits, *Chem. Phys. Lett.*, 1996, **210**, 343.

456. Q. Peng, Y. Dong, Z. Deng and Y. Li, *Inorg. Chem.*, 2002, **41**, 5249.

457 Q. Yang, K. Tang, C. Wang, Y. Qian and S. Zhang, *J. Phys. Chem. B*, 2002, **106**, 9227.

458. T. Nann and J. Riegler, *Chem. Eur. J.*, 2002, **8**, 4791.

459. L.-S. Li and P. Alivisatos, *Adv. Mater.*, 2003, **15**, 408.

460. D. Xu, D. Chen, Y. Xu, X. Shi, G. Guo, L. Gui and Y. Tang, *Pure. Appl. Chem.*, 2000, **72**, 127.

461. D. Xu, X. Shi, G. Guo, L. Gui and Y. Tang, *J. Phys. Chem. B*, 2000, **104**, 5061.

462. X. S. Peng, J. Zhang, X. F. Wang, Y. W. Wang, L. X. Zhao, G. W. Meng and L. D. Zhang, *Chem. Phys. Lett.*, 2001, **343**, 470.

463. Y.-W. Su, C.-S. Wu, C.-C. Chen and C.-D. Chen, *Adv. Mater.*, 2003, **15**, 49.

464. X.-Y. Wang, J.-Y. Zhang, A. Nazzal, M. Darragh and M. Xiao, *Appl. Phys. Lett.*, 2002, **81**, 4829.

465. Y. Xie, P. Yan, J. Lu, Y. Qian and S. Zhang, *Chem. Commun.*, 1999, 1969.

466. D. J. Péna, J. K. N. Mbindyo, A. J. Carado, T. E. Mallouk, C. D. Keating, B. Razavi and T. S. Mayer, *J. Phys. Chem. B*, 2002, **106**, 7458.

467. Z. Tang, N. A. Kotov and M. Giersig, *Science*, 2002, **297**, 237.

468. X. Peng, *Adv. Mater.*, 2003, **15**, 459.

469. L. Manna, D. J. Milliron, A. Meisel, E.C. Scher and A. P. Alivisatos, *Nat. Mater.*, 2003, **2**, 382.

470. D. J. Milliron, S. M. Hughes, Y. Cui, L. Manna, J. Li, L.-W. Wang and A. P. Alivisatos, *Nature*, 2004, **430**, 190.

471. M.-S. Mo, M.-W. Shao, H.-M. Hu, L. Yang, W.-C. Yu and Y.-T. Qian, *J. Cryst. Growth*, 2002, **244**, 364.

472. F. Gao, Q. Lu, X. Liu, Y. Yan and D. Zhao, *Nano Lett.*, 2001, **1**, 743.

473. Y. Liu, J. Cao, J. Zeng, C. Li, Y. Qian and S. Zhang, *Eur. J. Inorg. Chem.*, 2003, 644.

474. W. Wang, Y. Geng, Y. Qian, M. Ji and X. Liu, *Adv. Mater.*, 1998, **10**, 1479.

475. X.-H. Liao, H. Wang, J.-J. Zhu and H.-Y. Chen, *Mat. Res. Bull.*, 2001, **36**, 2339.

476. S.-H. Yu, L. Shu, J. Yang, Z.-H. Han, Y.-T. Qian and Y.-H. Zhang, *J. Mater. Res.*, 1999, **14**, 4157.

477. W. Zhang, Z. Yang, X. Huang, S. Zhang, W. Yu, Y. Qian, Y. Jia, G. Zhou and L. Chen, *Solid. State Commun.*, 2001, **119**, 143.

478. H. Wang, J.-J. Zhu, J.-M. Zhu and H.-Y. Chen, *J. Phys. Chem. B*, 2002, **106**, 3848.

479. G. Shen, D. Chen, K. Tang, F. Li and Y. Qian, *Chem. Phys. Lett.*, 2003, **370**, 334.

480. Q. Lu, F. Gao and S. Komarneni, *J. Am. Chem. Soc.*, 2004, **126**, 54.

481. A. L. Prieto, M. S. Sander, M. S. Martín-González, R. Gronsky, T. Sands and A. M. Stacy, *J. Am. Chem. Soc.*, 2001, **123**, 7160.

482. M. S. Sander, R. Gronsky, T. Sands and A. M. Stacy, *Chem. Mater.*, 2003, **15**, 335.

483. M. S. Sander, A. L. Prieto, R. Gronsky, T. Sands and A. M. Stacy, *Adv. Mater.*, 2002, **14**, 665.

484. F. L. Deepak, A. Govindaraj and C. N. R. Rao, *J. Nanosci. Nanotechnol.*, 2002, **2**, 417.

485. S. Wang and S. Yang, *Chem. Phys. Lett.*, 2000, **322**, 567.

486. J. Chen, S. Z. Deng, N. S. Xu, S. Wang, X. Wen, S. Yang, C. Yang, J. Wang and W. Ge, *Appl. Phys. Lett.*, 2002, **80**, 3620.

487. N. Pradhan and S. Efrima, *J. Phys. Chem. B*, 2004, **108**, 1964.

488. Q. Wu, N. Zheng, Y. Ding and Y. Li, *Inorg. Chem. Commun.*, 2002, **5**, 671.

489. C. Lan, K. Hong, W. Wang and G. Wang, *Solid. State Commun.*, 2003, **125**, 455.

490. Y. Wang, L. Zhang, C. Liang, G. Wang and X. Peng, *Chem. Phys. Lett.*, 2002, **357**, 314.

491. J.-P. Ge, J. Wang, H.-X. Zhang and Y.-D. Li, *Chem. Eur. J.* 2004, **10**, 3525.

492. C. Mao, D. J. Solis, B. D. Reiss, S. T. Kottmann, R. Y. Sweeney, A. Hayhurst, G. Georgiou, B. Iverson and A. M. Belcher, *Science*, 2004, **303**, 213.

493. S.-H. Yu and M. Yoshimura, *Adv. Mater.*, 2002, **14**, 296.

494. X. Wang, P. Gao, J. Li, C. J. Summers and Z. L. Wang, *Adv. Mater.*, 2002, **14**, 1732.

495. J. X. Ding, J. A. Zapien, W. W. Chen, Y. Lifshitz, S. T. Lee and X. M. Meng, *Appl. Phys. Lett.*, 2004, **85**, 2361.

496. Q. Xiong, G. Chen, J. D. Acord, X. Liu, J. J. Zengel, H. R. Gutierrez, J. M. Redwing, L. C. L. Y. Voon, B. Lassen and P. C. Eklund, *Nano Lett.*, 2004, **4**, 1663.

497. Q. Xiong, J. Wang, O. Reese, L. C. L. Y. Voon and P. C. Eklund, *Nano Lett.*, 2004, **4**, 1991.

498. P. V. Teredesai, F. L. Deepak, A. Govindaraj, A. K. Sood and C. N. R. Rao, *J. Nanosci. Nanotechnol.*, 2002, **2**, 495.

499. Q. Li, X. Gong, C. Wang, J. Wang, K. Ip and S. Hark, *Adv. Mater.*, 2004, **16**, 1436.

500. J. Hu, Y. Bando and D. Golberg, *Small*, 2005, **1**, 95.

501. Y. Jiang, X.-M. Meng, W.-C. Yiu, J. Liu, J.-X. Ding, C.-S. Lee and S. T. Lee, *J. Phys. Chem. B*, 2004, **108**, 2784.

502. M. Nath and C. N. R. Rao, *J. Am. Chem. Soc.*, 2001, **123**, 4841.

503. Y. Q. Zhu, W. K. Hsu, H. W. Kroto and D. R. M. Walton, *Chem. Commun.*, 2001, 2184.

504. Y. Z. Jin, W. K. Hsu, Y. L. Chueh, L. J. Chou, Y. Q. Zhu, K. Brigatti, H. W. Kroto and D. R. M. Walton, *Angew. Chem. Int. Ed.*, 2004, **43**, 5670.

505. M. Nath, S. Kar, A. K. Raychaudhari and C. N. R. Rao, *Chem. Phys. Lett.*, 2003, **368**, 690.

506. H. Liao, Y. Wang, S. Zhang and Y. Qian, *Chem. Mater.*, 2001, **13**, 6.

507. K. Tang, C. An, P. Xie, G. Shen and Y. Qian, *J. Cryst. Growth*, 2002, **245**, 304.

508. (a) M. Nath, A. Choudhury, A. Kundu and C. N. R. Rao, *Adv. Mater.*; **15**, 2098–2101. (b) M. Nath, A. Choudhury and C. N. R. Rao, *Chem. Commun.*, 2004, 2698.

509. Y. Jiang, Y. Wu, S. Yuan, B. Xie, S. Zhang and Y. Qian, *J. Mater. Chem.*, 2001, **16**, 2805.

510. C. An, Y. Jin, K. Tang and Y. Qian, *J. Mater. Chem.*, 2003, **13**, 301.

511. M. D. Hornbostel, S. Hillyard, J. Silcox and F. J. DiSalvo, *Nanotechnology*, 1995, **6**, 87.

512. J. H. Song, Y. Wu, B. Messer, H. Kind and P. Yang, *J. Am. Chem. Soc.*, 2001, **123**, 10397.

513. X. Duan and C. M. Lieber, *Adv. Mater.*, 2000, **12**, 298.

514. W. Shi, Y. Zheng, N. Wang, C. S. Lee and S. T. Lee, *Adv. Mater.*, 2001, **13**, 591.

515. W. S. Shi, Y. F. Zheng, N. Wang, C. S. Lee and S. T. Lee, *Appl. Phys. Lett.*, 2001, **78**, 3304.

516. H. Yu and W. E. Buhro, *Adv. Mater.*, 2003, **15**, 416.

517. Z. H. Wu, X. Y. Mei, D. Kim, M. Blumin and H. E. Ruda, *Appl. Phys. Lett.*, 2002, **81**, 5177.

518. X. Duan, J. Wang and C. M. Lieber, *Appl. Phys. Lett.*, 2000, **76**, 1116.

519. D. Routkevitch, A. A. Tager, J. Haruyama, M. Almawlawi Moskovits and J. M. Xu, *IEEE Trans.*, 1996, **43**, 1646.

520. X. Duan, Y. Huang, Y. Cui, J. Wang and C. M. Lieber, *Nature*, 2001, **409**, 66.

521. Y. Xiong, Y. Xie, Z. Li, X. Li and S. Gao, *Chem. Eur. J.*, 2004, **10**, 654.

522. M. S. Gudiksen, J. Wang and C. M. Lieber, *J. Phys. Chem. B*, 2001, **105**, 4062.

523. T. Mårtensson, P. Carlberg, M. Borgström, L. Montelius, W. Seifert and L. Samuelson, *Nano Lett.*, 2004, **4**, 699.

524. M. S. Gudiksen, J. Wang and C. M. Lieber, *J. Phys. Chem. B*, 2002, **106**, 4036.

525. J. Wang, M. S. Gudiksen, X. Duan, Y. Cui and C. M. Lieber, *Science*, 2001, **293**, 1455.

526. M. S. Gudiksen, L. J. Lauhon, J. Wang, D. C. Smith and C. M. Lieber, *Nature*, 2002, **415**, 617.

527. M. S. Gudiksen and C. M. Lieber, *J. Am. Chem. Soc.*, 2000, **122**, 8801.

528. W. S. Shi, Y. F. Zheng, N. Wang, C. S. Lee and S. T. Lee, *J. Vac. Sci. Technol. B*, 2001, **19**, 1115.

529. H. W. Seo, S. Y. Bae, J. Park, H. Yuang and S. Kim, *Chem. Commun.*, 2002, 2564.

530. S. Xu and Y. Li, *J. Mater. Chem.*, 2003, **13**, 163.

531. D. Kuang, A. Xu, Y. Fang, H. Ou and H. Liu, *J. Cryst. Growth*, 2002, **244**, 379.

532. C. Qian, F. Kim, L. Ma, P. Yang and J. Liu, *J. Am. Chem. Soc.*, 2004, 126, 1195.

533. P. Kluth, Q. T. Zhao, S. Winnerl, S. Lenk and S. Mantl, *Appl. Phys. Lett.*, 2001, **79**, 824.

534. C. A. Decker, R. Solanki, J. L. Freeouf, J. R. Carruthers and D. R. Evans, *Appl. Phys. Lett.*, 2004, **84**, 1389.

535. P. Zjou, D. Xue, H. Luo and X. Chen, *Nano Lett.*, 2002, **2**, 845.

536. Y. Huang, X. Duan, Y. Cui, L. J. Lauhon, K. H. Kim and C. M. Lieber, *Science*, 2001, **294**, 1313.

537. M. T. Björk, B. J. Ohlsson, C. Thelander, A. I. Persson, K. Deppert, L. R. Wallenberg and L. Samuelson, *Appl. Phys. Lett.*, 2002, **81**, 4458.

538. H. Kind, H. Yan, B. Messer, M. Law and P. Yang, *Adv. Mater.*, 2002, **14**, 158.

539. S.-M. Koo, A. Fujiwara, J.-P. Han, E. M. Vogel, C. A. Richter and J. E. Bonevich, *Nano Lett.*, 2004, **4**, 2197.

540. G. Zheng, W. Lu, S. Jin and C. M. Lieber, *Adv. Mater.*, 2004, **16**, 1890.

541. Y. Wu, J. Xiang, C. Yang, W. Lu and C. M. Lieber, *Nature*, 2004, **430**, 61.

542. (a) Y. W. Heo, L. C. Tien, Y. Kwon, D. P. Norton, S. J. Pearton, B. S. Kang and F. Ren, *Appl. Phys. Lett.*, 2004, **85**, 2274; (b) H. T. Ng, J. Han, T. Yamada, P. Nguyen, Y. P. Chen and M. Meyyappan, *Nano Lett.*, 2004, **4**, 1247.

543. N. I. Kovtyukhova, B. K. Kelley and T. E. Mallouk, *J. Am. Chem. Soc.*, 2004, **126**, 12738.

544. P. Nguyen, H. T. Ng, T. Yamada, M. K. Smith, J. Li, J. Han and M. Meyyappan, *Nano Lett.*, 2004, **4**, 651.

545. S. Tan, Z. Tang, X. Liang and N. A. Kotov, *Nano Lett.*, 2004, **4**, 1637.

546. S. Jin, D. Whang, M. C. McAlpine, R. S. Friedman, Y. Wu and C. M. Lieber, *Nano Lett.*, 2004, **4**, 915.

547. Z. Zhong, D. Wang, Y. Cui, M. W. Bockrath and C. M. Lieber, *Science*, 2003, **302**, 1377.

548. G. Y. Tseng and J. C. Ellenbogen, *Science*, 2001, **294**, 1293.

549. Y. W. Heo, L. C. Tien, D. P. Norton, S. J. Pearton, B. S. Kang, F. Ren and J. R. LaRoche, *Appl. Phys. Lett.*, 2004, **85**, 3107.

550. B. Lei, C. Li, D. Zhang, Q. F. Zhou, K. K. Shung and C. Zhou, *Appl. Phys. Lett.*, 2004, **84**, 4553.

551. Y. Huang, X. Duan and C. M. Lieber, *Small*, 2005, **1**, 142.

552. M. Law, D. J. Sirbuly, J. C. Johnson, J. Goldberger, R. J. Saykally and P. Yang, *Science*, 2004, **305**, 1269.

553. J. A. Zapien, Y. Jiang, X. M. Meng, W. Chen, F. C. K. Au, Y. Lifshitz and S. T. Lee, *Appl. Phys. Lett.*, 2004, **84**, 1189.

554. L. Gangloff, E. Minoux, K. B. K Teo, P. Vincent, V. T. Semet, V. T. Binh, M. H. Yang, I. Y. Y. Bu, R. G. Lacerda, G. Pirio, J. P. Schnell, D. Pribat,

D. G. Hasko, G. A. J. Amaratunga, W. I. Milne and P. Legagneux, *Nano Lett.*, 2004, **4**, 1575.

555. M. Adachi, Y. Murata, J. Takao, J. Jiu, M. Sakamoto and F. Wang, *J. Am. Chem. Soc.*, 2004, **126**, 14943.

556. L. A. Bauer, N. S. Birenbaum and G. J. Meyer, *J. Mater. Chem.*, 2004, **14**, 517.

557. J. Hahm and C. M. Lieber, *Nano Lett.*, 2004, **4**, 51.

558. A. K. Salem, J. Chao, K. W. Leong and P. C. Searson, *Adv. Mater.*, 2004, **16**, 268.

559. (a) W. F. Paxton, K. C. Kistler, C. C. Olmeda, A. Sen, S. K. St. Angelo, Y. Cao, T. E. Mallouk, P. E. Lammert and V. H. Crespi, *J. Am. Chem. Soc.*, 2004, **126**, 13424; (b) T. R. Kline, W. F. Paxton, T. E. Mallouk and A. Sen, *Angew. Chem. Int. Ed.*, 2005, **44**, 744.

Subject Index